AGRARIAN STUDIES 5

HOW DO
SMALL FARMERS
FARE?

Evidence from Village Studies in India

AGRARIAN STUDIES 5

HOW DO
SMALL FARMERS
FARE?

Evidence from Village Studies in India

Edited by
Madhura Swaminathan and Sandipan Baksi

 Tulika Books

Published by
Tulika Books
44, first floor, Shahpur Jat, New Delhi 110 049, India
www.tulikabooks.in

First edition (hardback) 2017

ISBN: 978-93-82381-97-6

Printed at Chaman Offset, Delhi 110 002

For
V. Namasivayam

Contents

Foreword

How Do Small Farmers Fare? Evidence from Village Studies in India is the fifth in the Agrarian Studies series published by the Foundation for Agrarian Studies in collaboration with Tulika Books, New Delhi. The volume is designed by M. V. Bhaskar of TNQ Books and Journals.

The book is an effort by the Foundation for Agrarian Studies to intervene in the debate on the efficiency and sustainability of small farms, and on the incomes and livelihoods of small farmers in India. It draws on empirical material collected through carefully designed and conducted household and farm economy surveys in 17 villages located in nine States of India. These surveys are part of the Project on Agrarian Relations in India (PARI) that is being constructed by the Foundation.

Much international scholarship on small farms and small farmers holds up small-scale farming as an example of efficiency and equity, as well as a solution to the problems of global hunger and environmental sustainability. The chapters in this book, which present meticulous empirical material on the crisis of small-scale farming in India, challenge that point of view. The authors demonstrate that a majority of small farmer households are unable to generate income adequate for a minimum standard of living. They also describe and analyse the stark and growing differentiation among rural cultivator households. Several chapters in the book examine policy measures to deal with problems of small farmers' incomes and livelihoods.

The book is an outcome of a two-year project on "Small-Scale Farming in Indian Agriculture" undertaken by the Foundation for Agrarian Studies with support from the Rosa Luxemberg Stiftung.

V. K. RAMACHANDRAN
General Editor, Agrarian Studies series

Preface

The United Nations declared 2014 as International Year of Family Farming, "in an effort to highlight the potential family farmers have to eradicate hunger, preserve natural resources, and promote sustainable development." The then Director General of the Food and Agriculture Organisation, José Graziano da Silva, claimed that the declaration was a recognition of the fact that "family farmers are leading figures in responding to the double urgency the world faces today: improving food security and preserving natural resources."

In this context, the Rosa Luxemburg Stiftung (RLS) invited the Foundation for Agrarian Studies (FAS) to take up a study on small-scale farming in India. The objective of the project was to examine the socio-economic characteristics and viability of small producers in different agro-ecological regions of India, locating them in the broader context of capitalist development of Indian agriculture. Given the limitations of macro-data on farm households in India, it was decided that a large part of the study will be based on primary data collected as part of the ongoing Project on Agrarian Relations in India (PARI) undertaken by FAS. This book is an outcome of the two-year research project. We warmly thank RLS for funding the writing of this book.

A core group of scholars was formed to design and implement the study, comprising V. K. Ramachandran, Venkatesh Athreya, T. Jayaraman, Madhura Swaminathan, R. Ramakumar, Aparajita Bakshi, Niladri Sekhar Dhar, Sandipan Baksi, Arindam Das, T. Sivamurugan, Shamsher Singh, and Biplab Sarkar. In addition to members of this core group, the contributors to the book include Pallavi Chavan, Kamal Kumar Murari, Deepak Kumar, A. Bheemeshwar Reddy, Tapas Singh Modak, Vijay Kumar, Sanjukta Chakraborty, Subhajit Patra, Ritam Dutta, and Rakesh Kumar Mahato. We are thankful to the FAS staff for their crucial support to the project. Special thanks are due to Divya S. Devadiga for help in preparing the manuscript, and to Pushpita Dhar for formating and proof-reading. We also thank Pinki Ghosh from FAS and Tauqueer Ali Sabri from RLS for their consistent administrative support.

The preliminary results of the research were presented at a conference held on 2–3 December 2016, in Thiruvananthapuram, Kerala. The conference was attended by senior academics, young scholars, and activists from the All

India Kisan Sabha and All India Agricultural Workers' Union. We thank the participants for their comments. Prior to the conference, draft chapters were sent to discussants, and we thank Abhijit Sen, A. Suresh, D. N. Reddy, Rajni Palriwala, Sheila Bhalla, Judith Heyer, V. Surjit, and Yoshifumi Usami for their critical feedback. We are grateful to Harshan, T. P., Deepak Johnson, S. Niyati, and Kaushik Bora for their help in organizing the conference.

We owe special thanks to Judith Heyer who read and commented on several draft chapters, and helped the editors bring the book together; and to Parvathi Menon who made the text more readable.

Finally, a big and sincere thanks to Indira Chandrasekhar of Tulika Books who has provided her inputs at all stages of the publication, and M. V. Bhaskar and the team at Authorcafe and TNQ Books and Journals for the final design of the book.

The structure of the book is as follows. Chapter 1 introduces the debates on small farmers and small farming, with an emphasis on classical Marxist literature. It also provides a working definition of small farmers, and outlines how detailed field data from PARI can speak to these issues and debates. Chapter 2 provides descriptions of the 17 PARI study villages, and introduces the PARI surveys which have been used for the purposes of this book. Apart from providing an agro-ecological, demographic, and socio-economic description of the villages, it discusses the importance of small farmers in terms of their numbers and area operated in each village. Chapter 3 elaborates on the nature of labour supply and labour use among small farmer households. Chapter 4 addresses questions of cropping pattern, yields, and returns from farming among small farmer households, including the debate on small versus large farmers. Chapter 5 deals with household incomes and diversification of incomes of small farmer households. Chapter 6 examines the components of costs of production as well as prices obtained for farm output. Chapter 7 investigates the use of fertilizers by small farmers in terms of quantity, cost, and efficiency. Chapter 8 analyses indebtedness among small farmers with a focus on access to crop credit. Chapter 9 examines the differential impact of climate change on small farmers in India. Access to education and basic amenities among small farmer households are taken up in Chapters 10 and 11, respectively. In Chapter 12 we bring together the main findings of the book, identify the major constraints faced by small farmers, and raise issues of policy support.

Madhura Swaminathan is grateful to the Indian Statistical Institute, Bengaluru, for the excellent research environment.

Bengaluru   MADHURA SWAMINATHAN
3 September 2017   SANDIPAN BAKSI

1

Small Farmers and Small Farming: A Definition

Venkatesh Athreya, Deepak Kumar, R. Ramakumar, Biplab Sarkar

This volume seeks to address some key questions concerning small farms and small farmers in the context of contemporary India. It draws on empirical material of exceptional quality collected through carefully designed and conducted household and farm economy surveys, mostly of the census type, in nearly 20 villages located in nine major States of India. In this chapter, we look at the importance of small farms and small farmers in India today; selectively review some of the literature; briefly discuss policy-related definitions of small farmers; outline the current agrarian context, highlighting the issue of agrarian distress; and indicate how the rich empirical material from the surveys conducted by the Foundation of Agrarian Studies (FAS) under the Project on Agrarian Relations in India (PARI) speak to some key issues concerning small farms and small farmers.

WHY STUDY SMALL FARMS AND SMALL FARMERS?

Oksana Nagayets (2005) notes:

> There are approximately 525 million farms worldwide, though small farm data are only available for 470 million. Of these, smallholders who operate plots of land of less than 2 hectares currently constitute 85 per cent. The overwhelming majority of these farms are located in Asia (87 per cent), while Africa is home to another 8 per cent, and Europe to approximately 4 per cent. . . . In Asia, China alone accounts for almost half the world's small farms (193 million), followed by India with 23 per cent. Other leaders in the region, in descending order, include Indonesia, Bangladesh, and Vietnam.

Small farms operated by households – which, as noted above, account for a high proportion of all family farms across the world – dominate the agrarian economy in the developing countries of Asia in terms of their share of all farms. They also account for a significant proportion of the area operated,

Table 1 *Small farms as a proportion of all farms, and extent of land under small farms as a proportion of extent of all farms* in per cent

Country	Share of number of farms	Share of extent of farm land
India, 2011	84.98	44.32
Pakistan, 2000	57.63	15.64
Nepal, 2002	92.44	68.72
Sri Lanka, 2002	45.25	5.36
Myanmar, 2003	56.92	19.03
China, 1997	97.91	–
Philippines, 2002	69.06	25.47
Indonesia, 2003	88.73	–
Vietnam, 2001	94.81	–
Thailand, 2003	64.50	–
Laos, PDR, 1998–99	73.50	42.82

Note: Small farms are farms whose extent is 2 hectares or less.
Source: Adapted from T. Haque (2016, p. 19, Table 1.1).

though the share of small farms in total area operated is substantially smaller in many countries than their share of the number of farms. Presented above are some data on the share of small farms, defined as being less than or equal to 2 hectares in extent, in area operated in some countries of Asia (Table 1). Farms of size less than 2 hectares accounted for 97.91 per cent of all holdings in China in 1997, 94.81 per cent in Vietnam in 2001, 92.44 per cent in Nepal, 88.73 per cent in Indonesia in 2003, and 84.98 per cent in India in 2011.

According to the Agricultural Census of 2010–11, there were a total of 138.35 million operational holdings in India. The total area operated was 159.59 million hectares and the average size of an operational holding was 1.15 hectares. The average size of all holdings of size 2 hectares or less – which constituted small and marginal holdings as per the official definition – was 0.60 hectare.[1] Holdings of size 2 hectares or less accounted for around 85 per cent of all holdings and 45 per cent of the total area operated. The number of persons who were part of small farmer households was close to half a billion.

It is obvious that small farms will continue to account for a large share of all farms across the world, and especially the developing countries, in the foreseeable future. The sheer numerical importance – in both absolute and relative terms – of both small farms and the small farmers who operate them makes it worthwhile and appropriate for social scientists to study small farms and small-scale farming.

---

[1] Available at https://factly.in/agricultural-land-holdings-statistics-india-account-for-close-to-a-third-of-the-total-agricultural-land, viewed on 8 January 2017.

THE PERSPECTIVE OF THE PRESENT STUDY[2]

While what the future holds for the small farmer may be a subject of much academic speculation, it cannot be denied that for a long time to come, small farmers will be present in the Indian as well as the global economy in large numbers. The present study seeks to offer some suggestions regarding policies of support to small farmers in the interregnum, even as we examine the hypotheses in the literature concerning small farmers against the evidence from FAS surveys conducted in several States of India since 2005.

In the voluminous literature on small farms and small farmers, there is a distinct tendency to romanticise small-scale farming. Various virtues are ascribed to the small farm: it is claimed to be more efficient than the large farm; it is said to be ecologically more worthy of preservation than the large farm; it is said to represent a higher moral economy. A recent study argues:

> Considerable research in the past several decades has indicated that the small-scale and family farming sector plays a key role for environmental sustainability and farmer livelihoods (e.g., Chappell *et al.* 2013), and, given the non-market values generated by agriculture (Sandhu *et al.* 2015), the true contribution to the global economy is likely much larger than the US$2.2 trillion figure. There is also consistent evidence that small-scale farms can be more productive per unit area (Barrett, Bellemare, and Hou 2010); may show enhanced stability and resilience (HLPE 2013; Holt-Gimenez 2002); generate more jobs and money within local economies (HLPE 2013; Lyson, Torres, and Welsh 2001); and harbour more agro-biodiversity and contribute to dietary diversity (HLPE 2013; Jarvis *et al.* 2008) – the latter being a key indicator of overall food security. (Graeub *et al.* 2016)

Small-scale farming has come to be accepted by some as a one-size-fits-all solution to problems in the countryside across much of the underdeveloped world. The support for and advocacy of small-scale farming come from (seemingly) antagonistic sources. While it is promoted by the World Bank, the United Nations and many of its constitutive organisations, and various other international and domestic NGOs, it also finds sympathisers amongst individuals and organisations, such as the popular Via Campesina, that oppose globalisation and the influence of transnational organisations.

The intellectual lineage of advocacy of small-scale farming goes back to the neo-populist theoretician Chayanov, who argued that a peasant household

---

[2] This section draws on an unpublished note prepared by the FAS team, led by Deepak Kumar, in 2015, entitled "Small-Scale Farming in Indian Agriculture."

carries out agricultural production for the purpose of consumption and employs only family labour. The objective of such production, he argued, was "labour–consumption" equilibrium – a balance between the utility of higher production and the disutility of greater drudgery by way of more intensive labour. Such household-based production was embedded in a larger theory of a "peasant economy," distinct from a capitalist economy. Even in conditions where capitalist farms go bankrupt, Chayanov argued, peasant families could continue production by working longer hours, selling at lower prices, and surviving without obtaining any net surplus. However, the concept of intensification of self-exploitation, which played a central role in Chayanov's work, hardly finds mention in the current advocacy for small-scale farming.

Chayanov's conception of a largely homogenous peasantry was in direct contradiction to the Marxian understanding of the social order in the countryside, which recognised the process of differentiation of the peasantry as an integral part of capitalist development. His ideas have been subjected to systematic criticism, most notably by Utsa Patnaik (1979). Chayanov's ideas have since been appropriated and altered substantially to feed two dominant trends of agrarian populism as identified by Henry Bernstein (2009): the technicist or neo-populist trend, and political populism. These roughly correspond to the positions of the international organisations and Via Campesina, respectively. Citing Gavin Kitching (1982), Henry Bernstein notes that:

> populist ideas are a response to the massive social upheavals that mark the development of capitalism in the modern world. Advocacy of intrinsic values and interest of the small producer . . . as emblematic of "the people" arises time and again as an ideology, and movement, of opposition to the changes wrought by the accumulation of capital. (Bernstein 2009, p. 68)

Terence Byres argues that neo-populism stands

> for a particular set of solutions, in the countryside, to the depredation visited upon labour, as capitalist transformation proceeds in circumstances of limited absorption of labour in manufacturing industry, and of a large and growing services sector and a continuing large informal sector . . . the problems of capitalist industrialisation are . . . conjured away by suggesting that an alternative lies in the countryside. (Byres 2004, p. 19)

Several strands of argument are invoked in support of this form of social organisation of production by political populists and technicist populists. We discuss them briefly below.

1. *Efficiency.* At the core of the argument in favour of small-scale farming in terms of its efficiency is the alleged inverse relationship between land productivity and size. It states that small farms are more efficient, defined in terms of yield per acre, than large farms. It is argued that this relationship holds true more or less universally. This assertion was also the basis of the debate in India on farm size and productivity based on findings from the Farm Management Studies. This argument, which continued through the 1960s, has seen a recent revival. Apart from the empirical challenge posed to this formulation (especially by the green revolution), it has also been theoretically rebutted by Terence Byres. He states that this argument posits

> the staying power, viability, and superiority of peasant agriculture *vis-à-vis* capitalist agriculture only by failing to acknowledge the existence and nature of capitalism. Methodologically, its crippling shortcoming is that it is a static approach in a dynamic context that does not and cannot capture relevant change and its contradiction. It is [an] . . . eminently well-intentioned but reactionary intervention inasmuch as it seeks to recreate a past that has never existed and institute an agrarian structure that contains the seeds of its own destruction. (Byres 2004, p. 41)

The body of empirical evidence from FAS surveys too does not support the hypothesis of an inverse relationship between farm size and output per unit of land.

2. *Social justice and equity.* The argument for social justice, while often well-intentioned, overlooks the conception of self-exploitation explicitly made by Chayanov. As we will see from the empirical material of the FAS surveys carried out for a decade now, most peasant farms and peasant households survive because they are based on intensive labour and considerable deprivation in respect of consumption, education, and health care. Peasant farms are overwhelmingly dependent on family labour, which is mostly a euphemism for intensive employment of the unpaid labour power of the women and children in a peasant household. This labour-intensive cultivation does not provide enough income (in cash or kind) to meet the consumption needs of the family. Incomes from crop cultivation can even be negative in many cases, as the FAS surveys show. The members of such households therefore often labour out on others' fields, and rely on other highly precarious forms of employment in the non-agricultural sector.

3. *Food security.* Supporters of small-scale farming claim that food production on small holdings ensures at least a basic minimum provisioning of food. Small-scale farming, then, is supported on the grounds that it provides food security in its four dimensions of availability, access, utilisation, and stability.

The evidence from FAS surveys, however, does not suggest that small farmer households are generally food-secure.

4. *Social solidarity and ecological sustainability.* Many advocates of small-scale farming claim that ecological sustainability and social solidarity are integral to the "peasant way" followed by the "people of the land." They argue that peasants (undifferentiated, in their understanding) follow ecologically sustainable cultivation practices and preserve bio-diversity. In their reading, small-scale farms offer a viable alternative to the devastating effects of capitalism on ecology. In addition, solidarity networks of mutual assistance are claimed to be integral to this way of life. These peasants are projected as an organised force against the onslaught of global neoliberal capitalism that seeks to endanger food sovereignty globally. This moral dimension of agrarian populism is a defence of a threatened and idealised way of life that involves anti-industrialism and anti-urbanism.

On the other hand, the FAS surveys suggest that, as against the assumption of social solidarity among an undifferentiated peasantry, an active process of differentiation of the peasantry is, and has been, the reality on the ground. Further, there is no evidence to show that small farmers follow cultivation practices that are significantly more ecologically sustainable than others. There is also little evidence of mutual assistance being a mainstay of peasant production.

While there may not be much ground for arguing that small-scale farming embodies the virtues claimed for it by its romantic advocates, it is important to recognise that any democratic agrarian policy/perspective should reckon with the fact that small farmers account for a substantial proportion of the rural/agrarian population, and require concrete policy support to stay viable in the context of the hostile assault on their livelihoods by neoliberal capitalist globalisation. The present study takes the view that just as it is necessary not to romanticise small-scale farming, it is equally important not to abandon the small farmer.

MARXIST CLASSICS ON THE SMALL FARMER UNDER CAPITALISM [3]

In this section, we briefly review the Marxist viewpoint on the development of capitalism in agriculture and its implications for small farmers, from which our perspective draws its inspiration.

In the Marxist understanding, a peasant under a capitalist mode of production faces a significantly different economic environment as compared to a peasant

---

[3] This section draws on Ramakumar (2016).

under a pre-capitalist mode of production. Capitalism revolutionises the forces of production and expands the "economic size" of the farm. As a result, the "isolated labour" of the peasant gets transformed into "social labour." In this context, the point made by Marx in *Capital*, Volume 3, is of relevance:

> Proprietorship of land parcels, by its very nature, excludes the development of social productive forces of labour, social forms of labour, social concentration of capital, large-scale cattle raising, and the progressive application of science. Usury and a taxation system must impoverish it everywhere. The expenditure of capital in the price of the land withdraws this capital from cultivation. An infinite fragmentation of means of production and isolation of the producers themselves.[4]

Reviewing Marx's views on cooperatives, Lenin made the following remarks in parentheses in his work, *Karl Marx*:

> (Co-operative societies, i.e., associations of small peasants, while playing an extremely progressive bourgeois role, only weaken this tendency, without eliminating it; nor must it be forgotten that these co-operative societies do much for the well-to-do peasants, and very little – next to nothing – for the mass of poor peasants; then the associations themselves become exploiters of hired labour.)[5]

It is important, however, to note that Marx's views on this issue continued to evolve and acquired much nuance later. Thus, Marx and Engels wrote in their discussion of the Polish question:

> The big agrarian countries between the Baltic and the Black seas can free themselves from patriarchal feudal barbarism only by an agrarian revolution, which turns the peasants who are serfs or liable to compulsory labour into free landowners, a revolution which would be similar to the French Revolution of 1789 in the countryside. (Engels 1848)

One can see here the germ of the idea of a "worker–peasant alliance" that emerged later in Marxist literature with regard to resolution of the agrarian question.

A dominant school of thought within Russia – the Narodniks – held the view that the *mir*, or the old Russian commune system, was essentially socialist in character and could be the platform on which Russian socialism could be built by means of a non-capitalist path, or by bypassing capitalism

---

[4] From Karl Marx, *Capital*, Volume 3, Chapter 47, available at https://www.marxists.org/archive/marx/works/1894-c3/ch47.htm

[5] Available at http://www.spartacist.org/english/wv/1076/marx.html

as an intervening mode of production. When the Narodniks put forward this formulation to Marx and Engels, they responded, cautiously at first but then rather firmly, in the negative. Marx and Engels were ready to appreciate that the *mir* could be a vehicle for social transformation of Russia without reliance on a capitalist path. However, there were two necessary conditions for this:

> The first condition necessary for this was an impulse from outside – a change in the economic system of Western Europe, destruction of the capitalist system in those countries where it had first arisen. (Engels 1893)

The second was a "popular revolution;" that is, as Trapeznikov (1981, p. 75) put it, "If popular revolution was victorious in Russia, the landlord-monarchical system abolished and private ownership of the instruments and means of production abolished . . . ." However, both conditions were not fulfilled for a prolonged period. In the intervening period, capitalist relations in Russia advanced rapidly. In such a circumstance of advancing forces of production, Engels wrote:

> As to the commune, it is only possible so long as the differences of wealth among its members are but trifling. As soon as these differences become great, as soon as some of its members become the debt-slaves of the richer members, it can no longer live. I am afraid that institution is doomed. But on the other hand, capitalism opens out new views and new hopes. Look at what it has done and is doing in the West. . . . There is no great historical evil without a compensating historical progress. (Engels 1893)

LENIN ON THE PEASANT QUESTION

Picking up the thread from Marx and Engels, Lenin fought a historic ideological battle against the Narodniks. To begin with, he recognised that the idea of equality embedded within Narodnik thought was indeed "progressive," and that its content was "historically real and historically legitimate." However,

> The mistake all the Narodniks make is that by confining themselves to the narrow outlook of the small husbandman, they fail to perceive the bourgeois nature of the social relations into which the peasant enters on coming out of the fetters of serfdom. They convert the "labour principle" of petty-bourgeois agriculture and "equalisation," which are their slogans for breaking up the feudal latifundia, into something absolute, self-sufficing, into something implying a special, non-bourgeois order. (Lenin 1907)

In other words, the Narodniks missed the fact that the economic principles under which peasant farming operates shift fundamentally under capitalism. With the advance of capitalism, the peasantry faced a qualitatively different material situation. Lenin also quoted Marx to underline his point.

One of the major results of the capitalist mode of production is that, on the one hand, it transforms agriculture from a mere empirical and mechanical self-perpetuating process employed by the least developed part of society into the conscious scientific application of agronomy, in so far as this is at all feasible under conditions of private property; that it divorces landed property from the relations of dominion and servitude, on the one hand, and, on the other, totally separates land as an instrument of production from landed property and landowner. The rationalising of agriculture, on the one hand, which makes it for the first time capable of operating on a social scale, and the reduction *ad absurdum* of property in land, on the other, are the great achievements of the capitalist mode of production. Like all of its other historical advances, it also attained these by first completely impoverishing the direct producers. (Marx, *Capital*, Volume 3, Chapter 37, 1894, p. 461, cited in Lenin 1914)

Lenin later wrote:

In agriculture, as in industry, capitalism transforms the process of production only at the price of the "martyrdom of the producer." Under capitalism, the small-holding system, which is the normal form of small-scale production, degenerates, collapses, and perishes. (Lenin 1914)

While this is a categorical statement, it is important to note that it describes a long-term tendency under capitalism and not an immediate outcome. The decline and demise of small peasant production under modern capitalism, while a continuing aspect of reality and inevitable in the historical long run, is also mediated and countered by numerous forces. Just as small industrialists do not cease to exist even under neoliberal capitalism, neither do small peasants, who are far more numerous, especially in agrarian countries devastated by centuries of colonial rule.

THE VIEWS OF KARL KAUTSKY

Karl Kautsky's classic work, *The Agrarian Question*, argued that large farms are superior to small farms with respect to development of the productive forces.

Although the individual large farm requires relatively less living and dead stock and less labour-power relative to its surface area with the same type of cultivation, it naturally always uses absolutely more than the individual small farm – meaning simply that the large farm can take much better advantage of the benefits of the division of labour than the small. Only large farms are able to undertake that adaptation and specialisation of tools and equipment for individual tasks which render the modern farm superior to the pre-capitalist. The same applies to breeds of animals. The dwarf-holder's cow is a dairy animal, a draught animal and breeding stock; there is no question of choosing a specific breed, or of adapting the stock and feed to specific requirements. Similarly, the dwarf-holder cannot delegate the various tasks on the farm to different individuals.

The ability of the large farm to practise this type of specialisation confers a number of advantages. The large-scale farmer can divide the work into those tasks requiring particular skill or care, and those merely involving the expenditure of energy. The first can be allotted to those workers who display particular intelligence or diligence, and who will be able to increase their skill and experience by concentrating completely, or mainly, on a particular task. As a result of the division of labour and the greater size of the farm, the individual worker will spend longer on each job, and will therefore be able to minimise the loss of time and effort associated with constant switching of tasks or workplaces. Finally, the large-scale farmer also has access to all the advantages of cooperation, of the planned collaboration of a large number of individuals with a common objective. (Kautsky [1899] 1988, p. 101)

At the same time, however, Kautsky also recognised that there were multiple reasons why small-scale farming may not be swept away by large-scale agriculture, at least in the short run. For instance, there may be important economic barriers to mechanisation becoming widespread even in large farms. Kautsky argued that if we assume that the major objective of mechanisation is to "save on wages, not labour-power," then, mechanisation in agriculture stood at a disadvantage vis-à-vis mechanisation in industry.

Whilst industry can use its machinery every day, most machines in agriculture are only required for a short period each year. Other things being equal, the labour-saving capacity of machinery is therefore much greater in industry. Given two machines, each of which can replace ten workers per day, of which one is used for ten days per year, and the other for 300, one will save 100 days labour a year, and the other 3,000. (*Ibid.*, p. 43)

As a result, "the lower the level of wages, the more difficult it is to introduce machines" (*ibid.*).

Kautsky further argued that there were many differences between agriculture and industry, and these differences may slow down the expansion of large farms in agriculture compared to large firms in industry. Therefore, a bigger farm is not necessarily better.

Under normal circumstances the large enterprise is always superior to the smaller in industry. Of course, even in industry every enterprise has its limits, beyond which it cannot go without risking profitability. The scale of the market, the size of the available capital, the amount of labour-power available, the supply of raw materials, and the limits of technology impose limits on every enterprise. However, within these limits, the larger enterprise is superior to the smaller.

In agriculture this is not wholly, or always, the case. In industry any expansion of the enterprise also represents an increasing concentration of productive forces, with all the advantages which this brings – savings in time, costs, materials, easier supervision and so on. By contrast in agriculture, other things being equal, any expansion of the enterprise means the same methods of cultivation being applied over a larger area – hence, an increase in material losses, increased outgoings on labour and means of production and greater delays associated with the transport of labour-power and materials. These are more significant in agriculture since most of its materials have a low ratio of value to volume – fertilizer, hay, straw, corn, potatoes – and methods are very primitive compared with industry. The larger the estate, the more difficult the supervision of individual workers which, under the wages system, is an important consideration. (*Ibid.*, p. 148)

Notwithstanding these nuanced arguments on why the small firm could survive for an extended period, Kautsky was firm that none of these barriers implied superiority of the small farm over the large farm. The process of agrarian transformation was dependent on multiple factors including the pace of technological development and the agro-ecological context.

The enormous advantages of the large farm more than outweigh the disadvantages of great distance – but only for a certain overall area. After a certain point, the advantages of the larger farm begin to be overtaken by the disadvantages of distance, and any further extension of the land area will reduce the profitability of the land.

It is impossible to specify in general when this point will be reached. It varies according to techniques, soil-type, and type of cultivation. A number of factors are currently moving the point upwards, such as the introduction of steam or electricity as motive power, or light railways. Others push in the opposite direction. The greater the number of animals and workers per given area of land, the greater the number of loads which have to be moved – machines and heavy implements, fertilizers, the harvest itself – the more noticeable the effects of longer distances. In general, the maximum size of a farm beyond which its profitability declines will be less the more intensive the type of cultivation, the more capital is invested in the soil: nevertheless, developments in technology can mean that this law is broken through from time to time. (*Ibid.*)

An important point that both Lenin and Kautsky highlight, and which is of great contemporary relevance, is related to the self-exploitation of the peasant in the small farm. This feature of small-scale farming is indeed the key to its prolonged survival despite the onslaught of capitalism. Quoting John Stuart Mill's discussion on the "almost incredible toil" of small peasants, Kautsky wrote:

> Small farms have two major weapons to set against the large. Firstly, the greater industriousness and care of their cultivators, who in contrast to wage-labourers work for themselves. And secondly, the frugality of the small independent peasant, greater even than that of the agricultural labourer.
>
> The small peasants not only flog themselves into this drudgery: their families are not spared either. Since the running of the household and the farm are intimately linked together in agriculture, children – the most submissive of all labour – are always at hand! And as in domestic-industry, the work of children on their own family's small peasant holding is more pernicious than child wage-labour for outsiders. (*Ibid.*, p. 110)

But this is a losing battle with the most reactionary and ruinous consequences for the peasant's family, especially women and children.

> The more agriculture becomes a science, and hence the more acute the competition between rational and small-peasant traditional agriculture, the more the small farm is forced to step up its exploitation of children, and undermine any education which the children might acquire.
>
> It takes a very obdurate admirer of small-scale land-ownership to see the advantages derived from forcing small cultivators down to the level of beasts of burden, into a life occupied by nothing other than work – apart from time set aside for sleeping and eating.
>
> Competing through lengthening working time always goes hand in hand with technical backwardness. The latter generates the former – and vice versa. An enterprise which cannot fight off the competition through technical innovation is forced to resort to the imposition of even greater demands on its workers. Conversely, an enterprise in which the workers can be pushed to their limits is much less exposed to the need for technical improvements than one in which workers place limits on their exertions. The possibility of prolonging working-time is a very effective obstacle to technical progress.

The greater care taken by peasants in their work is less ruinous for them than their drudgery and excessive frugality. Care plays a major role in agricultural production – greater than in industry for example. And workers working for themselves will clearly exercise more care than wage-labourers. Whilst this might not necessarily be an advantage in all types of large enterprise, it certainly is as far as large-scale capitalist farming is concerned. This should not be overstated, however. The other weapons in the small farm's arsenal – overwork, undernourishment, and accompanying ignorance – offset the effects of greater care. The longer the worker has to work, the lower the standard of diet, and the less time available for education, the less care ultimately exercised in work. And what is the point of taking great care if there is no time to clean the stall and livestock, if the draught animals – often simply a dairy cow – are just as overworked and underfed themselves.

The overwork of the small independent farmers and their families is not therefore a factor which should be numbered amongst the advantages of the small farm even from a purely economic standpoint, leaving aside any ethical or other considerations. (*Ibid.*, p. 112)

Lenin agreed with Kautsky in his review of Kautsky's book that:

The fundamental and main trend of capitalism is the elimination of small production by large-scale production both in industry and in agriculture. But this process must not be taken only in the sense of immediate expropriation. This elimination process also includes a process of ruination, of deterioration of the conditions of farming of the small farmers which may extend over years and decades. This deterioration manifests itself in overwork or underfeeding of the small farmer; in an increased burden of debt; in the deterioration of cattle fodder and the condition of the cattle in general; in the deterioration of the methods of cultivating and manuring the land; in the stagnation of technical progress, etc. (Lenin 1972, p. 248)

Small production in agriculture is doomed to extinction and to an incredibly crushed, oppressed position under capitalism. Being dependent on big capital and being backward compared with large-scale production in agriculture, small production can hold on only because of the desperately reduced consumption and laborious, arduous toil. The dispersion and waste of human labour, the worst forms of dependence of the producer, exhaustion of the strength of the peasant family, of peasant cattle and peasant land – this is what capitalism brings to the peasant everywhere. (*Ibid.*, p. 288)

"Small-scale production is compatible only with a narrow and primitive framework of production and society" (Lenin 1914).

On balance it is clear that, in the classical Marxist literature, the way forward – the democratic mode of advance – is for the working class to fight and overthrow capitalism in alliance with the peasantry, in countries where such an opportunity presented itself. It was to be the task of the working class-led democratic movement to defend the rights of the peasantry without succumbing to a reactionary outlook. In particular, democratic movements should seek to provide the small peasant the power of scale while ensuring that issues of material and cultural deprivation of peasant households (graphically described by Kautsky and Lenin) were addressed through collective action. It is noteworthy that, even while criticising the Narodniks for their romantic notions of peasant homogeneity, Lenin also put forward the concept of a worker–peasant alliance in the democratic revolution. In the course of the transition of the Russian revolution from February 1917 to October 1917, he put forward the slogan of distribution of seized land among the peasantry rather than its nationalisation, recognising the democratic character of the small peasantry in a historical context in which state power was with the working people. The recognition that petty peasant production must not be romanticised is not inconsistent with fighting for pro-small farmer public policies in specific circumstances.

THE MARXIST VIEW OF THE PEASANTRY IN THE INDIAN CONTEXT

The attitude towards the peasantry in the context of India's development, especially after the country won political independence, has been a matter of much discussion in Indian Marxist literature. Comprehensive land reform is essential to the completion of the democratic revolution in India. Achievement of the democratic revolution under a working class leadership in alliance with the peasantry, especially poor peasants and agricultural labourers as key rural classes in this process, is necessary for further democratic advance. Such a view envisages the continued presence of a large population of small and middle peasants for a long time. We need public policy that supports the peasantry, especially focusing on developing the productive forces among them.

This viewpoint is quite different from one that argues for support to small farmers on grounds of "efficiency," based on an alleged inverse relationship between farm size and productivity. In the Indian debate on the relationship between farm size and productivity, there is no consensus at all that small farms are more "efficient" than large farms. With advances in the productive forces over the years, especially in the decades of the "green revolution," the "inverse relationship" hypothesis has not had many takers.

From the early years of the new agricultural strategy, there has been official recognition of the preponderance of the small peasantry in India. Many policy measures address, in their formulation, the specific problems of "small" and "marginal" farmers. The establishment of the Small Farmers Development Agency (SDFA) and the Scheme for Marginal Farmers and Agricultural Labourers (MFAL) in 1971 was a case in point. Partly intended to quell what was viewed as rising peasant militancy in the late 1960s and early 1970s, public policies associated with the "green revolution," despite their overall orientation of "betting on the strong" in a context of entrenched land monopoly, also addressed – at least in terms of policy statements – the issues of small farmers. This was especially true of the long phase of social and development banking from the 1970s through the 1980s. The beginning of the 1990s, with the acceleration of neoliberal reforms, announced the arrival of an altogether different regime.

POLICY DEFINITIONS OF SMALL FARMERS IN INDIA

Small farms and small-scale farming are distinct concepts. The Marxist understanding of scale and of differentiation of cultivators is not based on the size of holdings and their physical extent alone. As V. K. Ramachandran has argued,

> a single size category of landholding may conceal considerable variations in the physical characteristics of land – variations, for instance, in the irrigation and drainage facilities available to the land, the type of soil and its fertility, land utilisation and cropping pattern and so on. (Ramachandran 1980, p. 2)

The analysis in this volume does not include farms that are owned or operated by corporations, cooperatives or other such organisations; it is confined to small farms that are also family farms. When 2014 was declared the International Year of Family Farming, the Food and Agriculture Organisation (FAO) of the United Nations defined family farms as a means of organising agricultural, forestry, fisheries, pastoral, and aquaculture production that is managed and operated by a family, and predominantly reliant on family labour including both women's and men's labour. The family and the farm are linked, co-evolve, and combine economic, environmental, social, and cultural functions (FAO 2013, p. 2).

Estimates suggest that family farms account for 98 per cent of all farms worldwide, 53 per cent of world food production, and 53 per cent of agricultural land (Graeub *et al.* 2016). In India and other South Asian countries, corporate farming is limited, and so family farming is the norm or near-universal. The other distinctive organisational form in South Asian agriculture is the plantation. Although some plantations may be family-owned, in general they grow crops for commercial purposes and by means of hired labour. Plantations are excluded from further discussion in this chapter as well as this volume.

Multiple criteria are needed to define small farms if account is to be taken of the type and scale of farming. However, a size-based criterion is often seen as practical. Thus, many national governments define small holders or small farmers in official documents in terms of size of landholding, though the precise measure or cut-off varies widely. The Agricultural Census and the Ministry of Agriculture, Government of India, use the following five-fold classification:

• Marginal, or below 1 hectare;
• Small, or between 1 and 2 hectares;
• Semi-medium, or between 2 and 4 hectares;
• Medium, or between 4 and 10 hectares; and
• Large, or above 10 hectares.

In India, Ministry of Agriculture and government policies/schemes related to farmers (such as the policy on crop insurance) identify small farmers as those operating less than or equal to 2 hectares (5 acres), a definition based solely on the extent of operational holding.

There are, however, two other definitions that are based on the extent of landholding.

> The Government of India has taken 10 hectares as the ceiling to define low income or resource-poor farmers for purposes of commitments to the World Trade Organisation (WTO). In India's Supporting Table Relating to Commitments on Agricultural Products in Part IV of the land Uruguay Round Schedule (WTO document G/AG/AGST/IND), landholders with less than 10 hectares are taken as low income or resource poor. (Sharma 2012)

This is in conformity with the bimodal definition of the High Level Panel of Experts of the FAO (HLPE 2013). The bimodal approach divides land distribution in the Agricultural Census of India into two different groups, small and large, where small farms indicate a size of less than 10 hectares.

Secondly, there is a set of policy definitions embedded in the land reform laws of different States of India. Land reform legislation specifies a ceiling in terms of the extent of land. Land here is usually demarcated by quality on the basis of irrigation or other characteristics – such as land on which plantation crops are grown – or on the basis of expected value of production from land or the revenue assessment on the land. In particular, the idea of a "standard acre" has been used in the context of land reform legislation. The norms vary across States. In Odisha, a standard acre refers to 1 acre of perennially irrigated land which is assured of water supply for at least three crops in a year, or 4 acres of dry land. In Tripura, a standard acre varies from 1 acre of lowland (*nal* or *lunga*) to 3 acres of upland (*tilla*).

Variations in ceilings across the country illustrate the diversity of farming systems. Land reform legislation is concerned with defining the upper limit of the extent of land that can be held by an individual or family, an issue that is distinct from defining a small farm.

To sum up, the policy definition of small farms in India is in general based on the extent of landholdings, with some provisions for incorporating criteria reflecting the quality of land (in terms of irrigation) but excluding other criteria that characterise the economic size of the farm as a unit.

DEFINITION OF SMALL FARMS IN INDIAN STUDIES

In the scholarly literature in India, while the extent of landholding has been used mostly to define small farms, further distinctions have been made based on the quality of land, or agro-ecological and physical features that can be used to characterise quality (Vyas *et al.* 1969). Scholars undertaking field studies have introduced variations in the extent of land to accommodate differences in quality of land. The most frequently made distinction is as between dry land and wet land, that is, on the basis of availability of irrigation, on the assumption that returns from crop cultivation are distinctly higher on irrigated than unirrigated land. Some studies, particularly of south India, use the categorisation of dry and wet villages.[6]

The PARI studies, based on detailed village surveys conducted by FAS, differentiate among cultivator households on the basis of socio-economic class: a category that takes account of ownership of the means of production, forms of labour employed, and household incomes.

WORKING DEFINITION FOR THIS BOOK

There can be no unique or strict definition of what constitutes a small farm or smallholder agriculture. In this volume, where data from a wide variety of villages are analysed, for purposes of simplicity and comparability, the starting

---

[6] See, for instance, Athreya, Djurfeldt, and Lindberg (1990), and Yanagisawa (2008).

definition of small farmers is based on the extent of operational landholding, with allowance for differences in irrigation.

First, in each village, households primarily dependent on non-agricultural incomes (even if they owned and operated cropland) were excluded. Among the remaining households, that is, those primarily dependent on crop cultivation and/or allied activities, we used a two-fold categorisation: small farms and large farms. Small farmer households operated up to 5 acres (2 hectares) of irrigated land or 15 acres (6 hectares) of unirrigated land. (In other words, we have assumed that 1 acre of irrigated land = 3 acres of unirrigated land.) All other cultivator households were termed "large" farmers. Landlords and capitalist farmers constitute a separate category.

Each author was requested to use this as a working definition for purposes of their study, but asked to evaluate it in the course of analysis. To put it differently, one of the research questions taken up in this volume is whether, and to what degree, categorisation by extent of landholding is correlated with other variables of interest such as yield and income.

Neoliberal Reforms and Agrarian Distress

The PARI surveys of the FAS have been carried out from 2006 onwards. A new policy regime emerged in India in 1991. While this new policy regime had continuities with the past, it had some altogether distinctive features (Ramachandran and Swaminathan 2000):

- reversal of land reform and acceleration, through legislation, of the takeover of agricultural land;
- changes in the policies of administered agricultural input costs and output prices, and sustained reduction in input subsidies;
- cutting back of public investment in rural physical and social infrastructure;
- moving towards the privatisation of public facilities for marketing and storing agricultural products;
- severe weakening of the institutional structure of social and development banking;
- lowering of barriers on trade in agricultural commodities and removal of quantitative restrictions on the import of agricultural products, resulting in considerable price volatility for agricultural outputs;
- weakening of the public infrastructure for storage and marketing;
- cutting back the public distribution system; and
- undermining national systems of research, and mechanisms for the protection of national plant and other biological wealth.

Relatively unfettered entry and exit of capital as finance has been a key feature of the country's policy regime since 1991. There has been a consequent emphasis on reducing the fiscal deficit (legislated as the Fiscal Responsibility and Budget Management Act in 2004), almost entirely through expenditure reduction. This policy compulsion has meant sustained attacks on state support to agriculture and the peasantry. It has led to unprecedented agrarian distress in large parts of the country, especially in the period from 1997 to 2003. While there has been some recovery in agriculture since then, the economic viability of small farms has been seriously challenged across the country. This context needs to be kept in mind in the ensuing discussions. The period over which the PARI surveys were carried out – 2006 to 2016 – witnessed some recovery in agriculture, but of an uncertain and non-uniform kind across time, space, crops, and classes.

How do the PARI data speak to issues concerning small-scale farming?[7] An important gap in the literature on small-scale agriculture is a critical analysis, based on empirical evidence, of many of the accepted theoretical assumptions about and the proclaimed benefits of small-scale farming. The present volume addresses some of these issues using the empirical material collected by the Foundation for Agrarian Studies from 2006.

The objectives of PARI are:

- To analyse village-level production, production systems and livelihoods, and the socio-economic characteristics of different strata of the rural population.
- To conduct specific studies of sectional deprivation in rural India, particularly with regard to Scheduled Caste and Scheduled Tribe populations, women, specific minorities, and the income-poor.
- To report on the state of basic village amenities and the access of rural people to the facilities of modern life.

PARI household data come from villages located in the following States: Andhra Pradesh, Bihar, Karnataka, Madhya Pradesh, Maharashtra, Punjab, Rajasthan, Telangana, Tripura, Uttar Pradesh, and West Bengal. In each of these States, the PARI team surveyed two or three villages in different agro-ecological regions. The villages cover a wide range of agro-ecological regions in the country. State-level mass organisations suggested the regions and the districts they would like to have studied, and helped FAS in the final selection of villages from a shortlist prepared by the Foundation.

---

[7] This section draws substantially on the description provided regarding PARI on the FAS website: http://fas.org.in/category/research/project-on-agrarian-relations-in-india-pari/

The PARI team generally conducts a census-type survey that covers every household and individual in each selected village; in five of the 22 villages surveyed so far, the team conducted sample surveys after initial house-listing surveys. A village-level questionnaire is also canvassed in each village. In addition, a village profile, based on existing sources of secondary data, is constructed.

The information gathered through the questionnaire covers the following (each of these items of information is further disaggregated in the questionnaire):

- Demographic data, including data on caste and religion
- Education levels
- Occupation and work status
- Ownership holdings and operational holdings of households
- Land sales and purchases
- Forms and terms of land tenure
- Cropping pattern and crop production
- Animal resources
- Costs of cultivation
- Ownership of assets
- Participation in selected government schemes
- Household electricity, sanitation, and water facilities
- Housing
- Incomes and earnings
- Patterns and levels of employment
- Forms of labour
- Indebtedness.

It is noteworthy that there are no official sources of serial data on household incomes in rural India. The National Sample Survey (NSS) provides regular data on monthly per capita household expenditure, and the Comprehensive Scheme for the Study of Cost of Cultivation of Principal Crops in India (CCPC) provides regular data on farm business incomes for selected crops. The PARI village data have information from all sources of tangible household income, under the following heads:

- Income from crop production
- Income from animal resources
- Income from agricultural and non-agricultural wage labour
- Income from salaries
- Income from business and trade, rent, interest earnings, pensions, remittances, scholarships, and all other sources.

Data on crop production and cost of cultivation, while based on the CCPC methodology, are somewhat more detailed in the PARI database, which includes household-wise data on the following variables:

- Value of hired human labour
- Value of hired bullock labour
- Value of owned bullock labour
- Value of owned machinery
- Value of hired machinery
- Value of seed, home-produced and purchased
- Value of insecticides and pesticides
- Value of manure, home-produced and purchased
- Value of fertilizers
- Irrigation charges
- Land revenue
- Marketing costs
- Miscellaneous expenses
- Rent paid for leased-in land
- Interest on working capital
- Depreciation of implements and machinery.

Both the level of detail and the quality of data collection make PARI data an exceptionally valuable base for investigating many issues concerning the agrarian economy, and rural socio-economic relations, structures, and processes. Specifically for the purposes of this volume, which focuses on small farms and small farmers in India, the detailed, itemised PARI data listed above can be used to examine questions concerning the socio-economic characteristics of small farms, and the difference in this regard between small and large farms. We can thus examine such issues as the productivity of small farms as against large farms; the economic viability of small farming; the multiple sources of household income for small farmers and their respective relative contributions; the educational characteristics of small farmer households; child labour; patterns of input use and the implications thereof for environmental sustainability; the extent of labour performed by men and women from small farmer households on their own holdings and elsewhere; the multiple modes of exploitation to which persons belonging to small farmer households are subjected; and so on.

The volume also makes suggestions for state policy that can enable small farmers to enhance their economic viability and lessen the extent of their deprivation with respect to specified parameters. Without going into the analyses and conclusions that emerge in the subsequent chapters, it can be stated here that they do not support the claims of efficiency, equity, food security, social solidarity, and environmental sustainability claimed for small farms by advocates of small-scale farming. Our analysis also brings out the need for stronger state support to enable small farmers to meet the challenges they now face.

This chapter draws from three background notes: Kumar (2016), Ramakumar (2016), and Sarkar (2016), prepared as part of the project on Small-Scale Farming in Indian Agriculture. We are grateful to V. K. Ramachandran for comments and suggestions, and to Aparajita Bakshi for a note on ceilings under land reform legislation.

REFERENCES

Athreya, V. B., Djurfeldt, G., and Lindberg, S. (1990), *Barriers Broken*, Sage Publications, New Delhi.

Barrett, Christopher B., Bellemare, Marc F., and Hou, Janet Y. (2010), "Reconsidering Conventional Explanations of the Inverse Productivity-Size Relationship," *World Development*, vol. 38, no. 1, January, pp. 88–97.

Bernstein, H. (2009), "V.I. Lenin and A.V. Chayanov: Looking Back, Looking Forward," *The Journal of Peasant Studies*, vol. 36, no. 1, pp. 55–81.

Byres, Terence J. (2004), "Neo-Classical Neo-Populism 25 Years On: *Déjà Vu* and *Déjà Passé*. Towards a Critique," *Journal of Agrarian Change*, vol. 4, pp. 17–44; available at doi:10.1111/j.1471-0366.2004.00071.x, viewed on 11 January 2017.

Chappell, M. J., Wittman, H. K., Bacon, C. M., Ferguson, B. G., García Barrios, L. E., and García Barrios, R. (2013), "Food Sovereignty for Poverty Reduction and Biodiversity Conservation in Latin America," available at http://dx.doi.org/10.12688/f1000research.2-235.v1, viewed on 11 January 2017.

Engels, Friedrich (1848), "The Frankfurt Assembly Debates: The Polish Question," *Neue Rheinische Zeitung*, no. 81, August 20; available at https://www.marxists.org/archive/marx/works/1848/08/09.htm, viewed on 22 May 2017.

Engels, Friedrich (1893), "Letter to Nikolai Danielson," October 17, 1888, in *Marx and Engels: Collected Works*, Volume 48, pp. 228–30; available at https://www.marxists.org/archive/marx/works/1893/letters/93_10_17.htm, viewed on 22 May 2017.

Food and Agriculture Organisation (FAO) (2013), "Master Plan," available at http://www.fao.org/fileadmin/user_upload/iyff/docs/Final_Master_Plan_IYFF_2014_30-05.pdf, viewed on 11 January 2017.

Graeub, Benjamin E., Chappell, M. Jahi, Wittman, Hannah, Ledermann, Samuel, Kerr, Rachel Bezner, and Gemmill-Herren, Barbara (2016), "The State of Family Farms in the World," *World Development*, vol. 87, November, pp. 1–15; available at http://dx.doi.org/10.1016/j.worlddev.2015.05.012, viewed on 11 January 2017.

Haque, T. (2016), "Sustainability of Small Farms in Asia – Pacific Countries: Challenges and Opportunities in Family Farming," Academic Foundation, New Delhi.

High Level Panel of Experts (HLPE) (2013), *Investing in Smallholder Agriculture for Food Security: A Report by the High Level Panel of Experts on Food Security and Nutrition of* the Committee on World Food Security, Rome, available at http://www.deza.admin.ch/ressources/resource_en_225682.pdf, viewed on 11 January 2017.

Holt-Gimenez, Eric (2002), "Measuring Farmers' Agro Ecological Resistance after Hurricane Mitch in Nicaragua: A Case Study in Participatory, Sustainable Land Management Impact Monitoring," *Agriculture, Ecosystems and Environment*, vol. 93, pp. 87–105; available at doi: 10.1016/S0167-8809(02)00006-3, viewed on 11 January 2016.

Jarvis, D. I. *et al.* (2008), "A Global Perspective of the Richness and Evenness of Traditional Crop-Variety Diversity Maintained by Farming Communities," *Proceedings of the National Academy of Sciences of the USA*, vol. 105, pp. 2101–03.

Kautsky, Karl ([1899] 1988), *The Agrarian Question*, translated from the German by Pete Burgess, Zwan Publications, London and Winchester.

Kitching, Gavin (1982), *Development and Underdevelopment in Historical Perspective: Populism, Nationalism and Industrialisation*, Routledge, New York.

Kumar, Deepak (2016), "Note on Small-Scale Farming in Contemporary Literature," unpublished note.

Lenin, V. I. ([1907] 1972), *Collected Works*, vol. 13, fourth English edition, Foreign Languages Publishing House, Moscow; available at http://www.marx2mao.com/Lenin/APFR07.html, viewed on 22 May 2017.

Lenin, V. I. (1914), *Karl Marx: A Brief Biographical Sketch and an Exposition of His Doctrine*, available at https://creandopueblo.files.wordpress.com/2011/09/lenin-karlmarx.pdf, viewed on 22 May 2017.

Lyson, T. A., Torres, Robert, and Welsh, Rick (2001), "Scale of Agricultural Production, Civic Engagement and Community Welfare," *Social Force*, vol. 80, pp. 311–27.

Nagayets, O. (2005), "Small Farms: Current Status and Key Trends," available at https://pdfs.semanticscholar.org/921f/bff5587afbb98550137bed8156771c9b2b88.pdf, viewed on 30 April 2017.

Patnaik, Utsa (1979), "Neo Populism and Marxism: The Chayanovian View of the Agrarian Question and Its Fundamental Fallacy," *The Journal of Peasant Studies*, vol. 6, issue 4, pp. 375–420.

Ramachandran, V. K. (1980), "A Note on the Sources of Official Data on Landholdings in Tamil Nadu," Data Series No. 1, Madras Institute of Development Studies, Chennai.

Ramachandran, V. K., and Swaminathan, Madhura, eds. (2002), *Agrarian Studies: Essays on Agrarian Relations in Less-Developed Countries*, Tulika Books, New Delhi.

Ramakumar, R. (2016), "Notes on Small Farmers and Small Farming: A Review of the Classical Marxist Literature," unpublished note.

Sandhu, Harpinder, Wratten, Steve, Costanza, Robert, Pretty, Jules, Porter, John R., and Reganold, John (2015), "Significance and Value of Non-Traded Ecosystem Services on Farmland," February 17, available at https://doi.org/10.7717/peerj.762, viewed on 11 January 2017.

Sarkar, Biplab (2015), "Productivity and Income Levels of Small Holders in India," background paper for the report prepared by M. S. Swaminathan Research Foundation (MSSRF) for assisting the World Food Programme (WFP) in achieving the Zero Hunger Challenge.

Sarkar, Biplab (2016), "Levels and Variations in Crop Income in a Small Farmer Economy: A Study of Three Villages of West Bengal," unpublished note.

Sen, A. (1962), "An Aspect of Indian Agriculture," *The Economic Weekly*, Annual Number, February.

Sharma, S. K. (2012), "Assessing the Extent of Low Income or Resource Poor Farmers in India (with Special Reference to Article 6.2 of AoA)," Centre for WTO Studies, June.

Trapeznikov, S. P. (1981), *Leninism and the Agrarian and Peasant Question*, vol. 1, Progress Publishers, Moscow.

Yanagisawa, H. (1996), *A Century of Change: Caste and Irrigated Lands in Tamilnadu 1860s–1970s*, Manohar Publishers and Distributors, New Delhi.

2

PARI Villages: An Introduction

T. Sivamurugan and Madhura Swaminathan

The Project on Agrarian Relations in India (PARI) began in 2006. One of the objectives of the Project is to analyse village-level production, production systems and livelihoods, and the socio-economic characteristics of different strata of the rural population. As of 2016, 25 villages from 11 States of the country, covering a wide range of agro-ecological regions in the country, have been studied under PARI. In this volume, we have used data on 17 villages from 9 States.

This chapter begins with a discussion of the methodology for classification of peasants and small farmers, followed by a description of the study villages in terms of location and selected socio-economic features. Each detailed village profile is prefaced by a thumbnail sketch of that village for ready reference. In the concluding section of the chapter, we provide some preliminary observations on the position of small farmers in the selected villages.

DEFINING RURAL CLASSES[1]

An important premise of the PARI surveys is that the rural population of India is differentiated, and the village provides a site from where to study the process of class differentiation in the Indian countryside. The surveys undertake a socio-economic classification of households in each village based on three classical criteria used to differentiate the peasantry, namely: control over the means of production; relative use of family and hired labour; and the surplus a household is able to generate within a working year. Based on these general criteria, households in the study villages are broadly categorised into the following five classes: landlords; capitalist farmers; peasants; manual workers; and households dependent on business, salaries or other sources of income. Within each village, the peasantry is further subdivided based on the specific conditions of the village.

---

[1] This entire section is an extract from Ramachandran (2011).

Landlords

Landlord households own a large part of the land in most villages. There is total absence of direct participation by these households in agricultural operations. Cultivation is conducted by hired labour or by tenants to whom land is leased out by the landlord households. This class traditionally controlled all aspects of social, economic, and political life in most villages.

Capitalist Farmers

Capitalist farmer households are similar to landlord households in terms of their land and asset ownership, and non-participation in agricultural operations. However, they are different from the former in that traditionally they did not belong to the class of landlords. These households invested the surplus they gained from non-agricultural sources in land. Agriculture was not their primary source of economic power. Many of them previously belonged to the class of rich peasants or upper middle class peasants, and often to the dominant caste group.

Manual Workers

At the bottom of the ladder of rural classes lie the manual workers. These households are characterised by their predominant dependence on wage incomes both in agricultural and non-agricultural work. This category includes both agricultural and non-agricultural workers due to the increasing difficulty in separating these two groups from the pool of rural manual workers. Farm servants, engaged in long-term work with a single employer, also belong to this class. Manual workers may have diverse sources of income, such as animal husbandry, domestic work and low-paid jobs in the private sector.

Peasants

The class of peasants lies between the classes of landlords and capitalist farmers on the one hand, and manual workers on the other. A fundamental characteristic of peasant households is their participation in all or some agricultural operations on the land. This class is itself differentiated, ranging from rich peasants to upper middle, lower middle and poor peasants. The broad criteria used to further categorise peasant households are: the extent of ownership of means of production (specifically land); the ratio of family labour and days of labouring out by members of the household (numerator), to the number of days of labour hired in (denominator); and net incomes. The exact criterion may vary from village to village, as specific cropping patterns, labour use patterns, caste configurations, and other socio-economic characteristics are kept in view while segregating the various classes of peasant households within a village.

Renters/Recipients of Remittances

These are households for whom the major sources of income are rents from agricultural/commercial land or remittances.

Business Households/Salaried Households

Business households are households for whom the major source of income is business activity other than crop production. Salaried households are households whose major source of income is salaries.

WHO IS A SMALL FARMER?

For this study, we have identified and defined "small farmers" as a category that is part of the peasantry, but whose position is distinct and therefore not interchangeable with that of the peasantry. In specific terms, a small farmer by our definition is a peasant with an operational holding of less than 2 hectares of irrigated land or 6 hectares of unirrigated land, or any combination thereof. In short, we use the extent of operational holding to demarcate two groups of farmers within the peasantry, so as to have some commonality with policy definitions in India and elsewhere that are usually based on extent of operational holdings.

The official definition of small and marginal farmers (all those with holdings less than 2 hectares) does not differentiate between irrigated and unirrigated land. Our categorisation does precisely that. We identify the type of land based on data on irrigation derived from the PARI field survey (which includes all sources of irrigation), and not based on revenue records that may be quite old and misleading. Our survey experience suggests that cultivators with 2 hectares of rainfed cropland are in no way similar to those with 2 hectares of irrigated land. Our assumption that 1 hectare of irrigated land is equivalent to 3 hectares of unirrigated land may not apply to every agro-ecological region of the country, but we make this assumption as a starting point for further study.

It should be noted that manual worker households, and households dependent primarily on incomes from business or salaries or remittances, whether or not they have operational holdings, are excluded from the analysis. Landlords and capitalist farmers, invariably with holdings of more than 2 hectares of irrigated land or its equivalent, are, however, included in our discussion as a separate category, in order to bring out the intra-village differences and inequalities among farming households.

VILLAGE PROFILES

We now turn to a description of the survey villages, which includes location, access to basic infrastructure, population size and composition, and major characteristics of agriculture. The villages are discussed in chronological order of the surveys, beginning with the three villages of Andhra Pradesh surveyed in 2005 and 2006, and ending with Tehang in Punjab surveyed in 2011. We also describe the basic socioeconomic classes in each village. The village profiles draw on earlier work including Ramachandran, Rawal, and Swaminathan (2001), Swaminathan and Rawal (2015), Swaminathan and Das (2017), and the PARI section of the Foundation for Agrarian Studies (FAS) website.

Appendix Tables 1 to 6 in this chapter give information on the location of each village, the agro-ecological zone to which it belongs, the cropping pattern, the number of households surveyed and the basic demographic structure of the village including caste composition, and the socio-economic categorisation of households. For most of the analysis, the broad caste groups used are Scheduled Castes (SCs), Scheduled Tribes (STs), Other Backward Classes (OBC), and Other Caste Hindus. Households belonging to other religions (Sikhs, Jains, Muslims, etc.) are grouped separately. Appendix Tables 7 to 9 give some data on small farmers in each village, in terms of number, land owned and land operated.

While each village is unique and studying a village gives us insights into specific processes, there is also a broad typology of villages that may help in the analysis that follows. One clear distinction is between rainfed and irrigated villages.

Bukkacherla in Anantapur district of Andhra Paradesh, Warwat Khanderao in Buldhana district of the Vidarbha region in Maharashtra, Zhapur in Kalaburagi district of Karnataka, and Rewasi in Sikar district of Rajasthan were all rainfed or dry villages. The share of irrigated area in total gross cropped area (GCA) was less than 10 per cent in Bukkacherla, Warwat Khanderao, and Zhapur. In Rewasi, this share was around 50 per cent, but the irrigation was from tubewells and depended on the monsoon. The survey of this village was conducted in a year of very poor rain and consequent crop failure, when the district was declared to be drought-affected.

The canal-irrigated villages (with surface and groundwater irrigation) were Ananthavaram in Guntur district of Andhra Pradesh, Nimshirgaon in Kolhapur district of Maharashtra, Alabujanahalli in Mandya district of Karnataka, and Tehang in Phillaur district of Punjab. Tank and groundwater irrigation were important in Kothapalle (Karimnagar district, Telangana), and in Siresandra (Kolar district, Karnataka) groundwater irrigation was important. The three

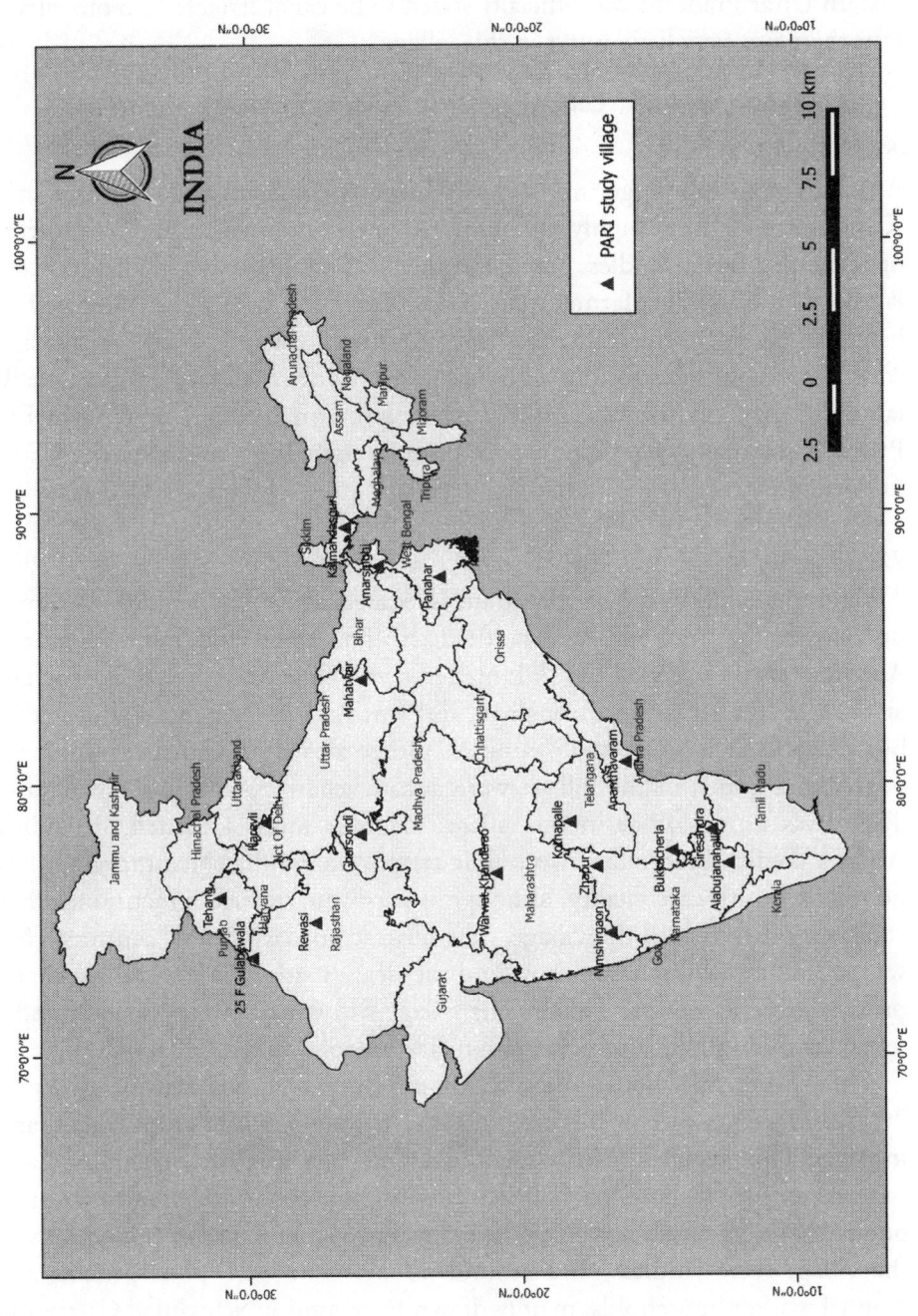

villages of West Bengal – Panahar, Amarsinghi, and Kalmandasguri – had a mix of surface and groundwater irrigation. While Mahatwar (Ballia district, eastern Uttar Pradesh) was officially stated to be canal-irrigated, in our survey year there was very little water in this village.

Ananthavaram, Guntur District, Andhra Pradesh, 2005–06

Brief sketch

Ananthavaram is a large, multi-caste village with canal irrigation from the Krishna river. It has a highly differentiated population with a high proportion of Scheduled Caste, landless manual worker households, and a small number of landlord and capitalist farmer households. The latter have agriculture as their base but have diversified incomes. Small farmers, owning 1 acre on average, constitute about one-third of all households. Many small farmers are actually landless tenant cultivators, subject to exploitation by high rent payments. Paddy is the main *kharif* (monsoon) crop in the village, and dairying is also important.

Detailed profile

The revenue village of Ananthavaram is located in Kollur *mandal* of Guntur district, in the State of Andhra Pradesh. The town and railhead nearest Ananthavaram is Tenali, 17 kilometres away. The village had a concrete, all-weather road passing through it, and was on a regular bus route with a bus available at least every 45 minutes. Other means of regular transport for people and goods in the village were autorickshaws, jeeps, and small vans. There was a post office in the village, a ration shop, a branch of Andhra Bank, a medical store, and two public telephone booths. Ananthavaram had no public health care facility, although two private medical practitioners had consultation rooms in the village. The nearest primary health centre (PHC) was at Kollur, 8 kilometres away, and the nearest sub-divisional hospital and private hospitals were at Tenali. The village had three primary schools and a secondary school, but no higher secondary school.

There were 667 households in Ananthavaram at our census survey of 2005, and we sampled 150 households for further details of employment and incomes. The average household size, at 3.6 persons, was low. Scheduled Caste (43 per cent) and Scheduled Tribe (6 per cent) households together constituted one-half of all households. A good majority (65 per cent) of households in the village were landless, and around 45 per cent were poor peasant and manual worker households, mainly drawn from among Scheduled Castes and Scheduled Tribes. Twenty-three households, constituting 4 per cent of all households, belonged to the class of landlords or capitalist farmers. This small proportion of the population controlled 41 per cent of all ownership holdings, thus reflecting the persistent inequality in landholdings.

The total geographical area of the village was 1,030 hectares, of which 86.5 per cent was irrigated land under cultivation – all cultivated land was irrigated. The agro-climatic zone under which the village falls according to the NARP (National Agricultural Research Project) classification is the Krishna–Godavari Zone. Canal irrigation (from the Varalapuram canal of Krishna river) and groundwater irrigation were the major sources of irrigation. Although official data suggest that almost the entire extent of cultivated land in the village was under canal irrigation, data from our census-type survey of December 2005 show that only 12 per cent of gross cropped area was solely under surface irrigation, 24 per cent was solely dependent on groundwater irrigation, and 55 per cent was under irrigation from both sources.

The major crops grown in Ananthavaram were paddy, maize, black gram, sesame, sugarcane, and betel-leaf. In the *kharif* (monsoon) season, paddy cultivation dominated the sown area of the village. The two most important crops of the *rabi* (winter) season in 2005 were maize and black gram. Sugarcane was cultivated through the year. Milch cattle were widely owned (58 for every 100 households) and incomes from livestock contributed significantly to household incomes.

Bukkacherla, Anantapur District, Andhra Pradesh, 2005–06

Brief sketch

Bukkacherla is a village in one of the most drought-prone districts of India. It had one traditional dominant landlord family. Small farmers accounted for 60 per cent of all cultivators and 30 per cent of all households in the village. They grew groundnut, pulses, and millets. Limited groundwater irrigation was available. Employment outside the village and dairying were two other sources of income in this village of risky agriculture.

Detailed profile

Bukkacherla village is located in Raptadu *mandal* of Anantapur district. The *mandal* headquarters, Raptadu, is 9 kilometres away, and Anantapur, the nearest town and railhead as well as the district headquarters, is at a distance of 15 kilometres. At the time of our survey, the approach road to the village was a *kaccha* (unmetalled) road and difficult to travel on during the monsoon. There was no bus to the village, though autorickshaws and vans from Gandlaparthy, a village 6 kilometres away, passed through depending on passenger demand. Nevertheless, several persons commuted to Anantapur, the district headquarters, for non-agricultural employment.

There was a post office in the village, a ration shop, and a pay telephone booth. There were no regular stores but there was a weekly market. There was a cooperative bank in Gandlaparthy, and the nearest commercial bank was in Anantapur. The nearest primary health centre (PHC) was at Raptadu. For any other medical service, private or public, people went to Anantapur. There was only one school in Bukkacherla, which provided education up to the seventh standard.

There were a total of 292 households resident in the village, and for the second survey, our sample size was 99 households. On average, there were four members per household. The single largest social group was that of Other Caste Hindus, accounting for nearly 44 per cent of all households. About one-third of all households belonged to Other Backward Classes (OBC) while Scheduled Castes (SCs) accounted for one-fifth of the households. A little less than 3 per cent of the households were Muslim. The main landlord family in Bukkacherla owned 280 acres, and had several non-agricultural sources of income including urban property. Members of this family controlled the panchayat, worked for the State bureaucracy, and were in State politics. "For this landlord family, land is not so much a source of income as of socio-political influence and power" (Ramachandran, Rawal, and Swaminathan 2010, p. 27). The village had a large number of peasant households and a relatively small section of manual workers. Only 15 per cent of all households owned no agricultural land.

The area of the village, according to the Census of India 2001, was 1,945 hectares. Of this, nearly 10 per cent (190 hectares) was not available for cultivation while a little over 2 per cent consistsed of culturable waste. The remaining land was under cultivation. Only 9 per cent of the area under cultivation (178 hectares) was irrigated. In terms of agro-ecology, the village is located in the Scarce Rainfall Zone of Rayalaseema as per the NARP classification. The main source of irrigation here is groundwater.

The village had a diversified cropping pattern. Groundnut, red gram, cowpea, green gram, sesame, pearl millet, sorghum, beans (intercropped with groundnut), and paddy were grown in the *kharif* season. Paddy, groundnut, fruits, and vegetables were grown in the *rabi* season, wherever water was available. There was widespread ownership of cattle, goats, and sheep.

Kothapalle, Karimnagar District, Telangana, 2005–06

Brief sketch

Kothapalle is a well-connected village, off a national highway with access to non-agricultural employment. It is a multi-caste, multi-occupation village. The village had both irrigated and unirrigated land, and cultivated paddy, maize, cotton, and other crops. Small farmers constituted 88 per cent of all cultivators but only 28 per cent of all households in the village.

Detailed profile

Kothapalle is located in Thimmapur (Lower Maner Dam Colony) *mandal* of Karimnagar district, in the newly formed State of Telangana. The village is 5 kilometres away from the *mandal* headquarters at Thimmapur. The road to Thimmapur is a *pucca* (metalled) road, constructed under the Pradhan Mantri Gram Sadak Yojana scheme. The nearest town is Karimnagar, 16 kilometers away. Kothapalle is situated off the main Hyderabad–Karimnagar highway, at a distance of 160 kilometres from Hyderabad, a fact that has major consequences for the village economy in terms of access to non-agricultural employment, especially construction activities. There is a bus stop in the village, and a bus passes through every 10 minutes. The village is well connected in terms of road transport, but the nearest railhead is at Kazipet, about 84 kilometres away. The village has a post office, a ration shop, and two pay telephone services. There is a weekly market and two medical stores. There is a primary health centre (PHC) at Thimmapur; a sub-divisional hospital at Nustlapur, 6 kilometres away; and a district hospital, and several private hospitals and nursing homes at Karimnagar. The nearest commercial bank branch is at Thimmapur.

There were 372 households in Kothapalle at the time of our census survey, and a sample of 101 households with less than four persons (3.9) per household was surveyed for further information. Other Backward Classes (OBC) constituted the largest proportion of households (40 per cent); Scheduled Castes (SCs) accounted for 32 per cent of households; and Other Caste Hindus, who were both economically and socially dominant in the village, accounted for a little less than 24 per cent of all households. The sole big landlord in the village had strong connections with Hyderabad, where he had a residence and where his children were educated. There was a significant section of manual workers in Kothapalle, accounting for 44 per cent of all households.

The area of the village, according to the Census of India, is 715.5 hectares, and around one-sixth of this area is not available for cultivation. There is no area under forest, but as much as 23 per cent of the total area is classified as culturable waste. Three-fifths of the village area is under cultivation as per the village records. Irrigated land accounts for a little over two-fifths of the village area, and about one-sixth is under unirrigated cultivation. In terms of agro-ecological categorisation as per the NARP classification, the village comes under the North Telangana region.

The major crops grown in Kothapalle were paddy, maize, cotton, and fodder crop in the *kharif* season; and maize and paddy in the *rabi* season. Oilseeds, fruits, and vegetables were grown round the year. The major sources of irrigation were canal irrigation (from the Lower Maner Dam), and groundwater irrigation from borewells and open wells. Almost all the households owned cattle, mainly because grazing land was available along the banks of the Lower Maner Dam.

Harevli, Bijnor District, Uttar Pradesh, 2006

Brief sketch

Harevli is a small, agriculturally advanced and irrigated village, with access to water from canals and tubewells. It grows paddy, wheat, and sugarcane. The village has a clear socio-economic hierarchy, with Tyagi landlords and rich peasants controlling a large part of the land. Landless manual workers, mostly belonging to Scheduled Castes (SCs), comprise a quarter of the population, and live in segregated and squalid conditions. Non-agricultural employment in nearby urban areas is an important source of income. Small farmers, constituting 45 per cent of all households, account for 43 per cent of the gross cropped area (GCA) but only 21 per cent of the irrigated area. The average family size, at six persons per household, is much larger than in villages of southern and western India.

Detailed profile

The revenue village of Harevli is located in Najibabad block of Bijnor district, Uttar Pradesh. The block headquarters, Najibabad, is a municipality and an important railway junction. The town nearest to Harevli is Mandavli, 4 kilometres away. Maujampur, also 4 kilometres away, is the nearest railway station. At the time of our survey in 2006, Harevli did not have a bus stop within the village, nor did it have a metalled approach road. It had just one school, a primary school for classes I to V. The nearest primary health centre (PHC) was at a distance of 4 kilometres. There was neither a bank nor a post office in Harevli.

There were a total of 109 households in Harevli at the time of our census survey. The average household size was 6.1. The sex ratio was low, at 864 females per 1,000 males. The single largest social group in the village comprised Scheduled Castes (SCs) or Dalits (37 per cent). Muslims comprised 12 per cent of all households, and the rest were divided almost equally between OBC and Other Caste Hindu (Brahmins and Tyagis) households. Landlords comprised 3 per cent of all households and rich peasants another 9 per cent. Peasant households together accounted for 61 per cent of all households. The class of manual workers – almost all of whom were landless and belonged to Scheduled Castes (Chamar and Balmiki) – comprised 24 per cent of households, and these households lived in a segregated settlement.

The total geographical area of Harevli was 505 hectares, of which 444 hectares were classified as area under cultivation. The area irrigated amounted to 364 hectares or nearly 82 per cent of the area under cultivation. The major sources of irrigation were government canals (surface water) and private tubewells (groundwater), both electrically operated and otherwise. The village belongs to the Bhabar and Tarai Agro-Ecological Zone as per the NARP classification.

The major crops grown in Harevli were paddy during the *kharif* season, and wheat and rapeseed in winter. Sugarcane was grown as an annual crop. A large majority of the households in the village owned milch animals. Members of manual worker households obtained non-agricultural employment in neighbouring urban centres.

Mahatwar, Ballia District, Eastern Uttar Pradesh, 2006

Brief sketch
Mahatwar is a small, Dalit-majority village in eastern Uttar Pradesh. It has limited groundwater irrigation, and grows wheat, maize, and other crops. Our survey of the village was conducted during a period of water shortage. Small farmers account for 93 per cent of all cultivators and 44 per cent of all households. There is a sizeable section of landless manual workers, employed in agriculture as well as non-agricultural tasks (including the digging of borewells).

Detailed profile
Mahatwar village, in Rasra tehsil of Ballia district, eastern Uttar Pradesh, is located on the side of the highway linking the towns of Rasra and Mau. The people of the village have access to bus and jeep services to nearby towns as well as to larger cities such as Varanasi. The town nearest to Mahatwar, located at a distance of 2 kilometres, is Pakwainar. It is also the nearest railway station. The village has a primary school and an *anganwadi* (crèche and pre-school) centre. The nearest health sub-centre is 6 kilometres away. There is neither a bank nor a post office within the village.

At the time of our survey, there were 156 households and 1,122 persons resident in the village. The average household size was high, at 7.2. Mahatwar is a multi-caste village with 10 different castes. Scheduled Caste (SC) households accounted for 60 per cent of all households. In fact, Mahatwar was selected twice for housing and related programmes under the Dr Ambedkar Gram

Vikas Yojana for Dalit-majority villages. There were no Scheduled Tribe (ST) and Muslim households in the village. The dominant landowners were Brahmin and Rajput families. There was a small section of landlords and rich peasants. Middle and poor peasants comprised 45 per cent of all households, and manual workers another 23 per cent.

The village records report the entire geographical area of 148 hectares as being available for cultivation, and all but 2 hectares as being irrigated. Irrigation is from groundwater sources, using tubewells energised by diesel or electricity. However, because of a water shortage during our survey year, the yields reported were low. The major crops grown in Mahatwar were paddy and maize during the *kharif* season, and wheat (sometimes intercropped with mustard) during the *rabi* season. Income from livestock was important for all sections of the population.

Non-agricultural employment accounted for around one-third of total wage employment per worker, on average. An important source of non-agricultural employment for male workers (particularly from the Chamar caste) was digging wells (or sinking borewells). Workers from Mahatwar also migrated to faraway places for employment. Home-based women workers were engaged in rolling and packing *bidis*.

Warwat Khanderao, Buldhana District, Maharashtra, 2007

Brief sketch

Warwat Khanderao is a dry, cotton-growing village, comprising 250 households belonging to many castes. The majority of cultivators in the village are small farmers, owning around 5 acres of land on average. The village is relatively remote in terms of its location, and access to non-agricultural sources of employment is limited.

Detailed profile

The revenue village of Warwat Khanderao is located in the Sangrampur block of Buldhana district in the Vidarbha region of Maharashtra. The town and railhead nearest to the village is Shegaon, 18 kilometres away. At the time of the survey in 2007, the village did not have a *pucca* (metalled) approach road, but there was a bus stop in the village. Warwat Khanderao has one primary school for classes I to V, but no high school or higher secondary school. The nearest primary health centre (PHC) is 4 kilometres away. A post office is located inside the village, and a branch of a cooperative bank on the outskirts. This village is less accessible than most other PARI study villages.

There were a total of 250 households with a total population of 1,308 in Warwat Khanderao at the time of our census survey. The average household size was 5.2 persons. It is a multi-caste village with Other Backward Classes (OBC) and Nomadic Tribes making up 69 per cent of all households. Muslims accounted for around one-fifth of all households and Scheduled Castes (SCs) for the remaining one-tenth. Warwat Khanderao had a relatively small proportion of landless households. While SC households owned very little land, Muslim households and those belonging to Nomadic Tribes did have small ownership holdings. The village had a small section of landlords and rich peasants (5 per cent), and a large section of small or poor peasants (37 per cent). There was also a section of manual worker households engaged in crop cultivation.

Warwat Khanderao belongs to the agro-ecological zone known as Western Maharashtra Plain, as per the NARP classification. Cotton (both Bt and non-Bt varieties, frequently intercropped with green gram and red gram) was the main monsoon crop, while sorghum, maize, sesamum, black gram, and other pulses were also grown. Wheat, groundnut, and sunflower were the main winter crops. There was very little irrigation in the village in 2006–07. Of the total number of operational holdings, 88 per cent was unirrigated and the remaining land was irrigated by tubewells.

Nimshirgaon, Kolhapur Distrist, Maharashtra, 2007

Brief sketch
Nimshirgaon is a large, well-connected, multi-caste village in the relatively advanced agricultural and industrial region of western Maharashtra. The village has canal irrigation from the Krishna river, supplemented by wells. It grows sugarcane, pulses, millets, and a variety of fruit, vegetables, and flowers. It has an active cooperative society, and the biggest landowner is the president of the panchayat as well as cooperative society. Small farmer households constitute a quarter of all households. Data for this village come from a sample survey.

Detailed profile
Nimshirgaon is a village in Shirol taluk of Kolhapur district in the sugarcane-growing region of western Maharashtra. It is connected to the highway by an all-weather road. The railway station, which bears the same name as the village, is 1 kilometre away and the nearest town is 10 kilometres away. The village has a post office, a ration shop, public telephones, two pharmacies, an office of the Kolhapur District Central Cooperative bank, and two cooperative societies. The nearest primary health centre (PHC) is 4 kilometres away, at Danoli. There is a registered medical practitioner in the village. It has two primary schools, a middle school, and a secondary school. There is a bus stop

in the village. Nimshirgaon is a village with better infrastructure than most other PARI villages.

At the time of our survey, there were 757 households in Nimshirgaon. We undertook a stratified sample survey of 137 households and later arrived at population estimates by applying appropriate weights. The average household size was 5.2. This was a multi-caste village. One-third of all households belonged to the Scheduled Castes (SCs) and the next largest group comprised Jain households. All the rich peasant and landlord households were Jain households. Manual workers comprised 40 per cent of all households. Manual workers and small peasants were caste-heterogeneous groups.

In terms of the NARP classification, Nimshirgaon belongs to the Western Maharashtra Plain Zone. The village was irrigated through a water supply system linked to the Krishna river. There were also hundreds of privately owned open wells, borewells, and tubewells in the fields. Sugarcane was the major crop, accounting for one-third of the cultivated area. Soybean, pulses, and millets were also cultivated. A wide variety of vegetables, fruits, and flowers (including grape, banana, tomato, and mango) were grown.

Since Shirol *taluk* is part of an industrial belt of western Maharashtra, workers from Nimshirgaon had access to non-agricultural employment in small factories of the region. For example, members of households resident in the village worked in small textile, rubber, sugar, pipe, and automobile spare parts factories.

25F Gulabewala, Sri Ganganagar District, Rajasthan, 2007

Brief sketch

25F Gulabewala is an irrigated village, with around 200 households newly settled along the Gang Canal. Gulabewala is distinctive for the virtual absence of small farmers. This is a village with a clear differentiation between capitalist farmers on the one hand, and landless manual workers on the other hand. It is also marked by high caste inequality. Scheduled Castes (SCs) comprise 60 per cent of all households, and almost all of them are landless manual workers. Jat Sikhs are the large, landowning "modern" farmers with an average farm size of 35 acres, mechanised farm operations, and intensive agriculture. Dairying is an important allied activity.

Detailed profile

25 F Gulabewala is a village in Karanpur tehsil of Sri Ganganagar district, Rajasthan. It is located about 25 kilometres from Sri Ganganagar town and is connected by an all-weather road. The nearest town and railhead is at Kesarisinghpur, 9 kilometres away. The village has two primary schools and

one secondary school, an anganwadi centre, a primary health centre (PHC), and a branch of the State Bank of Bikaner and Jaipur.

In 2007 a census survey of the village was undertaken as part of PARI. At that time there were 204 resident households in the village with a total population of 1,132 and an average household size of 5.5 persons. The population sex ratio was 972 females per 1000 males. Scheduled Castes (Hindu and Sikh) constituted 60 per cent of all households, and 36 per cent were Jat Sikhs. Land distribution in the village was extremely unequal: 65 per cent of all households were landless, which included almost all SC households.

Among the PARI villages, 25F Gulabewala is the most polarised village with a high correlation between class and caste. It is also a village in which the capitalist development of agriculture is a noteworthy feature. There was very little deployment of family labour among any of the cultivator households. Jat Sikh farmers owned 35 acres on average. All the landless SC households were manual worker households.

The village is irrigated by the Gang Canal project, and the entire cultivable area is irrigated. The main crops cultivated in 25F Gulabewala were wheat, rapeseed, cotton, cluster beans, and fodder crops. Agriculture in the village was characterised by highly mechanised farm operations, extensive use of hired labour, and high yields.

Rewasi, Sikar District, Rajasthan, 2010

Brief sketch
Rewasi is a multi-caste village with a small section of landlords (mainly Jats), a small section of landless workers, and a large peasantry. The village is located in a dry region, where agriculture is mainly rainfed and subject to the risk of monsoon failure. Small farmers account for over two-thirds of all cultivators and one-half of all households in Rewasi. They did not have access to any irrigation. In addition, the PARI survey year, 2009–10, was one of drought in the village and district; rearing of camels and goats provided a cushion against the drought. Remittances from workers in other parts of India and the world were an important addition to household incomes.

Detailed profile
The village of Rewasi in Sikar district of Rajasthan, surveyed by PARI in 2009–10, is situated at a distance of 31 kilometres from Sikar town. The railway station nearest to Rewasi is at Sikar. A metalled road connects the main habitation of the village with the Sikar–Salasar road. The nearest market is in Sewad Badi, 6 kilometres away, on the Sikar-Salasar road. There is a health sub-centre in the village that provides only first-aid facilities. People

need to travel to the primary health centre (PHC) in Phagalwa (9 kilometres) or to the district hospital in Sikar (31 kilometres) for other medical services. There is one primary school, one upper primary school, and one high school (privately owned) in the village.

At the time of the survey there were 219 resident households in Rewasi, with 5.9 members per household on average. It is a multi-caste village where Jats (classified as OBC) are the economically and politically dominant caste. Rajputs, the erstwhile landlords, no longer have a position of dominance in the village. The village also has Brahmins, Scheduled Castes (Meghwal) and Scheduled Tribes (Meena). Most households in the village owned land (only 4 per cent were landless). There were eight households in the category of landlords and rural rich (mainly Jats and Brahmins) at the top of the socio-economic hierarchy, who controlled 13 per cent of operated holdings and 19 per cent of irrigated holdings. Only 18 per cent of households were classified as manual workers, and the bulk of the population were classified as belonging to peasant households.

Rewasi is located in the Western Dry agro-climatic region of Rajasthan. Further, the monsoon failed in the reference year, 2009–10, and Sikar district was declared drought-affected. Pearl millet is the most important crop of the *kharif* season in Rewasi. In the *rabi* season, land irrigated by tubewells is sown with wheat, mustard, onions, and fenugreek. There are about 75 tubewells in the village. These irrigate about 41 per cent of the net sown area. Tubewells are used mostly in the *rabi* season. The *kharif* crop is mainly rainfed, even where landholdings are in the command area of tubewells. Unirrigated land is dependent entirely on scanty and uncertain rainfall for cultivation in the *kharif* season, and is not cultivated at all in the *rabi* season.

As in other dry and drought-prone regions, animal resources – cattle, camels, and goats – are an important source of household incomes. Another important aspect of the village economy is the high rate of migration to other cities in India and to countries of the Persian Gulf. Remittances from these migrants are an important source of income for many households.

Gharsondi, Gwalior District, Madhya Pradesh, 2008

Brief sketch

Gharsondi is a multi-caste village in Gwalior district, Madhya Pradesh. Although it has access to canal irrigation, there was a shortage of water in the survey year, and there was also a pest attack which destroyed the *kharif* (soybean) crop. A big Thakur landlord family controls a large part of the land

and other assets in the village. There is a small section of Jat Sikh capitalist farmers. Small farmers comprised about 50 per cent of all cultivators.

Detailed profile

The revenue village of Gharsondi is located in Bhitarwar tehsil of Gwalior district in western Madhya Pradesh. The nearest town as well as railway station for Gharsondi is Dabra, 21 kilometres away. There is a bus stop within the village, but no *pucca* (metalled) approach road. The nearest primary health centre (PHC) is 5 kilometres away. There is no bank in the village. There are four primary schools, a middle school and two high schools.

When Gharsondi was surveyed in 2008, there were 273 households in the village. Of these, three households refused to respond and seven households were found to be joint households during the survey. The PARI census therefore effectively covered 263 households. More than half the population of the village lived in large households with seven or more members. In terms of social composition, Other Backward Classes (OBC) accounted for the largest share of the population at 60 per cent, while Scheduled Castes (SCs) and Scheduled Tribes (STs) accounted for 10 per cent and 13 per cent respectively. Muslims constituted 5 per cent of the population. Households belonging to the Thakur caste accounted for hardly 2 per cent of the population, but were economically powerful. The major landlord family in the village was a Thakur household owning 150 acres. In addition, the village had Jat Sikh households, constituting 13 per cent of the population, which owned the largest share of landholdings. Landlord and big capitalist farmer households (Thakurs and Jat Sikhs) together accounted for 4 per cent of all households, 41 per cent of all ownership holdings, and 53 per cent of all assets. Gharsondi is characterised by high inequality in ownership of land and other assets.

In terms of agro-ecological classification, the village falls in the NARP Gird Zone (with mostly alluvial soil). Of 668 hectares of land in the village, 589 hectares were reported as under cultivation, of which all but 2 hectares was irrigated. Cultivable waste accounted for 40 hectares, while the remaining 39 hectares were not available for cultivation. In 2008, land in Gharsondi was irrigated by means of a canal from the Harsi dam and by privately owned tubewells. There was a water shortage in the canal in the survey year, so most of the land was under monocrops.

The major *kharif* crops in the village were soybean, sesame, and black gram – but a pest attack in 2008, the survey year, destroyed the soybean crop of most households. The main *rabi* crops in the village were wheat, rapeseed, chickpea, and lucerne grass. In 2008, the yield of *rabi* crops were very low because of inadequate water for irrigation.

Rich households in Gharsondi have diversified into business, trade, and real estate in the nearby town, while for the poor, the main sources of non-agricultural employment were the Mahatma Gandhi National Rural Employment Guarantee Scheme (MGNREGS), and work in the construction and transport sectors.

Alabujanahalli, Mandya District, Karnataka, 2009

Brief sketch

Alabujanahalli is a village of around 250 households in the rice-bowl of Karnataka, irrigated by the Cauvery river. It is a paddy and sugarcane-growing village, situated close to a major sugar factory of the region. The population is predominantly from the Vokkaliga caste (a Backward Class in Karnataka). Small farmers constitute more than 80 per cent of all cultivators and 50 per cent of all households. Dairying and sericulture are important allied activities.

Detailed profile

The village of Alabujanahalli is located in Maddur taluk, Mandya district of Karnataka. It is located at a distance of 25 kilometres from Mandya town and 95 kilometres from Bengaluru city. The nearest railway station is at Maddur, 15 kilometres away. There is a metalled approach road to the village but no public bus service. The nearest bus stop is at K. M. Doddi, a small market town 1.5 kilometres away and the site of a major sugar factory in Mandya district. There are private vans that ply to the village from K. M. Doddi. The nearest post office and bank branch are also at K. M. Doddi. The village has a primary school and milk collection centre. Although transport to Alabujanahalli is limited, it is a part of the urban periphery of Bengaluru and Mysore.

The PARI census survey of 2009 covered 248 households in Alabujanahalli, but since information was incomplete for five households, the study uses data on 243 households. The average household size in the village was 4.9 persons. Seventy per cent of the households belonged to the Vokkaliga caste, a major landowning caste, while many of the remaining households were Scheduled Caste (SC) households. There was only one Scheduled Tribe (ST) household. There was no traditional landlord in this village. The majority of households were peasants and manual workers, and there were two rich capitalist farmer families. There was a close association between caste and socio-economic class in the village: all the rich peasant households were Vokkaligas, while SC households had a concentration of poor peasants and manual workers.

This is a village belonging to the Cauvery-irrigated region of South Karnataka. Channels leading from a tank that is fed by canals from the

Krishnarajasagar dam irrigated the village. Tubewells were used for additional irrigation.

At the time of the survey, the major crops grown were sugarcane, rice, and finger millet. The village is very close to a major sugar factory. The sugarcane produced in the village was procured by the sugar factory as raw material, and the factory and its owners had a substantial hold on the village economy. A large number of households practised sericulture and livestock raising, which contributed significantly to household incomes.

Siresandra, Kolar District, Karnataka, 2009

Brief sketch
Siresandra is a small and relatively well-connected village, where 50 per cent of land is irrigated and there is a high proportion of small farmers. It has substantial sericulture and dairying activity. Small farmers account for 71 per cent of the entire village, suggesting that only a small section of households were engaged in manual labour and other activities.

Detailed profile
Siresandra is a revenue village located in Huttur block of Kolar taluk in the district of Kolar, Karnataka. The nearest town is Kolar, 20 kilometres away, and the nearest railway station is Kolar Gold Fields, 15 kilometres from the village. There is a metalled approach road to the village, and a bus stop is located inside the village. Siresandra has a primary school, *anganwadi*, and milk collection centre, but no post office or bank branch. The nearest primary health centre (PHC) is at Shapur, at a distance of 5 kilometres.

It is a small village with a geographical area of 265 hectares as per revenue records. Its population size is also small: 79 households at the time of our census survey of 2009, with an average household size of 5.9. There were only two major social groups in Siresandra in 2009, namely, Scheduled Castes (SCs) and Backward Classes. Backward Classes accounted for almost two-thirds of all households and the rest were SC households. There were no landlords or capitalist farmers in the village, only peasants and a small section of manual workers.[2] Only 11 per cent of households in Siresandra were landless. Four rich peasant households controlled 30 per cent of all operational holdings, and the rest was operated by medium and small peasants. As we will see later, this village did not have any cultivator household with more than 2 hectares of irrigated land.

[2] There was a landlord household in the next village with landholdings in Siresandra but, unfortunately, this family was excluded from our survey.

Cultivation in the village was mainly rainfed, supplemented by irrigation by means of borewells and drip irrigation. The area under cultivation was 114 hectares (43 per cent of the geographical area), of which 66 hectares were irrigated by borewells and drip irrigation. The cropping pattern included finger millet followed by vegetables and other crops (potato, tomato, carrot, cauliflower, beetroot, radish, fodder maize and fodder grass, and other vegetables, condiments, and tree crops). Finger millet was often intercropped with sorghum, red gram, and sesamum.

Sericulture and dairying were also important occupations in Siresandra, and contributed as much as crop production to household incomes.

Zhapur, Kalaburagi District, Karnataka, 2009

Brief sketch
Like Siresandra, Zhapur too is a small and relatively well-connected village, where Scheduled Castes (SCs) and Scheduled Tribes (STs) constitute the majority of the population. The village is almost entirely rainfed, with pulses and millets being the major crops. One traditional landlord family controls more than a third of all the agricultural land in the village. Small farmers constitute 24 per cent of all households and do not have any access to irrigated land. Stone quarries nearby are an important source of non-agricultural employment for small farmers and manual workers.

Detailed profile
Zhapur is a small village in Kalaburagi taluk of Kalaburagi district in Karnataka State. The village is 15 kilometres away from Kalaburagi town. The nearest railway station and primary health centre (PHC) are at Nandur, 5 kilometres away. There is a metalled approach road to the village, and a bus stop within the village. It does not have a bank branch or a post office. There is, however, an *anganwadi* centre and a primary school within the village. A new airport is to be located nearby, which means that the village will get better connected by road to the town.

The PARI survey of 2009 covered 113 households in Zhapur, of which four households had to be dropped because of incomplete schedules. This volume uses information from the remaining 109 households. The average household size in the village was 6.1 persons. The two main social groups were Scheduled Castes (SCs) and Lingayats (classified as a Backward Class in Karnataka). These two groups accounted for nearly 85 per cent of Zhapur's population (42 per cent each). Scheduled Tribe (ST) households accounted for another one-eighth. Zhapur had one traditional landlord family (now split into four households) that controlled 34 per cent of the ownership holdings and 47 per cent of all assets. There were two rich peasant households that controlled 13 per cent of ownership holdings and 20 per cent of operational holdings. At the other end of the socio-economic hierarchy, 57 per cent of the households were landless and 47 per cent were manual worker households, of whom 55 per cent belonged to the Scheduled Castes.

Zhapur falls in the Northeast Dry Zone as per the NARP classification. It has a geographical area of 628 hectares of which 84 per cent is under cultivation, most of it rainfed. Groundwater extraction is difficult on account of the rocky terrain.

The cropping pattern followed was that of a single mixed crop of rainfed cereals and oil seeds. Most cultivators grew red gram intercropped with maize, sesamum, pearl millet, and green gram. They also cultivated sorghum and safflower, either as pure crops or as mixed crops. In addition, Bengal gram was grown as a mixed crop with safflower.

Many workers from the village were employed as daily labourers in a stone quarry located partially on the boundaries of the village. This was a major source of non-agricultural employment for manual workers from Zhapur. It was also a factor contributing to the prevalence of child labour.

Panahar, Bankura District, West Bengal, 2010

Brief sketch

Panahar is a large, multi-caste village with almost 100 per cent groundwater irrigation (from tubewells), and good connectivity with nearby towns and markets. Most cultivators in this village grow three crops a year: *aman* paddy, potato, and *boro* paddy. Scheduled Castes (SCs) comprise 60 per cent of the population. Small and very small farmers constitute 95 per cent of all cultivators, with an average landholding of less than 1 acre. Many small farmers participate in wage labour and could be termed semi-proletarians. However the village is clearly differentiated with seven landlord families, and the largest landholding is 32 acres (very large by West Bengal standards).

Detailed profile

Panahar village (Deshra-Koalpara Gram Panchayat, Kotulpur block, Bankura district) falls in the Old Alluvial zone (NARP) of West Bengal. It is 3 kilometres from Kotulpur on the road that connects Kotulpur to Joyrambati and Arambagh. The boundaries of the village habitation are contiguous with the village habitations of two other villages, Koalpara and Palpuskarini. At the time of the 2010 PARI survey, there were a few shops and a primary school in Panahar; the nearest market, high school, and primary health centre (PHC) were located in Koalpara. Panahar is well-connected by road, as buses connecting Kotulpur, Bishnupur, and Bankura to Kolkata pass through the village.

At the time of the survey there were 248 households in Panahar, and the average household size was 4.2. Households in the village belonged to various castes and social groups. Scheduled Castes (SCs) comprised more than half of all households and Scheduled Tribes (STs) another 7 per cent. There were seven Muslim households (the biggest landowners were Muslims), and seven landlord families comprising 3 per cent of all households. The majority of households were of small peasants or semi-proletarians, that is, households cultivating small plots of land and engaging in wage labour as well.

Agricultural land in Panahar was primarily irrigated by tubewells. Although the village is in the command area of the Kangsabati Project, very little water for irrigation was received from it. In 2010, land irrigated by electrified tubewells was triple-cropped in this village. The major crops were an *aman (kharif)* crop of paddy, a winter crop of potato, and summer crops of *boro* paddy or sesame. Mustard, rapeseed, and wheat were also cultivated on a small scale.

For women, livestock rearing was an important economic activity, and for men, jobs in urban areas in the services sector were an alternative source of income.

Amarsinghi, Malda District, West Bengal, 2010

Brief sketch
Amarsinghi is a small, multi-caste village located near the market town of Samsi in Malda district of West Bengal. The biggest landowner of the village lives in a neighbouring village and was not interviewed during our survey. Amarsinghi is thus entirely a small farmer village, with an average farm size of 1.2 acres. Irrigation is by means of river lift and wells. Paddy and jute are the main crops.

Detailed profile
Amarsinghi is situated in Samsi gram panchayat, Ratua I block, Malda district, in the New Alluvial Plains zone of West Bengal. The nearest railway station is at Samsi, about 6 kilometres away. Samsi is also the nearest market centre, with a weekly market. Motorcycle vans frequently run between Samsi and Amarsinghi. An important feature of this village is that many residents of the adjoining village of Bandhaguri own land in Amarsinghi; so, in a way, the two villages are integrated.

The 2010 census-type survey conducted by the PARI team covered 127 households in Amarsinghi, and the average household size of the village was 4.3. The major social groups were Other Backward Classes (OBCs) including the *Tanti, Napit,* and *Goala* castes, and Scheduled Castes (SCs) comprising

45 per cent of the population. Amarsinghi did not have any large landlord or capitalist farmer household. As in Panahar, the peasantry of the village was subdivided into peasants and semi-proletarians. There was also a significant proportion (38 per cent) of manual workers.

The area of the village is 1.24 square kilometres (Census of India 2001). The village came under a river lift irrigation scheme and there were also privately owned tubewells, which were electrified in 2008.

A wide variety of crops were grown in Amarsinghi, the main crops being *aman* and *boro* paddy, and jute. Potato, mustard, and pulses (lentils and gram) were the other crops grown in the village. Different kinds of vegetables were grown in the village, although the acreage under vegetable cultivation was small.

Kalmandasguri, Koch Bihar District, West Bengal, 2010

Brief sketch

Kalmandasguri is a village that saw land reform under the Left Front Government of West Bengal in 1977. It lies in a relatively isolated and poorly connected part of northern Bengal. Muslims and Scheduled Castes (SCs) are the two main social groups, and they are the landowners in the village. This is an entirely small farmer village, with some irrigation. Paddy, jute, and, more recently, potato are the crops grown. Migration from the village for construction and other non-agricultural work is a noticeable feature. Fishing is also a major activity in the water bodies in and around the village.

Detailed profile

Kalmandasguri village (Bararangras gram panchayat, Cooch Behar II block, Koch Bihar district) falls in the Terai Teesta Zone of West Bengal. Koch Bihar, the district town, is 17 kilometres away. At the time of the 2010 survey, the nearest markets were in Bararangras (3 kilometres) and Pundibari (7 kilometres). The village did not have electricity and had no all-weather road. No public transport was available and the nearest bus stop was in Baudiardanga, where a bus passed through about five times a day. The nearest primary health centre (PHC) was in Thanesar (3 kilometres away). There was one primary school and one Children's Learning Centre (Shishu Shiksha Kendra) in Kalmandasguri. The nearest secondary school was in Baudiardanga.

At the time of the PARI survey of 2010, 147 households were resident in Kalmandasguri with an average household size of 4.5. In terms of social composition, Muslim households were numerically dominant (42 per cent). Scheduled Castes (Rajbanshis) comprised 33 per cent of all households, and another 7 per cent of households were Scheduled Tribes (Nagesis and Oraons).

Prior to land reforms, large tracts of land in Kalmandasguri were owned by a *jotedar* (landlord) household living in a neighbouring village (Basu 2015). The village was an active site of peasant mobilisation by the Krishak Sabha, and of implementation of land reform after 1977. Around one-half of its residents were beneficiaries of land reform. Today, Kalmandasguri is a small farmer or small peasant economy with the majority of peasants engaging in other self-employment or wage employment.

Around one-third of the net sown area in the village was irrigated, but a substantial part of the land in the village was double-cropped because of high rainfall. The major crops in the village were paddy and jute, and, more recently, potato.

Fishing was an important occupation among Muslim households in the village. Migration from the village for non-agricultural employment was substantial – both to nearby towns for jobs in construction, and to distant locations such as Delhi and Kerala.

Tehang, Jalandhar District, Punjab, 2011

Brief sketch

Tehang is a unique village in having been studied several times during the twentieth century. It is a village that directly experienced the effects of Partition in 1947. It is a large village in terms of population. Agricultural land in Tehang is irrigated by tubewells, with paddy and wheat in rotation being the most common crop cycle. Like 25F Gulabewala, this is a village that is highly differentiated by class and caste. Small farmers accounted for 7 per cent of all households and 8 per cent of operational holdings. Tehang is well connected to the larger region, and employment in nearby urban centres, as well as emigration to the United Kingdom, Canada, and other countries, plays an important role in the village economy.

Detailed profile

Tehang (Phillaur tehsil, Jalandhar district) is situated in the Doab region of Punjab (that is, the land between the rivers Beas and Sutlej), at a distance of about 5 kilometres from the town of Phillaur. The village was surveyed twice by the Board of Economic Inquiry, Punjab, in 1931 and later in 1962. It is what was termed a "refugee village." Before Independence about 80 per cent of the households resident in the village were Muslim; by the survey of 1962, there was not a single Muslim family in the village.

Tehang is well connected by means of public transport; it has a bus stop and frequent bus service. The nearest railway station is at Phillaur. There is one government primary school and higher secondary school, and one private higher secondary school in the village. It has branches of two commercial banks and one cooperative bank. There is one primary health centre (PHC) in the village.

The PARI census survey covered 681 households in Tehang with an average of 4.8 persons per household. Scheduled Castes (SCs) constituted 58 per cent of all households, and Jat Sikh households accounted for more than one-fifth of all households. There was only one Scheduled Tribe (ST) household in the village.

Tehang belongs to the Central Plain Zone of NARP. All cultivated land in the village was irrigated – mainly by electric tubewells, but also by canals. The main crops were paddy (*kharif*) and wheat (*rabi*). In addition, fodder crops, pulses, oilseeds, and sugarcane were cultivated as subsidiary crops.

Non-agricultural employment in construction and in factories (such as hosiery units) in neighbouring towns and cities was a vital source of income for landless households. There has been substantial emigration from Tehang to North America and Europe, and many migrants have settled abroad permanently. Other workers have migrated to countries of the Persian Gulf and Africa. Remittances from workers in other parts of India and the world were also an important component of household incomes.

SMALL FARMERS IN THE STUDY VILLAGES

We now turn to a preliminary description of small farmers and their position in the study villages (Appendix Tables 7 to 9).

First, small farmers were numerically significant in most of the study villages except 25F Gulabewala. The absolute number of small farmers was more than 100 in the villages of Warwat Khanderao, Rewasi, Alabujanahalli, and Panahar. Secondly, in terms of weight in the village production economy, there was a lot of diversity among the study villages. We can broadly categorise the study villages into three types based on the share of small farmers in cultivator households, and in gross cropped area (GCA) and operated area.

1. Villages where small farmers were more than 80 per cent of cultivators and accounted for more than 50 per cent of GCA: Amarsinghi, Kalmandasguri, Alabujanahalli, Mahatwar, Warwat Khanderao, Siresandra, Panahar, and Kothapalle.
2. Villages where small farmers were between 50 to 80 per cent of cultivators, and accounted for 30 to 50 per cent of GCA: Ananthavaram, Bukkacherla, Harevli, Rewasi, Nimshirgaon, Gharsondi, and Zhapur.

3. Villages where small farmers were less than 30 per cent of households and operated less than 10 per cent of GCA: 25F Gulabewala and Tehang. These two villages may be characterised as villages dominated by capitalist farmers.

The relation between small farmers' share in cultivator households and in operated area, rather than gross cropped area (see Table 1), presents an even more illuminating picture. Of the first group above, we see that in five villages – Panahar, Amarsinghi, and Kalmandasguri in West Bengal; Siresandra in Karnataka; and Kothapalle in Telangana – small farmers accounted for more than 50 per cent of operated area. These can be termed small farmer-dominated villages.

In a second subgroup comprising the three villages of Mahatwar (eastern Uttar Pradesh), Alabujanahalli (Karnataka) and Warwat Khanderao (Maharashtra), small farmers accounted for only 30 to 50 per cent of operated area. This suggests that there was a substantial section of larger farmers in these three villages as compared to the first subgroup. This can be confirmed from data on average land ownership among small farmers and all landowning households (Appendix Table 9).

Similarly, of the seven villages in the second subgroup, small farmers had less weight in terms of land operated in Ananthavaram, Harevli, Gharsondi, and Zhapur, than in Bukkacherla, Rewasi, and Nimshirgaon. This does not of course imply less overall inequality in the latter villages. Indeed, among all the villages studied, the second largest landowner (a landlord household) was in Bukkacherla, with ownership holdings of 280 acres.

The third subgroup, comprising 25F Gulabewala and Tehang in northwestern India, is clearly one where small farmers played a very minor role in the agricultural economy.

Table 1 *Grouping of villages by share of small farmers in total cultivators and in operated area*

Share of cultivators	Share of operated area			
	>50	>30–≤50	>10–≤30	≤10
>80	Amarsinghi, Kalmandasguri, Siresandra, Panahar, Kothapalle	Alabujanahalli, Warwat Khanderao, Mahatwar		
>50–≤80		Bukkacherla, Rewasi, Nimshirgaon	Ananthavaram, Harevli, Zhapur, Gharsondi	
≤50				Tehang, 25F Gulabewala

How Small is Small?

Although we have defined small farmers as those with less than 2 hectares of irrigated operated area or 6 hectares of unirrigated operated area, in reality, the extent of area operated and owned by small farmers varies tremendously across the villages. In West Bengal, small farmers on average owned less than 1 acre (although this is likely to be irrigated land). In Ananthavaram (Andhra Pradesh) it was even less, because most small farmers here were tenant cultivators. In Nimshirgaon (Maharashtra), Kothapalle (Telangana), Gharsondi (Madhya Pradesh), Siresandra and Alabujanahalli (Karnataka), and Harevli and Mahatwar (Uttar Pradesh), small farmers owned less than 3 acres on average. It was only in the rainfed villages of Zhapur (Karnataka), Bukkacherla (Andhra Pradesh), Rewasi (Rajasthan), and Warwat Khanderao (Maharashtra) that small farmers owned around 5 acres of land (unirrigated). It should be noted that although we allowed for 15 acres of dry land in our definition of small farmers, in the study villages small farmers owned only around 5 acres of dry land on average.

The coefficient of variation measures dispersion of ownership holdings from the mean holding in the study villages, and can be compared across villages. In all the study villages there was tremendous dispersion in the size of household ownership holdings.

In all but the three villages of Andhra Pradesh and Harevli in western Uttar Pradesh, the large majority of small farmers operated their own land. Last but not least, in all villages except Nimshirgaon (Maharashtra), the share of small farmers in irrigated area was less than their share in gross cropped area. The gap was large in Warwat Khanderao (Maharashtra), Zhapur (Kalaburagi), Rewasi (Rajasthan), and Bukkacherla (Andhra Pradesh). In other words, where irrigation was scarce, small farmers had disproportionately less irrigated land than large farmers.

In summing up, five of the study villages – namely Kothapalle (Telangana), Siresandra (Karnataka), and Amarsinghi, Kalmandasguri, and Panahar (all in West Bengal) – can be called small farmer-dominated village economies. At the other end of the spectrum, small farmers accounted for less than 10 per cent of irrigated area and gross cropped area in Tehang (Punjab), and less than 1 per cent in 25F Gulabewala (Rajasthan).

We are grateful to the Foundation for Agrarian Studies team, particularly Deepak Kumar, Tapas Singh Modak, and Rakesh Kumar Mahato, for providing inputs to this chapter. We have drawn on the collective work of the team, available on the FAS website, as well as V. K. Ramachandran's writing on PARI.

References

Foundation for Agrarian Studies (FAS) (2016), "Introduction to the Project on Agrarian Relations in India (PARI)," available at http://fas.org.in/category/research/project-on-agrarian-relations-in-india-pari/, viewed on 16 January 2017.

Ramachandran, V. K. (2011), "The State of Agrarian Relations in India Today," *The Marxist*, vol. 27, nos. 1–2, January–June.

Ramachandran, V. K. (2017), "Socio-Economic Classes in the Three Villages," in Madhura Swaminathan and Arindam Das (eds.), *Socio-Economic Surveys of Three Villages in Karnataka: A Study of Agrarian Relations*, Tulika Books, New Delhi.

Ramachandran, V. K., Rawal, Vikas, and Swaminathan, Madhura (eds.) (2010), *Socio-Economic Surveys of Three Villages in Andhra Pradesh: A Study of Agrarian Relations*, Tulika Books, New Delhi.

Swaminathan, Madhura, and Das, Arindam (eds.) (2017), *Socio-Economic Surveys of Three Villages in Karnataka: A Study of Agrarian Relations*, Tulika Books, New Delhi.

Swaminathan, Madhura, and Rawal, Vikas (eds.) (2015), *Socio-Economic Surveys of Two Villages in Rajasthan: A Study of Agrarian Relations*, Tulika Books, New Delhi.

Appendix Table 1 *List of study villages with location, year and type of survey, and number of households*

Village	Sub-district	District	State	Year of survey	Type	Number of households
Ananthavaram	Kollur	Guntur	Andhra Pradesh	2005–06	Census, Sample	664
Bukkacherla	Raptadu	Anantapur	Andhra Pradesh	2005–06	Census, Sample	292
Kothapalle	Thimmapur	Karimnagar	Telangana	2005–06	Census, Sample	370
Harevli	Najibabad	Bijnor	Uttar Pradesh	2006	Census	109
Mahatwar	Rasra	Ballia	Uttar Pradesh	2006	Census	156
Warwat Khanderao	Sangrampur	Buldhana	Maharashtra	2007	Census	250
Nimshirgaon	Shirol	Kolhapur	Maharashtra	2007	Sample	137
25F Gulabewala	Karanpur	Sri Ganganagar	Rajasthan	2007	Census	204
Rewasi	Sikar	Sikar	Rajasthan	2010	Census	219
Gharsondi	Bhitarwar	Gwalior	Madhya Pradesh	2008	Census	263
Alabujanahalli	Maddur	Mandya	Karnataka	2009	Census	243
Siresandra	Kolar	Kolar	Karnataka	2009	Census	79
Zhapur	Kalaburagi	Kalaburagi	Karnataka	2009	Census	109
Panahar	Kotulpur	Bankura	West Bengal	2010, 2015	Census, Sample	248
Amarsinghi	Ratua-I	Malda	West Bengal	2010, 2015	Census, Sample	127
Kalmandasguri	Cooch Behar- II	Koch Bihar	West Bengal	2010, 2015	Census, Sample	147
Tehang	Phillaur	Jalandhar	Punjab	2011	Census	681

Note: Villages are listed in order of State and survey year.

Appendix Table 2 *List of study villages with agro-ecological zones and type of irrigation*

Village	State	Agro-ecological zone	Type of irrigation
Ananthavaram	Andhra Pradesh	Krishna–Godavari Zone	Canal
Bukkacherla	Andhra Pradesh	Scarce Rainfall Zone of Rayalaseema	Groundwater
Kothapalle	Telangana	Northern Telengana Zone	Groundwater
Harevli	Uttar Pradesh	Bhabar and Tarai Zone	Canal
Mahatwar	Uttar Pradesh	Eastern Plain Zone	Groundwater
Warwat Khanderao	Maharashtra	Western Maharashtra Plain Zone	Unirrigated
Nimshirgaon	Maharashtra	South Konkan Coastal Zone	Groundwater
25F Gulabewala	Rajasthan	Irrigated North-Western Plain Zone	Canal and groundwater
Rewasi	Rajasthan	Transitional Plain Zone of Inland Drainage	Groundwater, rainfed
Gharsondi	Madhya Pradesh	Gird Zone	Limited canal and groundwater irrigation
Alabujanahalli	Karnataka	Southern Dry Zone	Canal
Siresandra	Karnataka	Eastern Dry Zone	Groundwater
Zhapur	Karnataka	North East Dry Zone	Unirrigated
Panahar	West Bengal	Old Alluvial Zone	Groundwater
Amarsinghi	West Bengal	New Alluvial Zone	Groundwater
Kalmandasguri	West Bengal	Terai Zone	Unirrigated
Tehang	Punjab	Central Plain Zone	Groundwater

Source: Agro-ecological zone as per National Agriculture Research Project (NARP) classification.

Appendix Table 3 *Total population, number of households, average household size, and sex ratio, study villages*

Village	State	Total population	Total number of households	Average household size	Sex ratio
Ananthavaram	Andhra Pradesh	543	150*	3.6	1011
Bukkacherla	Andhra Pradesh	398	99*	4	951
Kothapalle	Telangana	395	101*	3.9	1047
Harevli	Uttar Pradesh	660	109	6.1	864
Mahatwar	Uttar Pradesh	1122	156	7.2	1007
Nimshirgaon	Maharashtra	712	137*	5.2	904
Warwat Khanderao	Maharashtra	1308	250	5.2	994
25F Gulabewala	Rajasthan	1132	204	5.5	972
Rewasi	Rajasthan	1290	219	5.9	1129
Gharsondi	Madhya Pradesh	1838	263	7	876
Alabujanahalli	Karnataka	1182	243	4.9	1007
Siresandra	Karnataka	468	79	5.9	1000
Zhapur	Karnataka	667	109	6.1	968
Amarsinghi	West Bengal	542	127	4.3	1093
Kalmandasguri	West Bengal	663	147	4.5	985
Panahar	West Bengal	1048	248	4.2	1035
Tehang	Punjab	3285	681	4.8	980

Note: *Sample survey data.
Source: PARI data.

Appendix Table 4 *Caste and religious composition of study villages* in per cent

Village	State	Scheduled Caste (SC)	Scheduled Tribe (ST)	SC and ST	Muslim	All others
Ananthavaram	Andhra Pradesh	43	6	49	4	47
Bukkacherla	Andhra Pradesh	20	0	20	3	77
Kothapalle	Telangana	32	3	35	1	64
Harevli	Uttar Pradesh	37	0	37	12	51
Mahatwar	Uttar Pradesh	60	0	60	0	40
Nimshirgaon	Maharashtra	33	0	33	6	61
Warwat Khanderao	Maharashtra	10	0	10	21	69
25F Gulabewala	Rajasthan	60	0	60	0	40
Rewasi	Rajasthan	10	10	20	0	80
Gharsondi	Madhya Pradesh	10	13	23	5	72
Alabujanahalli	Karnataka	14	0	14	0	86
Siresandra	Karnataka	37	0	37	0	63
Zhapur	Karnataka	42	13	55	1	44
Amarsinghi	West Bengal	45	1	46	0	54
Kalmandasguri	West Bengal	33	7	40	42	18
Panahar	West Bengal	54	7	61	2	37
Tehang	Punjab	58	0	58	0	42

Source: PARI data.

Appendix Table 5 *Major crops grown by season in study villages*

Village	State	Kharif crop(s)	Rabi crop(s)	Other crop(s)
Ananthavaram	Andhra Pradesh	Paddy	Maize, Pulses	
Bukkacherla	Andhra Pradesh	Intercropped Groundnut	Paddy	
Kothapalle	Telangana	Paddy, Maize, Cotton	Paddy, Maize	
Harevli	Uttar Pradesh	Paddy	Wheat	Sugarcane
Mahatwar	Uttar Pradesh	Paddy, Wheat	Wheat	
Warwat Khanderao	Maharashtra	Cotton intercropped with Red and Green Gram		
Nimshirgaon	Maharashtra	Soybean, Groundnut		Sugarcane, Grape, Vegetables
25F Gulabewala	Rajasthan	Cotton, Cluster Bean, Fodder Crops	Wheat, Rapeseed	
Rewasi	Rajasthan	Pearl Millet intercropped with Cluster Bean, Green Gram	Wheat, Mustard, Onion, Fenugreek	
Gharsondi	Madhya Pradesh	Soybean, Soybean intercropped with Sesame, Paddy	Wheat, Rapeseed, Chickpea, Lucerne Grass	
Alabujanahalli	Karnataka	Paddy, Finger Millet	Paddy, Finger Millet	Sugarcane
Siresandra	Karnataka	Finger Millet intercropped with Sorghum, Red Gram and Sesame		Vegetable, Fodder Crops
Zhapur	Karnataka	Red Gram intercropped with Maize, Sesame, Pearl Millet, Green Gram		
Panahar	West Bengal	Paddy	Potato, Mustard, Rapeseed, Wheat, Paddy or Sesame	
Amarsinghi	West Bengal	Paddy, Jute	Paddy	
Kalmandasguri	West Bengal	Paddy, Jute	Vegetables, Sugarcane, Potato	
Tehang	Punjab	Paddy	Wheat	

Source: PARI data.

Appendix Table 6 *Distribution of households by socio-economic class in study villages* in number

Village	State	Small farmers	Large farmers	Landlords/Capitalist farmers	Manual workers	All others	All
Ananthavaram	Andhra Pradesh	48	16	11	25	50	150
Bukkacherla	Andhra Pradesh	36	16	4	20	23	99
Kothapalle	Telangana	28	1	3	41	28	101
Harevli	Uttar Pradesh	45	21	3	26	14	109
Mahatwar	Uttar Pradesh	69	1	4	36	46	156
Warwat Khanderao	Maharashtra	106	13	3	76	52	250
Nimshirgaon	Maharashtra	33	12	3	62	27	137
25F Gulabewala	Rajasthan	2	36	20	114	32	204
Rewasi	Rajasthan	107	38	8	39	27	219
Gharsondi	Madhya Pradesh	71	59	12	70	51	263
Alabujanahalli	Karnataka	113	23	2	73	32	243
Siresandra	Karnataka	56	4	–	13	6	79
Zhapur	Karnataka	26	10	4	51	18	109
Panahar	West Bengal	144	1	7	63	33	248
Amarsinghi	West Bengal	54	–	–	48	25	127
Kalmandasguri	West Bengal	68	–	–	55	24	147
Tehang	Punjab	49	75	–	–	557*	681

Note: *The detailed socio-economic classification for Tehang village is not complete. Sample data are reported for sample survey villages (not population estimates).
Source: PARI data.

Appendix Table 7 *Share of small farmer households in cultivating households and all households in study villages in number and per cent*

Village	State	Number of small farmer households	Share of small farmer households in all cultivator households	Share of small farmer households in all households
Ananthavaram	Andhra Pradesh	48	64	32
Bukkacherla	Andhra Pradesh	36	64	36
Kothapalle	Telangana	28	88	28
Harevli	Uttar Pradesh	45	65	41
Mahatwar	Uttar Pradesh	69	93	44
Warwat Khanderao	Maharashtra	106	87	42
Nimshirgaon	Maharashtra	33	69	24
25F Gulabewala	Rajasthan	2	3	1
Rewasi	Rajasthan	107	70	49
Gharsondi	Madhya Pradesh	71	50	27
Alabujanahalli	Karnataka	113	82	47
Siresandra	Karnataka	56	93	71
Zhapur	Karnataka	26	65	24
Amarsinghi	West Bengal	54	100	43
Kalmandasguri	West Bengal	68	100	46
Panahar	West Bengal	144	95	58
Tehang	Punjab	49	38	7

Source: PARI data.

Appendix Table 8 *Share of small farmers in cultivating households, in gross cropped area (GCA), in operational holdings, and irrigated area in study villages* in per cent

Village	State	Share of small farmers in total cultivators	Share of small farmers in GCA	Share of small farmers in operational holdings	Share of small farmers irrigated area to total irrigated area
Ananthavaram	Andhra Pradesh	64	49.3	35.9	35.8
Bukkacherla	Andhra Pradesh	64	47.9	35.8	25.5
Kothapalle	Telangana	88	68.1	53.7	67.2
Harevli	Uttar Pradesh	65	43.1	21.7	21.4
Mahatwar	Uttar Pradesh	93	63.4	48.3	48.5
Warwat Khanderao	Maharashtra	87	54.4	41.5	15
Nimshirgaon	Maharashtra	69	41.8	41.9	44.5
25F Gulabewala	Rajasthan	3	1.5	0.2	0.2
Rewasi	Rajasthan	70	44.2	32.1	26.3
Gharsondi	Madhya Pradesh	50	33.1	11.8	11.8
Alabujanahalli	Karnataka	82	61	48.8	48.7
Siresandra	Karnataka	93	75.3	67.9	65.9
Zhapur	Karnataka	65	46.7	19.5	0
Amarsinghi	West Bengal	100	79.7	87.4	87.3
Kalmandasguri	West Bengal	100	69.9	77.4	76.3
Panahar	West Bengal	95	75.8	58.8	58.1
Tehang	Punjab	38	7.7	7.8	7.7

Note: GCA = gross cropped area.
Source: PARI data.

Appendix Table 9 *Basic statistics on land ownership in the study villages*

Village	State	Small farmers			All landowning households		
		Mean	Maximum	CV	Mean	Maximum	CV
Ananthavaram	Andhra Pradesh	1.0	4.0	0.9	2.7	40	1.6
Bukkacherla	Andhra Pradesh	5.5	17.0	3.7	8.4	280	2.2
Kothapalle	Telangana	2.0	10.0	2.2	2.4	34	1.6
Harevli	Uttar Pradesh	2.1	9.6	2.3	5.2	43	1.5
Mahatwar	Uttar Pradesh	1.3	10.3	1.5	1.6	37	2.5
Warwat Khanderao	Maharashtra	4.8	15.0	3.2	5.7	85	1.5
Nimshirgaon	Maharashtra	2.5	9.0	1.7	2.8	50	1.4
25F Gulabewala	Rajasthan	9.4	12.8	4.8	35.0	287	1.1
Rewasi	Rajasthan	4.8	15.4	3.1	6.0	45	1.0
Gharsondi	Madhya Pradesh	2.7	5.0	1.4	7.8	150	2.1
Alabujanahalli	Karnataka	2.2	7.0	1.4	2.6	25	1.2
Siresandra	Karnataka	4.2	12.0	2.6	5.1	48	1.3
Zhapur	Karnataka	4.9	12.0	3.0	9.8	60	1.4
Amarsinghi	West Bengal	1.2	4.1	0.8	0.8	4	1.0
Kalmandasguri	West Bengal	1.3	5.0	1.0	1.0	5	1.0
Panahar	West Bengal	0.9	5.2	1.1	1.4	32	2.3
Tehang	Punjab	2.8	7.5	1.5	6.0	57	1.1

Note: Landless households are excluded from all households.
Source: PARI data.

3

Labour in Small Farms:
Evidence from Village Studies

Niladri Sekhar Dhar, with Subhajit Patra

The specific characteristics of small farming have left a sharp imprint on the labour process within India's agrarian economy. This chapter sets out the relevant data and draws conclusions on the special characteristics of small farming within the overall labour process, but with a focus on crop production as the central means of livelihood for small farmers. We examine the following issues. First, the supply of labour, which in turn is related to the size and composition of small farmer households. Secondly, the extent to which small farmer households continue to exploit family labour to produce their subsistence needs, and the extent to which, in the process of producing marketable surpluses, they deploy hired labour. In this section, we also discuss separately aspects of female labour and child labour in crop production. Thirdly, we examine the impact of different types of wage contracts and the mechanisation of agricultural tasks on total labour use, especially on female labour use. Lastly, we discuss the extent to which small farmers labour out in multiple spheres of production to maintain a decent standard of living.

The data for this chapter come from village surveys undertaken by the Foundation for Agrarian Studies (FAS) as part of its Project on Agrarian Relations in India (PARI). We have used data from 15 study villages here.

One feature of official databases in India that deal with rural workers is the changing definitions that have been applied to the concept of "work" itself. While these changes in definition reflect the emergence of new conceptual frameworks, they make inter-temporal comparisons more difficult than they would have been if the definitions had remained constant. The definitions and concepts of "work" in the Census of India from 1951 to 2001 concentrate specifically on the transition from "work" as an "earning" criterion, to work as "gainful" and an "economic activity" (see Agarwal 1985; Krishnamurty 1984; and Nath 1968). In this chapter we have adopted a broad definition of the term "worker," to mean any individual participating in any economic activity except housework for at least one day.

In India, as in many other developing countries, the majority of workers are dependent on agriculture and related activities. Among rural workers, those from small farmer households are engaged in own cultivation and also participate in the wage labour market. Small farmer households exhibit a very high degree of division of labour in performing farm activities and household chores.

LABOUR SUPPLY

In an agrarian economy, the supply of labour in any production process is an outcome of the interplay of economic, social, and demographic factors. In this section different indicators are used to understand the process by which members of small farmer households supply labour in their own production process and also production processes in other spheres.

Average Size of Household in the Study Villages

Mukherjee and Krishnaji (1995) reached the conclusion that big landholdings are correlated with large family sizes as these remain joint households, while small landowners and agricultural labourer households form nuclear households.

The PARI village-level data confirm the hypothesis that the probability of partition of households increases with reductions in the level of ownership of productive assets. In other words, the poorer the household, the smaller it is likely to be. Across all the study villages, the average size of household of small farmers ranged from four to seven persons, and was largely concentrated between four and five persons. On the other hand, the average size of household of landlord and rich farmer households ranged from three to ten, and was largely concentrated in the five to eight range. In the villages studied in Andhra Pradesh and in Nimshirgaon in Maharashtra, the economically better-off households moved out of the joint family norm because of economic and demographic transitions over the last two generations. Landlord and rich peasant households realised the value of investment in modern technical and high income-generating higher education, which in turn resulted in migration and smaller household size of their village residences. In the other study villages, landlord and rich peasant households continued with undivided households to make the best of economic opportunities available in the village and neighbouring towns (Ramachandran, Rawal, and Swaminathan 2010).

In the case of manual worker households, the average household size ranged from three to seven, and in most cases it was concentrated between three and five. As landholding size – the primary productive asset of small farmer

Table 1 *Number of members per household, study villages*

Village	State	Landlords and Large farmers	Manual workers	Small farmers
Ananthavaram	Andhra Pradesh	4	3	4
Bukkacherla	Andhra Pradesh	5	3	4
Kothapalle	Telangana	3	4	4
Harevli	Uttar Pradesh	8	5	6
Mahatwar	Uttar Pradesh	6	7	7
Gharsondi	Madhya Pradesh	9	6	7
Rewasi	Rajasthan	8	5	5
Nimshirgaon	Maharashtra	7	5	5
Warwat Khanderao	Maharashtra	10	5	5
Alabujanahalli	Karnataka	8	4	5
Siresandra	Karnataka	14	5	6
Zhapur	Karnataka	8	6	7
Amarsinghi	West Bengal	NA	4	4
Kalmandasguri	West Bengal	NA	4	5
Panahar	West Bengal	8	4	4

Note: NA = not applicable.
Source: PARI survey data.

households – decreases, the probability of separation or partition increases, thus leading to smaller household sizes.

In all the study villages, the average household size of small farmer households was less than that of landlords and large farmers. A closer look reveals that small farmer households resembled manual worker households in terms of average size of household.

Average Number of Workers per Household and Quality of Occupations

The number of workers per household was positively correlated with household size across all socio-economic classes. Among landlord and rich peasant households, those with high working age members were in a better position to take advantage of diversified economic activities. Working members of these households primarily took on supervisory roles in their own farm production processes, and diversified their sources of income into remunerative business, salaried jobs, and other non-farm activities.

On the other hand, the members of small farmer households worked on their own farms, and also participated in manual wage work in agricultural and non-agricultural activities. Very few people who were engaged in remunerative business and salaried activities belonged to small farmer households. The

Table 2 *Number of workers per household, study villages*

Village	State	Landlords and Large farmers	Manual workers	Small farmers
Ananthavaram	Andhra Pradesh	2	2	2
Bukkacherla	Andhra Pradesh	3	2	2
Kothapalle	Telangana	2	2	2
Harevli	Uttar Pradesh	3	2	3
Mahatwar	Uttar Pradesh	2	3	3
Gharsondi	Madhya Pradesh	4	3	3
Rewasi	Rajasthan	4	2	3
Nimshirgaon	Maharashtra	4	2	3
Warwat Khanderao	Maharashtra	4	3	3
Alabujanahalli	Karnataka	3	3	3
Siresandra	Karnataka	NA	3	4
Zhapur	Karnataka	4	3	3
Amarsinghi	West Bengal	NA	2	2
Kalmandasguri	West Bengal	NA	2	3
Panahar	West Bengal	3	2	2

Note: NA = not applicable.
Source: PARI survey data.

occupations of members of small farmer households are discussed in detail later in this chapter.

Dependency Ratio

The data suggest that the worker to non-worker ratio for persons aged 15 years and above was substantially higher for small farmer households than for landlord and large farmer households, and manual worker households. For landlords and large farmers, the ratio of worker to non-worker varied from 0.4:1 to 2.1:1 across the study villages, and for manual worker households, the ratio ranged from 1.5:1 to 3:1. For small farmer households, the ratio of worker to non-worker ranged from 1.4:1 to 4.3:1. Except for Rewasi (Sikar district, Rajasthan) and Nimshirgaon (Kolhapur district, Maharashtra), in all the other villages the ratio of worker to non-worker for landlords and large farmers was lower than the same ratio for small farmers and manual workers.

The ratios indicate that small farmer households involved more family members in economic activities. In Ananthavaram (Guntur district, Andhra Pradesh), Bukkacherla (Anantapur district, Andhra Pradesh), Siresandra (Kolar district, Karnataka), Warwat Khanderao (Buldhana district, Maharashtra), and Zhapur (Kalaburagi district, Karnataka), the ratio of worker to non-worker was more than 3:1. The average number of household members in

Table 3 *Ratio of worker to non-worker for persons aged 15 years and above, study villages*

Village	State	Landlords and Large farmers	Manual workers	Small farmers
Ananthavaram	Andhra Pradesh	0.8	1.8	3.3
Bukkacherla	Andhra Pradesh	1.5	1.9	3.1
Kothapalle	Telangana	1.9	3	2.4
Harevli	Uttar Pradesh	1.1	1.7	1.8
Mahatwar	Uttar Pradesh	0.4	1.8	2.5
Gharsondi	Madhya Pradesh	1.4	2.7	2.7
Rewasi	Rajasthan	1.9	1.5	1.5
Nimshirgaon	Maharashtra	2.1	1.8	2
Warwat Khanderao	Maharashtra	1.5	2.9	3.6
Alabujanahalli	Karnataka	1.2	2.8	2.1
Siresandra	Karnataka	1.9	3	4.3
Zhapur	Karnataka	1.4	2.6	3.1
Amarsinghi	West Bengal	NA	2.9	1.4
Kalmandasguri	West Bengal	NA	2.6	2.7
Panahar	West Bengal	0.7	2	2

Note: NA = not applicable.
Source: PARI survey data.

these villages was also among the lowest in the study villages, except in the case of Zhapur. Though these villages exhibited similar patterns, the reasons for the existence of these patterns were different. In the case of Ananthavaram, the participation of more household members in economic activities was crucial for two reasons. First, as rents were exorbitantly high, to earn a minimum requirement for household consumption, workers from small farmer households had to participate in wage labour markets. Secondly, workers from small farmer households also spent a substantial amount of labour time rearing animals, which contributed a major share of the total household income.

In Warwat Khanderao and Siresandra, the reasons for high worker to non-worker ratios were twofold. First, a large extent of land was under the cultivation of labour-intensive crops: cotton in the case of Warwat Khanderao, and vegetables and fruits in Siresandra. Secondly, the average extent of land operated by small farmers was higher in these villages than in the other study villages.

Child Labour

Let us now consider the aspect of child labour, that is, the probability that persons below 15 years participated in self-cultivation. This probability was invariably higher in the case of small farmer households than in the case of

Table 4 *Proportion of population below 15 years of age that participated in work, study villages* in per cent

Village	State	Landlords and Large farmers	Manual workers	Small farmers
Ananthavaram	Andhra Pradesh	–	10.8	14.3
Bukkacherla	Andhra Pradesh	–	–	8.1
Kothapalle	Telangana	–	7.3	0
Harevli	Uttar Pradesh	4.5	5.7	22.2
Mahatwar	Uttar Pradesh	–	7.8	12.6
Gharsondi	Madhya Pradesh	6.8	13.5	8.8
Rewasi	Rajasthan	7.1	1.1	2.2
Nimshirgaon	Maharashtra	17.3	0.3	21
Warwat Khanderao	Maharashtra	–	8.7	2.2
Alabujanahalli	Karnataka	4	8.7	3
Siresandra	Karnataka	–	–	5.7
Zhapur	Karnataka	–	17.2	9.5
Amarsinghi	West Bengal	–	6.1	8.1
Kalmandasguri	West Bengal	–	1.4	5
Panahar	West Bengal	–	–	6.5

Source: PARI survey data.

landlord and large farmer households. In only one-third of the study villages did anyone below 15 years of age belonging to landlord and large farmer households participate in work. In the case of small farmer households, in 14 of the 15 villages, at least one child participated in work. The proportions were very high in Harevli (Bijnor district, Uttar Pradesh) at 22.2 per cent, Nimshirgaon (Kolhapur district, Maharashtra) at 21 per cent, and Ananthavaram (Guntur district, Andhra Pradesh) at 14.3 per cent. In four other villages, the proportion was between 8 and 9 per cent.

To sum up, a large number of children below 15 years of age were not a part of the work force, especially in landlord and large farmer households. Most of them were enrolled in educational institutions. A substantial number of children below 15 years of age in small farmer and manual worker households participated in economic activities both in family production systems and as manual workers. Moreover, the incidence of children below 15 years engaged in work was higher among small farmer households than among manual worker households across most of the study villages. What emerged as a surprise was that the prevalence of child labour was higher in prosperous villages like Ananthavaram, Harevli, and Nimshirgaon. In these villages, children from small farmer households were involved in labour-intensive and time-bound agricultural tasks like harvesting, threshing, making bundles, winnowing cereal crops, and cutting sugarcane. Transplanting and weeding were the other two tasks for which child labourers were often used.

FEMALE WORK FORCE IN THE RURAL PRODUCTION SYSTEM

The supply of female labour from small farmer households depends on a complex set of features of reproduction and production. FAO (2010) broadly defines "feminisation" of the work force in agriculture as women's increasing presence (or visibility) in the agricultural labour force, whether as agricultural wage workers, independent producers, or unremunerated family workers. FAO (1999) showed that even though the proportion of the labour force engaged in agriculture declined over the 1990s, the proportion of women working in agriculture increased, particularly in the developing countries. Bennett (1992), using census data from 1961, concluded that there was a clear trend towards feminisation of the agricultural labour force across all States of India. The degree of feminisation was observed to be higher in the south Indian States than elsewhere (Duvvury 1989a).

Scholars have identified many reasons for the increase in female participation in agricultural work. Duvvury (1989a) examined structural changes in Indian agriculture, and identified irrigation, shifts in cropping pattern, and green revolution technologies as major determinants of changes in female work participation in agriculture. One reason for the increased feminisation of agricultural work was that male workers moved to non-agricultural occupations, according to da Corta and Venkateshwarlu (1999). According to Kelkar (2009), the increase in the number of women in agricultural production appears to be linked to a variety of factors, including male-specific migration, an increase in the number of households headed by women, and the introduction of labour-intensive cash crops.

To understand feminisation we need at least two data points to establish a time-series. However, this research does not have time-series data, and so we cannot comment on changes over time in the female work force in different production processes of the rural economy. Here, we have tried to discuss the gender dimension of labour supply of small farmer households with respect to other farming and non-farming households in the study villages. We have used two indicators, namely, the number of female workers as a proportion of all workers, and the number of female workers as a proportion of all females aged 15 years and above.

In official statistics, the definition of work leaves out a wide range of housework performed by household members, as for example the tending of animals, which is not considered an economic activity. This distinction between outside work and household activities is crucial, for in a country like India, the bulk of the work done by individuals is considered to be household activities. In other words, the dominance of the non-market, domestic subsistence production system leads to undercounting and non-inclusion of work force members, especially female work force members. Women in the working age population in rural production systems suffer from all three processes of exclusion from labour market statistics, namely, unemployment, underemployment, and exclusion from the labour force.

We have adopted a broad definition of work to understand women's participation in the rural production process. In the production processes of the study villages, women were by and large engaged in home farm activities and in maintaining animal resources. Apart from this, women from small farmer and manual worker households also participated in the wage labour market. If any woman reported participation in the home farm or animal upkeep, we have considered that as work. However, housework is not considered an economic activity in this study, and women engaged solely in housework are not considered workers.

The number of female workers as a proportion of all workers ranged from 25.9 per cent to 48.2 per cent across the study villages. In two of the 15 villages, the number of female workers as a proportion of all workers was less than 30 per cent; and in the other 13 villages, the proportion was more than 35 per cent. This indicates a generally lower female participation than male participation. The female work participation rate of manual worker households was higher than that of small farmer households, except in the case of Rewasi (Sikar district, Rajasthan) and Zhapur (Kalaburagi district, Karnataka). Female work participation was much lower for landlord and large farmer households. The women of landlord and large farmer households were primarily engaged in housework, which is outside the definition of work considered here. The landlord and large farmer households generally belonged to 'upper caste' Hindu groups, so cultural and social factors played a pivotal role in their decision to not participate in work outside the home.

The number of female workers as a proportion of all females in the age-group of 15 years and above ranged from 30.1 per cent to 73.5 per cent. In seven out of the 15 villages, the proportion was more than 60 per cent. Moreover, in six villages, the proportion ranged from 45 per cent to 60 per cent. The reasons for these high proportions varied.

The engagement of female members of small farmer households in non-market rural economies was the major reason for the undercounting of women's participation in economic activity. Women in small farmer households participated more significantly in home production, and were therefore more

Table 5 *Proportion of female workers to all workers, study villages* in per cent

Village	State	Landlords and Large farmers	Manual workers	Small farmers
Ananthavaram	Andhra Pradesh	15.1	38.6	38.2
Bukkacherla	Andhra Pradesh	34	40.5	40.7
Kothapalle	Telangana	35.3	50.6	42
Harevli	Uttar Pradesh	10.3	33.3	34.6
Mahatwar	Uttar Pradesh	–	39.8	36.4
Gharsondi	Madhya Pradesh	25.4	37.6	35.4
Rewasi	Rajasthan	44	38.4	48.2
Nimshirgaon	Maharashtra	26.3	36.3	30.7
Warwat Khanderao	Maharashtra	37.1	45.5	44.9
Alabujanahalli	Karnataka	20.3	43.4	34.9
Siresandra	Karnataka	37.9	45.8	42.2
Zhapur	Karnataka	32.7	38.1	42.9
Amarsinghi	West Bengal	–	44.4	25.9
Kalmandasguri	West Bengal	–	40.1	38.2
Panahar	West Bengal	10	36.2	35.7

Source: PARI survey data.

Table 6 *Proportion of female workers in all females of age-group 15 and above, study villages* in per cent

Village	State	Landlords and Large farmers	Manual workers	Small farmers
Ananthavaram	Andhra Pradesh	12.8	46.2	62.7
Bukkacherla	Andhra Pradesh	47.2	53.6	66.2
Kothapalle	Telangana	50	73.9	60.6
Harevli	Uttar Pradesh	13.7	44.2	45.1
Mahatwar	Uttar Pradesh		50.6	55.2
Gharsondi	Madhya Pradesh	30.6	55.5	56.7
Rewasi	Rajasthan	53.8	42.5	51.7
Nimshirgaon	Maharashtra	40	49.2	47.5
Warwat Khanderao	Maharashtra	44.8	65.6	73.5
Alabujanahalli	Karnataka	23.2	61	47.5
Siresandra	Karnataka	55	68.8	71.2
Zhapur	Karnataka	43.2	54	70.6
Amarsinghi	West Bengal		61.8	30.1
Kalmandasguri	West Bengal		58.9	60.7
Panahar	West Bengal	8.7	47.4	45.7

Source: PARI survey data.

likely to be invisible in the estimation of workers. Women in landlord and rich farmer households did not participate in farm-related activities because of their socio-economic status in the rural society. On the other hand, the participation of women of manual worker households was more visible as they directly participated in the wage labour market.

The rural production system was still dominated by male workers. However, the recognition of different non-market activities in the definition of work resulted in higher work participation rates for female workers, as compared to results provided by secondary data sources.

LABOUR USE IN CROP PRODUCTION

In most of the study villages, small farmer households accounted for a large portion of the total labour use in crop production. In some villages, landlords and capitalist farmers were few, so the major crop production activities were undertaken by small farmers. For instance, small farmer households were numerically dominant in terms of total labour use in crop production in the villages of West Bengal (Amarsinghi, Kalmandasguri, and Panahar). Small farmer households accounted for 70 and 86 per cent, respectively, of total labour use in crop production in Panahar and Amarsinghi. In Mahatwar (Uttar Pradesh) and Siresandra (Karnataka), small farmer households were responsible for 68 and 72.4 per cent, respectively, of total labour use. In the other villages, small farmer households accounted for around half of the total labour used in crop production.

Use of Family Labour and Hired Labour

In Mahatwar (Ballia district, Uttar Pradesh), Rewasi (Sikar district, Rajasthan), Gharsondi (Gwalior district, Madhya Pradesh), and Harevli (Bijnor district, Uttar Pradesh), 70 per cent and more of the total labour deployed by small farmer households was family labour. The ratio of family labour to hired labour use in small farmer households in the study villages varied between 0.4 and 7.8. In six villages the ratio was less than 1, which implies that in these villages hired labour was more dominant than family labour on small farms. The ratio of family labour to hired labour for landlord and large farmer households was less than 1 in all the study villages. Exploitation of family labour was more prominent on small farms. However the extent of this varied, and in some villages the use of hired labour dominated even on small farms.

Small farmer households in the majority of the study villages, especially small farm-dominated villages, used more of the labour of family members than of hired labourers to produce crops. In the case of cultivation of short-duration

Table 7 *Composition of family and hired labour use in crop production in the study villages* in per cent

Village	State	Family labour	Hired labour – daily rate	Hired labour – piece rate	Exchange labour
Ananthavaram	Andhra Pradesh	12	42	38	8
Bukkacherla	Andhra Pradesh	30	56	12	2
Kothapalle	Telangana	39	55	1	5
Harevli	Uttar Pradesh	44	15	35	6
Mahatwar	Uttar Pradesh	73	11	13	3
Gharsondi	Madhya Pradesh	44	18	36	2
Rewasi	Rajasthan	74	18	4	4
Nimshirgaon	Maharashtra	43	41	10	6
Warwat Khanderao	Maharashtra	36	43	20	1
Alabujanahalli	Karnataka	44	37	17	2
Siresandra	Karnataka	48	51		1
Zhapur	Karnataka	48	47		5
Amarsinghi	West Bengal	41	33	22	4
Kalmandasguri	West Bengal	57	22	21	0
Panahar	West Bengal	38	31	22	9

Source: PARI survey data.

crops, specific time-constrained agricultural tasks were performed by hiring wage labour. Though family labour and hired labour were the main forms of labour used in crop production, the use of exchange labour was also observed in the study villages. Of the total human labour use in crop production, a little under 10 per cent was provided through exchange of labour between farming households. Here, too, the incidence (of exchanging labour between farming households) was more prominent among small farmer households.[1]

Hired Labour: Extensive Use of Piece-Rate Contracts

The proliferation of piece-rated wage contracts is almost universal in developed agricultural regions of India (Duvvury 1989a; Ramachandran 1990; Harriss 1991; Gidwani 2001). While the increasing incidence of piece-rated work under different types of contracts has been well captured by village studies

[1] A unique form of exchange labour was observed in Khakchang village, Dasda block, North district of Tripura. Forest land was distributed for cultivation through lottery among households. All operations, starting from clearing the forest and preparing the land until post-harvest operations, were performed by all working members who participated in the lottery. This practice made sure that no household lagged behind in carrying out crop operations, and the spirit of community was preserved by sharing land and labour (Shamsher Singh, personal communication).

Table 8 *Composition of family and hired labour use by type of farm household, study villages* in per cent

Village	State	Small farmers		Landlords and Large farmers	
		Family labour	Hired labour	Family labour	Hired labour
Ananthavaram	Andhra Pradesh	25	64	8	87
Bukkacherla	Andhra Pradesh	38	60	24	76
Kothapalle	Telangana	47	51	5	67
Harevli	Uttar Pradesh	70	28	30	62
Mahatwar	Uttar Pradesh	86	11	3	97
Gharsondi	Madhya Pradesh	73	26	30	67
Rewasi	Rajasthan	74	20	72	24
Nimshirgaon	Maharashtra	52	48	22	60
Warwat Khanderao	Maharashtra	39	59	18	81
Alabujanahalli	Karnataka	50	48	26	72
Siresandra	Karnataka	53	45	31	0
Zhapur	Karnataka	55	39	40	0
Amarsinghi	West Bengal	37	59	–	–
Kalmandasguri	West Bengal	57	43	–	–
Panahar	West Bengal	45	52	12	61

Source: PARI survey data.

at various points of time, large-scale surveys have failed to monitor or gauge this change in labour-hiring in agriculture. Many village studies have reported that the prevalence of piece-rated contracts is increasing. In a study of villages in North Arcot district, Tamil Nadu, Harriss (1991) observed that there had been a shift away from standard daily payments to employment on the basis of various piece-rated contracts between 1973–74 and 1982–83. Other studies have pointed out that piece-rated contracts are widespread in areas dominated by high yielding varieties, in operations such as paddy transplanting and cotton picking (Heyer 2014; Duvvury 1989a; Ramachandran 1990).

The rise in piece-rated work in the 1990s occurred particularly in areas of capitalist agriculture because of an expansion in cropping intensity that involved a shorter time-period for the completion of agricultural operations (Gidwani 2001). Employers resorted to piece-rated work because workers had an inbuilt incentive to work faster for better remuneration. However, while piece rates reduced "effort-shirking," they allowed "quality-shirking." Ramachandran (1990) has argued that piece-rated operations are effective when the quality of work can be determined by the eye. He cited Marx, who stated that "since the quality and intensity of work are here controlled by the form of wage itself, superintendence of labour in great part becomes superfluous." In a piece-rated work regime, both employer and employee minimise the disutility related to supervision (*ibid.*). When there is a shift from time-rates to piece-rates in an agricultural operation, the employer hands the task of supervision to a person who has emerged from the ranks of the workers, a head labourer who is also negotiator, to supervise the entire operation (*ibid.*). The opportunity for direct confrontation between the dominant landowner class and the economically weaker workers is reduced, as the task of supervision previously done by the landowner is now conducted by a member of the working group (Breman 1985; Gidwani 2001).

The choice of labour contract is also determined by variations in labour demand that arise from differences in the growth cycle of a crop, by an individual farmer's perception about the time and precision required to perform specific operations, and by the utilisation of new machines to achieve optimal levels of timeliness and reduce crop losses (Hayami and Kikuchi 2000; Toquero and Duff 1985).

In all the PARI study villages, a significant proportion of farm labour was hired on a piece-rate basis. The hiring of labour on piece rates was not confined only to capitalist and large farmers. Small farmers also hired a significant proportion of labourers on piece rates in most of the study villages.

In Rewasi (Sikar district, Rajasthan), Siresandra (Kolar district, Karnataka), Zhapur (Kalaburagi district, Karnataka), Kothapalle (Karimnagar district, Telangana) and Nimshirgaon (Kolhapur district, Maharashtra), only a small proportion of the total hired labour was contracted on piece-rate wages. The reasons for the low incidence of piece rates were different across villages. In Rewasi and Zhapur, the cropping pattern (coarse cereals, pulses, and oilseeds) and low levels of production were the primary reason. In Nimshirgaon and Siresandra, the cropping pattern of landlords and capitalist farmers was dominated by cultivation of fruit and vegetables, which required precision in performing tasks and where the use of machinery was almost negligible. To ensure quality and avoid shirking in these kinds of crop production, the employers preferred daily wage contracts to piece-rated contracts.

The hiring of labour on piece-rated contracts was prominent in villages where crops like paddy, wheat, soybean, and sugarcane were cultivated extensively. In the case of paddy, the tasks of transplanting, harvesting, and threshing were contracted out on piece-rates. In sugarcane cultivation, weeding, cane-cutting, making bundles, and loading and transporting of cane to the sugar factory were operations mostly done on piece rates. In wheat cultivation, the dominant piece-rate operations were harvesting and threshing. In all these villages these operations were mostly done by using hired workers; the use of family labour was very limited even among small farmer households.

Table 9 *Proportion of total hired labour use by form of wage contract, study villages* in per cent

Village	State	Per cent of hired labour in total labour use	Hired labour – daily rate	Hired labour – piece rate
Ananthavaram	Andhra Pradesh	80	53	47
Bukkacherla	Andhra Pradesh	68	82	18
Kothapalle	Telangana	55	99	1
Harevli	Uttar Pradesh	50	29	71
Mahatwar	Uttar Pradesh	24	45	55
Gharsondi	Madhya Pradesh	53	33	67
Rewasi	Rajasthan	22	82	18
Nimshirgaon	Maharashtra	51	80	20
Warwat Khanderao	Maharashtra	63	68	32
Alabujanahalli	Karnataka	54	69	31
Siresandra	Karnataka	51	100	0
Zhapur	Karnataka	47	100	0
Amarsinghi	West Bengal	55	60	40
Kalmandasguri	West Bengal	43	51	49
Panahar	West Bengal	53	58	42

Source: PARI survey data.

In Harevli (Bijnor district, Uttar Pradesh), 71 per cent of hired labour was contracted out to labour gangs. Workers in sugarcane cultivation were hired on piece-rates. In Gharsondi (Gwalior district, Madhya Pradesh), 67 per cent of hired labour was contracted out on a piece-rate basis. In this village, harvesting and post-harvesting operations of wheat, paddy, and pulses were contracted out on piece rates. In Ananthavaram (Guntur district, Andhra Pradesh), transplanting was mostly contracted out on piece rates. Harvest and post-harvest operations were entirely done by hired workers, mostly on piece rates. The group of operations comprising harvesting of the standing crop, stacking, and transporting up to the threshing ground was given on contract to gangs of workers. Threshing was mechanised, and the employers contracted out this operation to threshing gangs that came with tractor-driven threshers and performed the operation. Almost all the threshing was done on piece-rated contracts.

Rationale for Using Piece-rated Contracts

Agricultural tasks like sowing, transplanting, harvesting, post-harvest operations, and in some cases weeding, are considered to be highly labour-absorbent and must be completed within a very short period. At the same time, these tasks require a large number of labourers at the peak of the agricultural season when the labour market situation is tight. It becomes very difficult for landowners to organise and supervise a large number of workers, and coordinate the tasks in the most efficient way. Employers therefore resort to trading off precision against efficiency and speedy completion of tasks, and hence opt for piece-rates.

Piece-rates have become near-universal in the above-mentioned agricultural operations, which implies that almost all cultivators equally employed gang labourers to perform these operations. Given wide variations in the cropping pattern and scale of operation, differences in the intensity of crop production were observed among cultivators of different capacities. For instance, capitalist farmers and large farmers always require a large number of workers to complete operations within a stipulated time. Hence timeliness was an important factor that guided the decision to hire labour at piece rates by capitalist and large farmers. On the other hand, the acreage of land cultivated by small farmers was very small, and their crop cycles were paddy–maize, paddy–pulses, paddy–wheat, paddy–oilseeds, and other cereal crops. Therefore, after harvesting the first crop, small farmers had ample time for sowing the second crop. For small farmers the choice of form of labour did not stem from the timeliness of performing operations. Nor did it stem from the opportunity cost incurred in supervising hired workers, as small farmers operated small areas of land.

The question then arises as to why small farmers hired gang labourers to perform these operations, which inevitably resulted in higher costs? The answer might be institutionalisation of piece-rated contracts in the labour market and higher labour productivity in piece work. It is also true that the choice of piece-rated contracts stems from the scarcity of workers willing to work on daily rates (as all able-bodied workers try to secure more remunerative piece-rated tasks) during the peak season, thus forcing small farmers to hire labourers at more costly piece rates. Evidently, capitalist farmers, rich peasants, and other non-working peasants enjoy the advantage of piece rates by relieving themselves from arranging large contingents of wage workers, and setting themselves free from supervision and ensuring timely completion of agricultural tasks. Small farmers, on the other hand, find themselves in a disadvantageous position as they face a scarcity of daily wage workers and hence are forced to incur a higher cost by employing gang labourers.

Gender Division of Labour and Impact of Piece-rated Wages

The gender division of work is a crucial feature of the labour process in rural India. Within agricultural tasks, transplanting, weeding, and non-mechanised post-harvest operations are considered "female-specific" jobs. On the other hand, land preparation-related activities such as ploughing, levelling the land by using animals or machines, and plant protection are often considered male-specific jobs. Harvesting and threshing generally bring both sexes to the field. Technical changes in agriculture, new job opportunities in the non-agricultural sector, changes in cropping patterns, and the social burden of domestic labour on female workers play an important role in determining the gender division in crop production (Ryan and Ghodake 1984; Ramachandran 1990).

Ramachandran (1990) showed that in specific regions, although technical changes in agriculture altered the combination of male and female labour-time, the pattern of allocation of specific tasks between male and female workers remained unaltered. The introduction of tools and machinery in harvest and threshing also had a definite impact on the gender division of labour. The use of mechanical reapers and threshing machines not only reduced the demand for total labour, but also changed the composition of male and female labour-time in favour of males (Paris 1998). Within the labour process, the type of crop is often a determinant of the gender division of labour. Also, the gender division of labour is not static. In certain areas, over time, because of altered social and economic circumstances, tasks undertaken by men have become female-dominated tasks (Paris 1998).

The gender division of labour suggests that male workers contribute more substantially to household crop production than female workers. Across all the PARI study villages, the contribution of men in household production varied from 55 per cent (in Warwat Khanderao) to 93 per cent (in Alabujanahalli). In small farmer households, however, the contribution of females in household crop production was greater than in landlord and large farmer households.

However, the pattern was very different in the case of deployment of hired labour in crop production. In 11 out of 15 villages, the hired labour market was dominated by female workers. The deployment of female hired workers in crop production was uniform across different socio-economic classes. However, in Alabujanahalli (Mandya district, Karnataka), Mahatwar (Ballia district, Uttar Pradesh), Nimshirgaon (Kolhapur district, Maharashtra), Siresandra (Kolar district, Karnataka), and Warwat Khanderao (Buldhana district, Maharashtra), small farmer households deployed more female hired labour as compared to landlords and large farmers.

It can be argued that the proliferation of piece-rated contracts is obscuring the gender division in agricultural operations, making it difficult to identify the gender composition of labour for agricultural tasks that were contracted out. The labour gang participating in agricultural tasks consisted of both male and female workers. The composition of male and female workers was determined by the nature of the task and the gang leader's ability to constitute the group.

As the employer interacted with the gang leader to give the contract and also to pay wages, he was not in a position to identify the gender composition of the labour gang in the PARI village studies.[2] Analysis of the gender division of labour-time remained somewhat inconclusive, as little information was available on gender division of labour-time in piece-rated operations. However, the study of employment indicates that "increased occurrence of piece-rated contracts, accompanied in some cases by mechanisation, has tended to displace female workers from crop operations in which women traditionally worked (or dominated)" (Ramachandran, Rawal, and Swaminathan 2010).

To recapitulate, in all the study villages a substantial portion of the labour was hired on a piece-rate basis. Proliferation of piece rates was prominent in villages where paddy, wheat, soybean, and cotton were cultivated. Transplanting of paddy, harvesting and threshing of paddy and wheat, and various operations in the cultivation of sugarcane were piece-rated in the study villages. The hiring of labour on piece rates helped cultivators whose farming was characterised by large-scale operations and the need to complete certain tasks in a very short period. The introduction of piece-rated wages relieved large employers of the job of supervision, which now became the job of the labour contractor. Though the institutionalisation of piece-rated operations was cost-effective for landlords and large farmers, small farmers were forced to choose the more costly piece-rated contracts because of the scarcity of able-bodied workers on daily rates.

Use of Machine Labour

In India, the use of machinery in crop production is still restricted mainly to land preparation for all crops, and for harvest and post-harvest operations of cereal crops. Machines are put to very limited use in the cultivation of horticultural crops. The use of machinery to perform agricultural tasks in the study villages was determined by the nature of crops grown and the affordability of machinery by the farmers concerned. Most of the machinery was owned by landlords and by capitalist farmers. Small farmers rented machines on an hourly or piece-rated basis to carry out agricultural tasks.

In Gharsondi (Gwalior district, Madhya Pradesh), Harevli (Bijnor district, Uttar Pradesh), Rewasi (Sikar district, Rajasthan), Ananthavaram (Guntur district, Andhra Pradesh), Panahar (Bankura district, West Bengal), Nimshirgaon (Kolhapur district, Maharashtra), and Warwat Khanderao (Buldhana district, Maharashtra), there were significant hours of machine use in crop production. In these villages, the cropping pattern was dominated

---

[2] It is not clear how other studies have resolved this methodological issue.

by the production of paddy, wheat, and oil seeds, and land preparation operations were entirely done by tractors and power tillers. Except in Zhapur (Kalaburagi district, Karnataka) and Warwat Khanderao, where small farmers used animal-drawn ploughs and also hired animals out for wages to other villages, more than 50 per cent of the plots were cultivated by tractors. In five of the study villages, namely, Amarsinghi (Malda district, West Bengal), Kalmandasguri (Koch Bihar district, West Bengal), Nimshirgaon, Rewasi, and Mahatwar (Ballia district, Uttar Pradesh), more than 70 per cent of the total machine hours in crop production were utilised for land preparation. This also indicates that very few actual agricultural operations were mechanised. In Gharsondi, Harevli, Panahar, and Siresandra, more than 60 per cent of the total machine hours were utilised for land preparation. Harvesting and post-harvesting operations were primarily done by human labour. The threshing of cereal crops was done by mechanised threshers. Paddy harvesting and threshing in Panahar, and wheat harvesting and threshing in Harevli were mainly done by machines. Among other agricultural tasks, the spraying of chemicals was done by using mechanical sprayers.

The use of machines was spread across all sections of farmers. It is pertinent to note that while small farmers used the largest share of machine hours in crop production in the study villages, the average hours of machine use were higher for landlords and large farmers than for small farmers. This was primarily because of the larger land area operated by landlords and large farmers. In Nimshirgaon, however, landlords and large farmers used fewer machine hours, as only 36 per cent of land preparation, harvesting, and post-harvest and other operations were performed by machines – a mere 24 per cent of the total machine hours. The low machine use can be attributed to the cropping patterns of landlords and large farmers in this village, which was dominated by grape, sugarcane, tomato, and other fruits and vegetables. Except in the case of sugarcane cultivation, the use of machines was low due to the non-availability of appropriate machines. On the other hand, small farmers grew sorghum and other cereal crops, and they used machines for land preparation and for harvest and post-harvest operations.

Mechanisation has multiple impacts on the farming practices of small farmer households. Mechanisation of physically demanding tasks like ploughing certainly reduces human drudgery. It also allows small farmer households to save family labour and use it in other economic activities that contribute to household income. The labour displacement argument might not be very strong in the case of the study villages, as mechanisation was expended in operations like ploughing, which did not impact significantly on total labour absorption in crop production. Interestingly, the release of male labour from

Table 10 *Use of machine hour by type of operation, study villages* in per cent

Village	State	Share of total machine hours used		
		Land preparation	Harvest and post-harvest operations	Other operations
Ananthavaram	Andhra Pradesh	39	16	45
Bukkacherla	Andhra Pradesh	50	14	37
Kothapalle	Telangana	24	18	57
Harevli	Uttar Pradesh	67	32	1
Mahatwar	Uttar Pradesh	85	13	2
Gharsondi	Madhya Pradesh	69	23	8
Rewasi	Rajasthan	77	21	2
Nimshirgaon	Maharashtra	73	24	3
Warwat Khanderao	Maharashtra	34	53	13
Alabujanahalli	Karnataka	37	48	1
Siresandra	Karnataka	61	34	4
Zhapur	Karnataka	40	60	0
Amarsinghi	West Bengal	98	0	2
Kalmandasguri	West Bengal	94	0	6
Panahar	West Bengal	64	35	2

Source: PARI survey data.

power-intensive operations opened up the opportunity for them to diversify their sources of income within and outside the village. On the other hand, mechanisation of female-specific and control-intensive operations would have a different impact on female employment. In the study villages, absorption of female labour and mechanisation of control-intensive operations has been very limited. So displacement of female labour did not take place in these villages. In the case of small farmer households, the use of machinery for control-intensive operations was even lower as compared to landlords and large farmers. So labour use in control-intensive, female-specific agricultural tasks was high in the study villages.

Labouring Out by Small Farmers

Self-cultivation was not the only source of employment and income for small farmer households. Labouring out at agricultural and non-agricultural tasks was also an important component of their household income. Along with manual worker households, members of small farmer households participated heavily in the wage labour market.

In Mahatwar (Ballia district, Uttar Pradesh), Rewasi (Sikar district, Rajasthan), Kothapalle (Karimnagar district, Telangana), and Zhapur (Kalaburagi district, Karnataka), non-agricultural work was an important source of employment for small farmer households. The stone quarry in Zhapur (Kalaburagi district, Karnataka), well-digging in Mahatwar, and work under the state-sponsored employment programme MGNREGS were the main sources of wage employment in the non-agricultural sector. In Kothapalle, the lower participation of males in agricultural work was due to the availability of non-agricultural employment opportunities in the district town, 16 kilometres away, and tapping and sale of toddy in the village. In all the other villages the non-agricultural sector played a secondary role in the wage labour market, and agriculture was the main source of wage employment. In almost all the study villages the main source of non-agricultural wage employment other than those mentioned above was construction-related work, followed by work in services like laundry, transportation, and technical jobs (electrical work, plumbing, welding, etc.).

The participation of small farmer households in the wage labour market was high in the study villages. Except in Rewasi, Gharsondi (Gwalior district, Madhya Pradesh), and Nimshirgaon (Kolhapur district, Maharashtra), at least one member from more than 50 per cent of small farmer households participated in some kind of wage work. In Ananthavaram (Guntur district, Andhra Pradesh), at least one member in 86 per cent of small farmer households participated in wage work. In Warwat Khanderao (Buldhana district, Maharashtra), at least one member in 71 per cent of small farmer households participated in wage work.

In the case of Nimshirgaon, the reasons for low participation in wage work – one member from only 36 per cent of small farmer households – were as follows. First, small farmers of this village cultivated soybean and groundnut in the *kharif* season on 70 per cent of the operated land, and they also cultivated highly remunerative crops like sugarcane, fruit and vegetables. These crops yielded enough cash income for them to sustain their consumption demand. Secondly, most small farmers in this village cultivated their own land, giving them the opportunity to retain the surplus generated in crop production. Thirdly, crop production in this village was highly irrigated with multiple sources of irrigation, which made cultivation less vulnerable.

In Ananthavaram, on the other hand, small farmers had to participate in the wage labour market for the following reasons. First, most small farmer households in this village were tenant cultivators. In 2005–06, around 67 per cent of the total operated land was leased-in land. Therefore, the surplus generated by small farmers was appropriated by the landowning class in the village. Rents were high, paid both in kind and cash. Secondly, the cropping pattern of small farmer households was dominated by paddy–maize and paddy–pulses crop cycles. Interestingly, the cropping pattern of small farmers in this village was markedly different from that of landlords and rich capitalist farmers. Apart from the paddy–maize crop cycle, capitalist farmers produced remunerative crops like sugarcane, and fruits and vegetables. Their cropping pattern was highly labour-absorbent, making the demand for hired labour high. Thirdly, the production of paddy, maize, and pulses provided small farmers with food, and they could produce very little marketable surplus from the total produce. Because of this particular cropping pattern small farmer households were unable to secure the cash income required to maintain their consumption needs, and so they participated in the wage labour market.

In Warwat Khanderao, the majority of small farmers cultivated cotton as the main crop on their own land. The relatively low income from crop production resulted in low rates of labour absorption in the *rabi* season, as very little land was under wheat, pulses, and oilseeds owing to lack of irrigation in the *rabi* season. The low income from cotton cultivation and relatively lean *rabi* crop season compelled workers of small farmer households in this village to participate in the wage labour market.

In Mahatwar (Ballia), it was yet another situation. The average area of land operated by small farmer households here was 1.3 acres. The average annual income from crop production in this village, dominated by a paddy–wheat crop cycle, was low. Moreover, workers from small farmer households had the opportunity to access non-agricultural employment within and around the village. With large household sizes and therefore more workers per household, small farmer households in Mahatwar could participate in a remunerative rural non-agricultural wage labour market. The coming together of these three factors in Mahatwar – namely, small operational landholdings, low incomes from crop production, and a small agricultural wage labour market – resulted in small farmer households seeking work in the non-agricultural sector.

Summary

The above analysis points to two important features of the labour market in the study villages. First, a very large number of workers from small farmer households worked as manual workers. Indeed, "the catchment area for hired workers extends well beyond the class of manual workers (that is, the class whose major income is from payment for hired manual work), and covers wide sections of the peasantry" (Ramachandran 2011). However, the level of participation of members of small farmer households differed across villages.

Secondly, the engagement of small farmer households in the wage labour market would suggest that a large section of them were impoverished. These impoverished small farmers constituted the class of the semi-proletariat in the study villages. The reasons for their impoverishment lay in the relatively small area of operated land, the high rents paid by tenants, the nature of crops produced, high input costs, lack of access to means of production, low incomes from crop production, and, in some cases, unavailability of non-agricultural wage employment. The institutionalisation of piece-rated agricultural tasks and use of machines in crop production were also factors that forced small farmer households to earn cash incomes, so that they in turn could pay cash wages to hired workers and cash rents for machines by participating in the wage labour market.

Labouring Out by Small Farmer Households in Agricultural Operations

Except in Mahatwar (Ballia district, Uttar Pradesh), Kothapalle (Karimnagar district, Telangana), Zhapur (Kalaburagi district, Karnataka), and Rewasi (Sikar district, Rajasthan), agricultural wage employment was the main source of outside employment for small farmer households. In 11 out of the 15 study villages, agricultural wage employment provided more than 45 per cent of the total number of days of wage work obtained by small farmer households.

In crop production, two kinds of labour-intensive tasks were prevalent. First, there were tasks that had to be completed within a short duration. Timeliness in performing these tasks had a high correspondence with the growth of the crop and the level of output. Transplanting paddy and harvesting of crops are examples of such tasks. Secondly, there were tasks which could be completed over a comparatively longer period of time, like land preparation, cotton picking, weeding in sugarcane cultivation, etc., for which labour was also hired. In both these cases, the high demand for labour encouraged members of small farmer households to participate in the wage labour market, as jobs were easy to obtain.

Another important motivation for small farmers to hire out their labour power in transplanting, harvesting, and post-harvesting operations, was the high wage rate for such operations as compared to the wages for weeding, applying fertilizer, and crop protection. Task-specific wage rates were the same for small farmer households and manual worker households. As discussed earlier, employers hired labour for transplanting, harvesting, and post-harvesting operations on piece-rate contracts, and workers had an inbuilt incentive to work faster for better remuneration. In 13 out of 15 study villages, workers of small farmer households took the opportunity to increase their daily wage earnings by obtaining employment in the above-mentioned agricultural tasks through the intensification of labour. The scope for using family labour in production on family-owned farms and the high demand for labour in specific operations allowed the workers of small farmer households to be selective in their choice of agricultural tasks. Members of manual worker households, however, did not have the advantage of choosing only better-paid agricultural tasks (see Appendix Table 1). The participation of workers from manual worker households in all agricultural tasks brought down the average daily wage earnings. However, female workers from small farmer households did not have much of an advantage over female wage workers from manual worker households. So the decision of small farmer households to supply wage labour was determined by the time used in their own production units, sufficient knowledge about labour demand in lean and peak phases of the crop production season, and the existence of differential wage rates across agricultural tasks.

The participation of members of small farmer households in the wage labour market also helped them to recruit workers to perform agricultural tasks, especially in tight labour market situations during peak seasons. Knowledge of the labour market made the search for workers easier and deployment of labour effective for small farmer households.

PLURIACTIVITY OF SMALL FARMER HOUSEHOLDS

The average number of occupations at the household level for small farmer households in the study villages varied from three to five. The three most important occupational sources of income for small farmer households were crop production, wage earnings, and animal resources. In terms of time spent on different occupations, crop production and animal resources were the most prominent. Some households also received some transfer earnings in the forms of pensions, scholarships, and remittances. However, the contribution of these sources to total income was minuscule.

Here we present the total number of days of work spent on own farms and the total number of days of wage work at the household level. Estimating the labour time spent on animal resources by members of small farmer households was methodologically a difficult task. As all members of households that owned animals spent some labour time on activities related to maintaining them, we were unable to record the exact time spent by different members of each household at different points in the day.

Raut (2004) has given an estimate of time-use on animal resources and the role of women in rearing animals. In our PARI studies, we found that in almost all small households with animal resources, female members reported

Table 11 *Average number of sources of income of small farmer households, study villages* in numbers

Village	State	Average number of sources of income
Ananthavaram	Andhra Pradesh	3
Bukkacherla	Andhra Pradesh	3
Kothapalle	Telangana	3
Harevli	Uttar Pradesh	4
Mahatwar	Uttar Pradesh	4
Gharsondi	Madhya Pradesh	4
Rewasi	Rajasthan	3
Nimshirgaon	Maharashtra	3
Warwat Khanderao	Maharashtra	3
Alabujanahalli	Karnataka	4
Siresandra	Karnataka	5
Zhapur	Karnataka	4
Amarsinghi	West Bengal	5
Kalmandasguri	West Bengal	5
Panahar	West Bengal	5

Source: PARI survey data.

the rearing of animals as their secondary or tertiary occupation (Swaminathan and Usami 2016). Children and older members of small farmer households also actively participated in animal care.

As mentioned earlier wage employment was the second most important source of income and employment for small farmer households. The PARI survey data show that 20 per cent (in Nimshirgaon, Kolhapur district, Maharashtra) to 51 per cent (in Harevli, Bijnor district, Uttar Pradesh) of all wage workers came from small farmer households. Numerically, the participation of workers from small farmer households ranked next to the participation of workers from manual worker households in the wage labour market. Small farmer households contributed 12 per cent of the total number of days of employment generated in the village-specific wage labour market in Nimshirgaon and Kothapalle (Karimnagar district, Telangana), and 54 per cent in Harevli. Further disaggregation suggests that members of small farmer households participated in both agricultural and non-agricultural wage employment. For instance, in Mahatwar (Ballia district, Uttar Pradesh), workers from small farmer households participated more actively in the non-agricultural wage labour market than in the agricultural wage labour market. Of the total number of days of employment generated in the wage labour market, 91 per cent was in the non-agricultural sector, where workers dug wells in neighbouring villages. In all the other study

Table 12 *Average number of days of wage employment obtained by workers in agricultural and non-agricultural work, by small farmer and manual worker households, study villages* in 8-hour day

Village	State	Small farmer households			Manual worker households		
		Agriculture	Non-Agriculture	Total	Agriculture	Non-Agriculture	Total
Ananthavaram	Andhra Pradesh	87	99	186	97	168	265
Bukkacherla	Andhra Pradesh	90	74	164	94	197	292
Kothapalle	Telangana	31	39	70	102	63	165
Harevli	Uttar Pradesh	313	26	339	221	55	276
Mahatwar	Uttar Pradesh	22	218	240	36	225	261
Gharsondi	Madhya Pradesh	45	50	96	117	97	221
Rewasi	Rajasthan	20	83	103	34	121	155
Nimshirgaon	Maharashtra	60	55	114	127	98	229
Warwat Khanderao	Maharashtra	106	29	135	170	58	230
Alabujanahalli	Karnataka	68	21	89	240	56	295
Siresandra	Karnataka	200	0	200	168	0	168
Zhapur	Karnataka	135	128	263	84	338	422
Amarsinghi	West Bengal	3	57	60	10	63	74
Kalmandasguri	West Bengal	72	75	125	93	120	196
Panahar	West Bengal	84	74	159	117	143	260

Source: PARI survey data.

villages, agricultural wage work provided more employment to small farmer households as compared to non-agricultural wage work. Nimshirgaon and Kothapalle, with very low participation of small farmer households in wage labour, were exceptions.

Across all the study villages a large section of small farmer households received a considerable amount of income from animal resources. The data reveal that in 11 out of 15 villages more than 85 per cent of small farmer households derived some income from animal resources. The dependence on animal resources was strikingly high in Ananthavaram (Guntur district, Andhra Pradesh), Kothapalle, and Gharsondi (Gwalior district, Madhya Pradesh), where 68 per cent, 51 per cent, and 33 per cent, respectively, of total household income was earned from animal resources. A significant proportion of household income was derived from animal resources in other study villages as well. Hence, the labour time spent by the family to maintain animal resources has been a crucial component in the exploitation of family labour by small farmer households.

To understand the degree of self-exploitation of family labour in different spheres of production, we have furnished the number of days of self-employment and wage employment at the household level. In 13 out of 15 study villages, the number of days of labour deployment in own crop production was relatively low compared to the number of days of wage employment obtained by small farmer households. In Harevli and Mahatwar, small farmer households obtained four to five months of work in family crop production. In all the other villages, the number of days of employment generated in family cultivation was three months or less. To utilise unspent labour time and increase household income, members of small farmer households participated heavily in the wage labour market. For instance, on average, a small farmer household performed 240 days of wage work in Mahatwar and 339 days of wage work in Harevli. In the other villages, a small farmer household received, on average, around four months of wage employment to increase household income.

In a small-scale farming system, small farmer households exploited family labour and wage labour to meet personal and productive consumption needs. The richer sections of small farmer households hired more wage labour in their production process and sold less labour power in others' production than the poorer sections of small farmer households. At the same time, small farmers as wage workers were exploited by landlords and rich farmers in the process of surplus generation.

The selling of small farmers' labour power to others' crop production indicates that self-exploitation of labour of small farmers within the family's

Table 13 *Number of days of employment obtained in own and wage work by small farmer households, study villages* in 8-hour day

Village	State	Self-employment	Wage employment	Total
Ananthavaram	Andhra Pradesh	39	186	225
Bukkacherla	Andhra Pradesh	75	164	239
Kothapalle	Telangana	96	70	166
Harevli	Uttar Pradesh	135	339	474
Mahatwar	Uttar Pradesh	159	240	399
Gharsondi	Madhya Pradesh	46	96	142
Rewasi	Rajasthan	87	191	87
Nimshirgaon	Maharashtra	76	114	190
Warwat Khanderao	Maharashtra	56	135	191
Alabujanahalli	Karnataka	126	89	215
Siresandra	Karnataka	94	103	197
Zhapur	Karnataka	88	263	351
Amarsinghi	West Bengal	53	60	113
Kalmandasguri	West Bengal	80	127	207
Panahar	West Bengal	53	159	212

Source: PARI survey data.

production process failed to meet their personal and productive consumption needs. This could only be achieved by hiring out their labour power in the rural labour market. Small farmer households therefore would appear to be doubly exploited: in the family production sphere, as well as in the production sphere of landlords and rich capitalist farmers where they work for wages. This was a striking feature of those study villages where employment opportunities in the non-farm sector were limited.

SUMMARY

A unique feature of small farmer households is the supply of labour to the production process by all members of the household. From our study we observed that even though the work of members of small farmer households was spread both within and outside crop production, it was in crop production that their labour was concentrated. Members of small farmer households participated in multiple occupations in order to supplement their incomes. They worked as hired workers on land owned by others and also as non-farm workers. Labouring-out constituted a high proportion of the total number of days of labour worked by the household, indicating the impoverishment of small farmer households.

With respect to the gender dimensions of labour supply, three observations can be made. First, male workers still dominated the production systems of the rural economy. However, female members of small farmer and manual worker households had a significant presence in the work force. Secondly, the participation of female workers varied across study villages for a variety of reasons. In some villages it had to do with the status of operational landholdings of small farmer households; in others it was the particular cropping pattern and low levels of income from crop production. Thirdly, female work participation was still determined by cultural and social as well as economic factors. The participation of female members of landlord and rich farmer households was relatively low as they belonged to the 'upper' castes and enjoyed better economic status.

It was also observed that the incidence of child labour was more prevalent among small farmer households than other households across the study villages. Children of small farmer households were primarily engaged in own production, but, in some cases, also participated in wage work.

A very important feature of labour absorption in agriculture was that the number of days of hired labour far outstripped the total number of days of family labour utilised in crop production. Family labour was a major form of labour for small farmer and manual worker households. However, the cultivation of labour-absorbing crops like cotton, fruits and vegetables, sugarcane, and other short-duration crops in tight labour market conditions forced all farmers, including small farmers, to hire wage labour.

Although the ratio of family labour to hired labour was higher for poorer sections of small farmer households, practices of hiring labour on piece rates have emerged, as a result of which even small farmers hired workers extensively for specific operations. The practice of employing large groups of workers organised by contractors or head-labourers who were paid at piece rates has become almost universal in agricultural tasks such as transplanting, harvesting, and threshing rice, and also in the cultivation of maize, sugarcane, groundnut, and other cereal crops.

Our evidence shows that the gender division of labour applied to specific crop operations and not to all crops. Farming practices in each crop involved some operations that are traditionally male-specific and others that are female-specific. At any given time, the gender division of labour is determined by a combination of factors including cropping pattern, type and degree of mechanisation, and institutional arrangements (such as contract labour). It is, of course, also clear that food crops in these villages were not subsistence crops, and that a significant share of the tasks in non-cereal cultivation, for example, in betel-leaf, sugarcane, and fruits and vegetables farming, were performed by women. It was also observed that in villages where piece rates were the dominant form of wage contract, the deployment of female hired labour on daily wage contracts was low.

An important feature of the labour market in the study villages was that a very wide section of workers, especially from small farmer households, worked as manual workers. Small farmer households rarely obtained income from remunerative employment in other spheres of production. Mechanisation and the institutionalisation of piece-rated wage contracts along with other technological innovations in crop production created a situation where small farmer households reallocated their labour time and sought wage employment in the village-specific wage labour market. Small farmer households employed wage labour and machinery along with family labour to generate surpluses to maintain personal and productive consumption. Since family-owned production invariably failed to provide sufficient income to meet personal and production consumption, they participated in the village-specific rural labour market to sell their labour. Therefore, small farmers constituted a class of the semi-proletariat in village economies. On the one hand they tended to persist with own-farm production involving low remuneration, and on the other hand the impoverishment caused by low remuneration from crop production forced them to opt for wage employment.

CONCLUSIONS

This chapter discusses the labour process on small farms using data collected from 15 villages across seven States of India. In the agrarian economy of villages a large number of households comprised small farmers, and their major source of labour for the production process was family labour including child labour. We observed that small farmers tended to hire labour for time-bound agricultural tasks and utilised machine power for specific mechanised operations. The analysis also suggests that small farmer households participated in multiple occupations, specifically wage work, to supplement their income. The probability of participation of poorer sections of small farmer households in the wage labour market was high, indicating the impoverishment of these households.

Village-level data confirmed that labour in the village production system came from small farmer and manual worker households. In particular, the participation of female workers from small farmer households was determined by the cropping pattern, level of household income, and cultural and social norms. The incidence of child labour in the wage labour market was more prominent among small farmer households.

On aspects of labour use in crop production for small farmers vis-à-vis landlords and large farmers, the village-level data showed that the use of hired labour, especially at piece rates, accounted for a significant proportion in total labour use. Our evidence also suggests that the gender division of labour is dynamic, and is determined by cropping pattern, level of mechanisation, and institutional arrangements for piece-rated work.

We observed that a large section of small farmer households regularly participated in the rural wage labour market, particularly in the agricultural wage labour market. This points to the fact that own-crop production failed to provide the means to meet the production and consumption requirements of small farmer households. This situation has led to the emergence of a class of the semi-proletariat, whose class position changes between that of the 'peasant' (in a good production year) and the 'proletariat' (in a bad production year).

REFERENCES

Agarwal, Bina (1980), "Tractorisation, Productivity and Employment: A Reassessment," *Journal of Development Studies*, vol. 16, no. 3, pp. 375–86.

Agarwal, Bina (1985), "Work Participation of Rural Women in the Third World: Some Data and Conceptual Biases," *Economic and Political Weekly*, vol. 20, no. 51/52, pp. A155–57, A159–A161, A163–64.

Agarwal, Bina (1993), "The Gender and the Environment Debate: Lessons from India," *Feminist Studies*, vol. 18, no. 1.

Barker, R., Herdt, Robert W., and Rose, Beth (1985), "The Rice Economy of Asia, Resource for the Future," International Rice Research Institute, Washington D. C.

Bennett, L. (1992), "Women, Poverty and Productivity in India," Economic Development Institute Seminar Paper, No. 43, World Bank, Washington D. C.

Breman, Jan (1985), *Of Peasants, Migrants and Paupers: Rural Labour Circulation and Capitalist Production in West India*, Oxford University Press, New Delhi.

Breman, Jan (1996), *Footloose Labour: Working in India's Informal Economy*, Cambridge University Press, New Delhi.

da Corta, Lucia, and Venkateshwarlu, Davaluri (1999), "Unfree Relations and the Feminisation of Agricultural Labour in Andhra Pradesh," *The Journal of Peasant Studies*, vol. 26, nos. 2 and 3, pp. 71–139.

Dhar, Niladri Sekhar (2012), "On Days of Employment of Rural Labour Households," *Review of Agrarian Studies*, vol. 2, no. 2, available at http://www.ras.org.in/on_days_of_employment_of_rural_labour_households, viewed on 16 January 2017.

Dhar, Niladri Sekhar, with Kaur, Navpreet (2013), "Features of Rural Underemployment in India: Evidence from Nine Villages," *Review of Agrarian Studies*, vol. 3, no. 1, available at http://www.ras.org.in/features_of_rural_underemployment_in_india, viewed on 16 January 2017.

Duvvury, Nata (1989a), "Women in Agriculture: A Review of the Indian Literature," *Economic and Political Weekly*, vol. 24, no. 43, pp. WS96–WS112.

Duvvury, Nata (1989b), "Work Participation of Women in India: A Study with Special

Reference to Female Agricultural Labourers, 1961 to 1981," in A. V. Jose (ed.), *Limited Options: Women Workers in Rural India*, ARTEP, World Employment Programme, International Labour Organisation, India.

Food and Agricultural Organisation (FAO), International Fund for Agricultural Development (IFAD), and International Labour Office (ILO) (2010), "Gender Dimensions of Agricultural and Rural Employment: Differentiated Pathways Out of Poverty Status, Trends and Gaps," Rome.

Gidwani, V. (2001), "The Cultural Logic of Work: Explaining Labour Deployment and Piece-Rate Contracts in Matar Taluka, Gujarat – Parts 1 and 2," *Journal of Development Studies*, vol. 38, no. 2, pp. 57–108.

Harriss, J. (1985), "What Happened to the Green Revolution in South India? Economic Trends, Household Mobility and the Politics of an 'Awkward Class'," DEV Discussion Paper No. 175, School of Development Studies, University of East Anglia, Norwich.

Harriss, J. (1991), "Population, Employment and Wages: A Study of North Arcot Villages, 1973–1983," in Peter B. R. Hazell and C. Ramasamy (eds.), *The Green Revolution Reconsidered: The Impact of High-Yielding Rice Varieties in South India*, The Johns Hopkins Press, London.

Hayami, Yujiro, and Kikuchi, Masao (2000), *A Rice Village Saga: Three Decades of Green Revolution in the Philippines*, International Rice Research Institute, Philippines.

Heyer, Judith (2014), "Dalit Households in Industrialising Villages in Coimbatore and Tiruppur, Tamil Nadu: A Comparison across Three Decades," in V. K. Ramachandran and Madhura Swaminathan (eds.), *Dalit Households in Village Economies*, Tulika Books, New Delhi.

Kelkar, Govind (2009), "The Feminisation of Agriculture in Asia: Implications for Women's Agency and Productivity," UNIFEM South Asia Regional Office, available at www. agnet.org/library/eb/594/, viewed on 16 January 2017.

Krishnaji, N. (1980), "Agrarian Structure and Family Formation: A Tentative Hypothesis," *Economic and Political Weekly*, vol. 15, no. 13, pp. A38–43.

Krishnamurty, J. (1984), "Changes in the Indian Work Force," *Economic and Political Weekly*, vol. 19, no. 50, pp. 2121–28.

Mukherjee, Chandan, and Krishnaji, N. (1995), "Dynamic of Family Size and Composition: A Computer Simulation Study with Reference to Rural India," *The Journal of Peasant Studies*, vol. 22, no. 2, pp. 279–99.

Nath, Kamla (1968), "Women in the Working Force in India," *Economic and Political Weekly*, vol. 3, no. 31, pp. 1205–13.

Paris, T. R. (1998), "The Impact of Technologies on Women in Asian Rice Farming," in Prabhu L. Pingali and Mahabub Hossain (eds.), *Impact of Rice Research*, International Rice Research Institute, Philippines.

Ramachandran, V. K. (1990), *Wage Labour and Unfreedom in Agriculture: An Indian Case Study*, Clarendon Press, Oxford.

Ramachandran, V. K. (2011), "Classes and Class Differentiation in India's Countryside," paper presented at the International Workshop on Advancing Knowledge in Developing Economies and Development Economics: Towards the Understanding of Institutions in Development, Hitotsubashi University, September 23–24.

Raut, K. C. (2004), "Estimation of Woman Labour in Animal Rearing Activities," *Journal of Indian Society of Agricultural Statistics*, 57 (special volume).

Ryan, G. James, and Ghodake, R. D. (1984), "Labour Market Behaviour in Rural Villages in South India: Effects of Season, Sex, and Socioeconomic Status," in Hanse P. Binswanger, and Mark R. Rosenzweig (eds.), *Contractual Arrangements, Employment, and Wages in Rural Labour Markets in Asia*, Yale University Press, New Haven.

Ryan, G. James, Ghodake, R. D., and Sarin, R. (1979), "Labour Use and Labour Markets in Semi-Arid Tropical Rural Villages of Peninsular India," in *Proceedings of the International Workshop on Socio-Economic Constraints to Development of Semi-Arid Tropical Agriculture*, Hyderabad, India.

Swaminathan, M., and Usami, Y. (2016), "Women's Role in the Livestock Economy," *Review of Agrarian Studies*, vol. 6, no. 2, available at http://www.ras.org.in/womens_role_in_the_livestock_economy, viewed on 16 January 2017.

Toquero, Z. F., and Duff, B. (1985), "Physical Losses and Quality Deterioration in Rice Post Production Systems," IRRI Research Paper Series No. 107, International Rice Research Institute, Philippines.

Appendix Table 1 *Average wage earnings of workers from agricultural wage work, 2006–15* in Rs

Village	State	Small farmers		Manual workers	
		Female	Male	Female	Male
Ananthavaram	Andhra Pradesh	57	108	51	108
Bukkacherla	Andhra Pradesh	54	103	59	82
Kothapalle	Telangana	42	129	39	75
Harevli	Uttar Pradesh	34	49	38	47
Mahatwar	Uttar Pradesh	35	35	29	33
Gharsondi	Madhya Pradesh	76	83	61	67
Rewasi	Rajasthan	156	164	157	153
Nimshirgaon	Maharashtra	51	133	54	80
Warwat Khanderao	Maharashtra	55	86	53	70
Alabujanahalli	Karnataka	75	186	73	150
Siresandra	Karnataka	72	134	65	85
Zhapur	Karnataka	47	102	50	103
Amarsinghi	West Bengal	96	113	90	91
Kalmandasguri	West Bengal	90	128	84	116
Panahar	West Bengal	99	103	94	93

Note: No deflator is used here, as the purpose of the table is not to compare wage earnings of workers across time and space.
Source: PARI survey data.

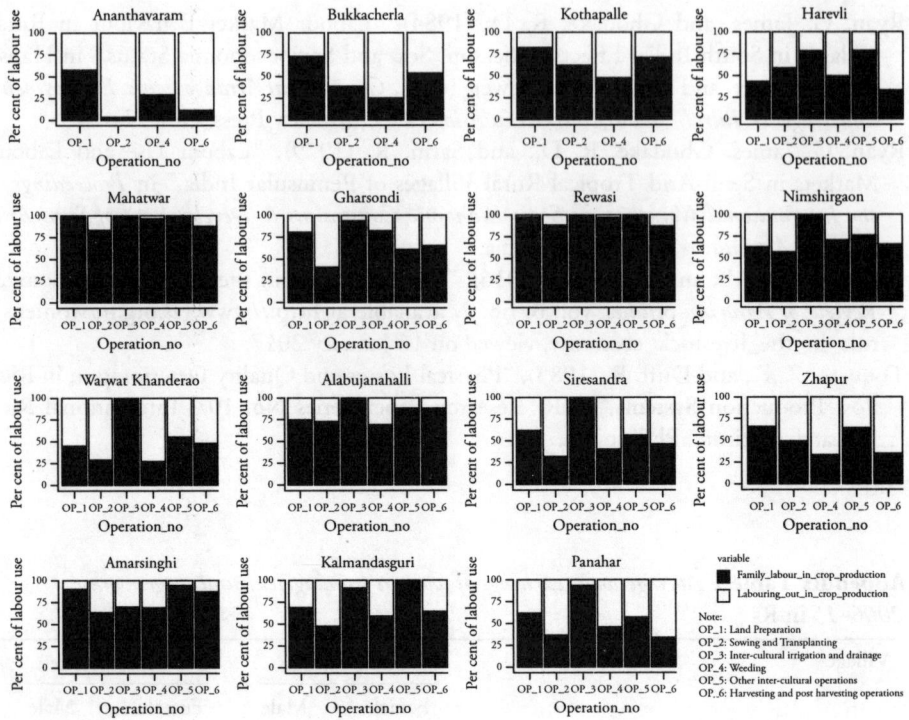

Figure 1 *Proportion of family labour use and labouring out of small farmer households by agricultural operation, PARI study villages*

4

Cropping Pattern, Crop Productivity, and Incomes from Crop Production

Arindam Das and Madhura Swaminathan

INTRODUCTION

This chapter examines two sets of issues concerning the small farmer economy, drawing on data from 17 villages across nine States belonging to different agro-ecological regions. First, do small farmers adopt a distinct cropping pattern in respect of preference for intercropping as opposed to monocropping, and preference for foodgrains (subsistence-related) as opposed to non-foodgrain or purely commercial crops? The second issue concerns the productivity and profitability of farming per unit of operational holding among small farmers, both in absolute terms and in relation to large farmers operating a larger extent of land. This issue draws attention to the "efficiency," or inverse farm size and productivity (profitability) argument. Small farmers' "efficiency" as compared to larger farmers has been debated in the existing literature (and has been dealt with in Chapter 1 of this volume).

The data on which this chapter is based has been drawn from the archives of the Project on Agrarian Relations in India (PARI) of the Foundation for Agrarian Studies (FAS). For a list of the study villages and basic features of the village economy, we refer the reader to Chapter 2 of this volume. The definition of small farmer used here, as noted in Chapter 2, is a cultivator with an operational holding of less than 2 hectares of irrigated land or 6 hectares of unirrigated land, or any combination thereof. In other words, we focus on households whose main source of income is from cultivation, and exclude households with small holdings but other major sources of income (such as salaries or business).

An important feature of the PARI data is that they provide a consistent method for estimating incomes from crop production (see FAS 2015). For each crop, we have computed the gross value of output (GVO). This is simply the value of total production – of the main product and all by-products. The costs of cultivation considered here are the costs paid out by the cultivator, or "Cost A2" as defined by the Commission for Agricultural Costs and Prices

(CACP). Cost A2 covers all paid-out costs incurred by the cultivator (see Appendix 1). In it, no value is imputed to family labour or to the rental value of owned land. Net income or crop income is defined as GVO minus Cost A2. For one section, we imputed the value of family labour (FL) and calculated Cost A2 + FL, and then calculated GVO net of Cost A2 + FL.

This chapter uses data on 17 villages surveyed by the PARI team between 2006 and 2011. To allow for some comparability, we have used the Consumer Price Index for Rural Labour of 2010–11 at the State level as a deflator. All numbers are at 2010–11 prices. As discussed in Chapter 2, the survey year (or reference year) was a bad rainfall year in some of the villages (Rewasi in Rajasthan, Gharsondi in Madhya Pradesh, and Mahatwar in Uttar Pradesh), which in turn affected water availability – a fact that has to be kept in mind while interpreting the data. At the same time, we provide an illustration of adjusting the data to reflect a normal year (Table 10). The major variables discussed in this chapter – yields, costs, and incomes – vary from year to year, and it is useful to remember that our data capture a snapshot and not a long-term view of returns from farming.

In terms of landholdings, we have used two variables: operational holding and gross cropped area. Operational holding refers to the total extent of area operated by a household; gross cropped area refers to the total area operated during a year. So, 1 acre of operated area, if sown with two crops, will be equivalent to 2 acres of gross cropped area.

CROPPING PATTERN

The cropping pattern of any region is based, first, on agro-ecological factors such as soil condition, rainfall, and temperature. Secondly, the selection of crops and cropping pattern depend on access to inputs, capital, irrigation, technology, and market conditions.

In the literature on small farmers, two additional factors of importance have been raised. First, small farmers, it is argued, may have distinct preferences – such as a preference for cultivating for their own consumption (even if not entirely for subsistence), or a preference for intercropping, crop diversity, and biodiversity in general. Such preferences are likely to be reflected in the extent of area sown to food crops, the extent of area intercropped rather than monocropped, and the diversity of crops grown.

A second important factor, given less importance in the mainstream literature, is that cropping patterns depend on resource availability (Rawal 2016; Swaminathan and Rawal 2015). Small farmers are likely to have less area than large farmers under crops that require a large initial investment and have a long gestation period. Hence, crops like betel-leaf, grape or sugarcane may be less prominent in the cropping pattern of small farmers. In other words, choices of cropping pattern are driven by access to resources, such as water or irrigation. In all the study villages, the landlords and capitalist farmers had access to the best quality land in addition to owning large extents of land. The situation among small farmers was more diverse, as shown in Chapter 2.

As Appendix Table 1 shows, in most of the villages small farmers had a lower proportion of area under *rabi* crops and annual crops (both usually possible only with access to irrigation) than large farmers. To take an example, in Ananthavaram village of Guntur district, Andhra Pradesh, 1 per cent of the gross cropped area (GCA) of small farmers was sown with annual crops as compared to 13 per cent of large farmers. The important annual crops in this village were sugarcane and betel-leaf, both requiring irrigation and high initial investment.

We now examine the share of area under single crops or in monocropping, and that under intercropping in the total GCA of small farmers and large farmers (Table 1).

Intercropping as a system of crop production was more prevalent in dry or rainfed villages than in irrigated villages, and could be seen as a way of reducing the risks of crop failure. Across the study villages, intercropped area was low in

Table 1 *Area under intercropping as share of total GCA in kharif, study villages* in per cent

Village	State	Intercropped area as per cent of total GCA	
		Small farmers	Large farmers
Ananthavaram	Andhra Pradesh	0	5
Bukkarcherla	Andhra Pradesh	75	75
Kothapalle	Telangana	14	5
Harevli	Uttar Pradesh	2	5
Mahatwar	Uttar Pradesh	0	0
Nimshirgaon	Maharashtra	9	17
Warwat Khanderao	Maharashtra	76	55
Gharsondi	Madhya Pradesh	12	8
25F Gulabewala	Rajasthan	0	0
Siresandra	Karnataka	43	18
Zhapur	Karnataka	68	34
Alabujanahalli	Karnataka	0	0
Rewasi	Rajasthan	43	24
Amarsinghi	West Bengal	0	0
Kalmandasguri	West Bengal	0	0
Panahar	West Bengal	0	0
Tehang	Punjab	5	3

canal-irrigated, high-productivity (growing paddy/wheat/sugarcane) villages such as Ananthavaram (Guntur district, Andhra Pradesh), Harevli (Bijnor district, Uttar Pradesh), Alabujanahalli (Mandya district, Karnataka), and Tehang (Jalandhar district, Punjab), irrespective of size of area operated or farmer category. In other words, in areas of intensive agriculture, intercropping was not common.

Intercropping was widely prevalent in dry villages – in Bukkacherla (Anantapur district, Andhra Pradesh), Warwat Khanderao (Buldhana district, Maharashtra), and Zhapur (Kalaburagi district, Karnataka). The proportion of GCA under intercropping in the *kharif* season among small farmers was of the order of 75 to 76 per cent in Bukkacherla and Warwat Khanderao (Table 1). In Warwat Khanderao cotton was intercropped with pulses and sorghum, whereas in Bukkacherla groundnut was intercropped with pulses and millets. In Siresandra (Kolar district, Karnataka), half the area under cultivation was used for millet (finger millet), vegetables, and mulberry. Mulberry cultivation was very important and was used for large-scale sericulture by all sections of farmers. In Rewasi (Sikar district, Rajasthan), monocropping covered more than half the area on account of pearl millet cultivation. In Mahatwar (Ballia district, Uttar Pradesh), the farmers preferred monocropping rather than intercropping for paddy cultivation.

In rainfed villages, the intercropped area accounted for a larger share of the total operated area (in the *kharif* season) among small farmers as compared to large farmers.[1] In Warwat Khanderao, for example, intercropping was practised on 76 per cent of the total GCA by small farmers and 55 per cent by large farmers. This was primarily on account of differences across categories of farmers in the adoption of Bt cotton. The practice of growing Bt cotton as a standalone crop (not intercropped) was mainly adopted by cultivators with larger landholdings (Rawal and Swaminathan 2011, p. 108)

Are there distinct differences between small farmers and large farmers in respect of the cultivation of food crops? It has been argued that small farmers prefer to cultivate food crops for subsistence or own consumption. At the same time, larger farmers may also wish to have homegrown grains. In Table 2, we report the share of operated area sown with foodgrains (that is, cereals and pulses) in total operated area by category of farmer.

In ten villages, small farmers allocated a higher share of their cultivated area to foodgrain production than large farmers, though the difference in many cases was small. This was so in a rainfed village like Bukkacherla in Anantapur district, Andhra Pradesh (94 per cent of GCA among small farmers and 89

[1] There was much less intercropping in the *rabi* season, so we have not reported the results here.

Table 2 *Share of foodgrains (cereals plus pulses) in total GCA by category of farmer, study villages*

Village	State	Small farmers	Large farmers
Ananthavaram	Andhra Pradesh	98	80
Bukkarcherla	Andhra Pradesh	97	89
Kothapalle	Telangana	88	52
Harevli	Uttar Pradesh	47	27
Mahatwar	Uttar Pradesh	91	93
Nimshirgaon	Maharashtra	22	28
Warwat Khanderao	Maharashtra	88	70
Gharsondi	Madhya Pradesh	61	59
25F Gulabewala	Rajasthan	25	33
Siresandra	Karnataka	57	23
Zhapur	Karnataka	94	87
Alabujanahalli	Karnataka	50	46
Rewasi	Rajasthan	85	68
Amarsinghi	West Bengal	70	N.A
Kalmandasguri	West Bengal	50	N.A
Panahar	West Bengal	64	69
Tehang	Punjab	78	90

Notes: (i) Foodgrains here include all cereals and pulses, even where either cereals or pulses were intercropped with other non-foodgrain crops.
(ii) The data for Nimshirgaon are from a sample.

per cent among large farmers) as well as in a canal-irrigated, paddy-growing village such as Ananthavaram in Guntur district, Andhra Pradesh (98 per cent among small farmers and 80 per cent among large farmers). In all three villages of West Bengal, a substantial section of small farmers allocated land to jute and potato cultivation, bringing down the share of foodgrains in total GCA. Most of these farmers harvested at least one crop of paddy as well.

The actual proportion of GCA devoted to foodgrain cultivation varied from 47 to 98 per cent across the study villages (excluding Nimshirgaon, Kolhapur district, Maharashtra, where the data came from a sample). In six villages, area sown with foodgrains exceeded 85 per cent of the land. In Siresandra, the area under foodgrains was 57 per cent among small farmers. This was largely on account of the widespread cultivation of mulberry trees for sericulture. In Harevli, the proportion of area sown with foodgrains was 47 per cent, and this was on account of cultivation of sugarcane.

To put it differently, differences across farmer categories within a village in terms of share of foodgrains in operated area were less than differences across villages. Nevertheless, small farmers had sown half or more of their gross cropped area with foodgrains in 14 of the 17 study villages. This proportion was over 85 per cent in seven villages. The area under food crops was high in all the rainfed villages. The two villages in which only a quarter of the GCA was sown with food crops were 25F Gulabewala in Sri Ganganagar district, Rajasthan (with only two small farmer households growing cotton), and Nimshirgaon, a village with multiple crops including fruits and vegetables. The third village with less than half of GCA under food crops was Harevli in western Uttar Pradesh, where sugarcane was popular.

To sum up, we see a distinct pattern of intercropping among small farmers in villages in arid and semi-arid regions. Commercial crops such as cotton or groundnut were intercropped with cereals. This can be seen as a risk-reducing strategy in relatively unfavourable areas. Secondly, small farmers in a majority of the villages had relatively more area sown with foodgrains and less area under purely commercial crops like sugarcane as compared to large farmers, though the differences were not large. However, more area under food crops was not an indicator of subsistence production or lack of market participation among small farmers.

CROP PRODUCTIVITY

For a long time, the discussion on small farmers was centred around the debate on the inverse relationship between farm size and productivity. When we plotted productivity of paddy and wheat (on y axis) against size of operational holding (on x axis), there was no clear pattern to the scatter plot in terms of positive or inverse correlation between extent of operated area and productivity, except for a positive relation for wheat in Gharsondi (Gwalior district, Madhya Pradesh).

Table 3 reports estimates of yields for paddy and wheat in villages where these crops were grown. Paddy yields were above 4 tonnes per hectare in all the villages except Mahatwar. In terms of variations across villages, wheat and paddy yields in Tehang (Jalandhar district, Punjab) were way ahead of the other villages. Excluding Mahatwar, which suffered a shortage of water during the survey year, paddy yields among small farmers were 4.5 tonnes per hectare or less in three villages, but almost 5 tonnes in Ananthavaram village (Guntur district, Andhra Pradesh) and 5.5 tonnes in Tehang. In the case of wheat, yields in Tehang were almost twice those in Harevli and Mahatwar, in Uttar Pradesh (note, however, that the surveys were separated by five years).

To sum up, yields obtained by small farmers were not necessarily lower than those obtained by large farmers in the case of paddy. In the case of wheat, small farmers obtained lower yields than large farmers in two villages and higher yields in the other two, though the differences were small.

Table 3 *Crop productivity for selected crops, study villages* in kilograms/hectare

Village	State	Paddy (*kharif*)		Wheat	
		Small farmers	Large farmers	Small farmers	Large farmers
Tehang	Punjab	5526	5952	4644	4721
Anathavaram	Andhra Pradesh	4954	4414		
Kothapalle	Telangana	4868	4270		
Harevli	Uttar Pradesh	4534	5416	2220	2788
Bukkacherla	Andhra Pradesh	4459	4775		
Alabujanahalli	Karnataka	4143	4574		
Gharsondi	Madhya Pradesh	4331	3976		
Mahatwar	Uttar Pradesh	1658	2200	2465	2304
Amarsinghi	West Bengal	4290			
Panahar	West Bengal	4300	4740		
Rewasi	Rajasthan			2465	2655

Note: Yields of *rabi* or *boro* paddy were higher in Amarsinghi (5.7 tonnes) and Panahar (4.5 tonnes) for small farmers.

INCOMES FROM CROP PRODUCTION

Absolute Incomes per Hectare

We now turn to costs and returns from crop cultivation per unit of operational holding. How did this vary across regions and farmer categories? The data in Tables 4, 5, and 6 show the GVO, Cost A2 and net income per hectare of operated area by farmer category for 17 villages.

Let us start with the GVO or gross value of output per hectare of operated area (Table 4).[2] The first striking but unsurprising finding was that the GVO varied hugely across villages representing different agro-ecological conditions and cropping patterns. The GVO per hectare of operational holding for small farmers ranged from about Rs 16,000 in Zhapur (Kalaburgi district, Karnataka), a rainfed village, and Rewasi (Sikar district, Rajasthan), a drought-affected village, to over Rs 85,000 in Ananthavaram (Guntur district, Andhra Pradesh), Nimshirgaon (Kolhapur district, Maharashtra), Alabujanahalli (Mandya district, Karnataka), Amarsinghi (Malda district, West Bengal), and Tehang (Jalandhar district, Punjab). Low GVO per hectare was observed in rainfed villages and villages prone to frequent drought, with most farmers growing only one crop a year. Bukkacherla in Anantapur (Andhra Pradesh) is located in the second most drought-prone district of the country. Rewasi

[2] There is no distinction made here between irrigated and unirrigated land.

Table 4 *Average GVO from crop production per hectare of operational holding by farmer category, study villages, 2010–11 prices* in Rs per hectare

Village	State	Small farmers GVO	Large farmers GVO
Ananthavaram	Andhra Pradesh	92409	281341
Bukkacherla	Andhra Pradesh	23075	36087
Kothapalle	Telangana	43906	31150
Mahatwar	Uttar Pradesh	35571	32448
Harevli	Uttar Pradesh	59704	76341
Nimshirgaon	Maharashtra	85200	76566
Warwat Khanderao	Maharashtra	41648	51866
Gharsondi	Madhya Pradesh	40799	43061
25F Gulabewala	Rajasthan	14806	50133
Siresandra	Karnataka	53996	111121
Zhapur	Karnataka	16999	23482
Alabujanahalli	Karnataka	112258	118340
Rewasi	Rajasthan	16016	29737
Amarsinghi	West Bengal	108706	–
Kalmandasguri	West Bengal	78764	–
Panahar	West Bengal	89313	108541
Tehang	Punjab	94848	131351

experienced a bad monsoon and severe water shortage during the survey year, and the entire district was declared to be drought-affected. Warwat Khanderao (Buldhana district, Maharashtra) was somewhat of an exception, as it was a largely unirrigated village that cultivated a more remunerative crop, namely cotton including Bt cotton, as compared to pigeon pea in Zhapur (Kalaburagi district, Karnataka) or groundnut in Bukkacherla.

Villages with higher GVO per unit of area were irrigated villages. In four of these villages – Ananthavaram, Alabujanahalli, Amarsinghi, and Tehang – the GVO per hectare was of the order of Rs 90,000 to Rs 110,000 a year. Ananthavaram, irrigated by canals from the Krishna river, was a village where two crops were grown in a year. In Tehang there was tubewell irrigation with commercial cultivation of paddy and wheat. In Alabujanahalli, a village in the Cauvery irrigation system, high-yielding paddy and sugarcane were grown. And in Amarsinghi, newly irrigated by tubewells, there was multiple cropping of paddy, potato, and other crops.

These data bring out the variations in gross returns from crop production among small farmers across agro-ecological regions (that is, with varying soil type, irrigation, and cropping pattern). Thus it is difficult to talk of the incomes of small farmers across India or its geographic regions because of

Table 5 *Average Cost A2 of crop production per hectare of operational holding by farmer category, study villages, 2010–11 prices* in Rs per hectare

Village	State	Small farmers	Large farmers
		Cost A2	Cost A2
Ananthavaram	Andhra Pradesh	93703	145066
Bukkacherla	Andhra Pradesh	19407	35792
Kothapalle	Telangana	32662	23472
Mahatwar	Uttar Pradesh	26967	21293
Harevli	Uttar Pradesh	45331	42921
Nimshirgaon	Maharashtra	41991	43844
Warwat Khanderao	Maharashtra	17618	26483
Gharsondi	Madhya Pradesh	24338	24835
25F Gulabewala	Rajasthan	4413	24215
Siresandra	Karnataka	35594	55714
Zhapur	Karnataka	13544	15922
Alabujanahalli	Karnataka	72268	72480
Rewasi	Rajasthan	11762	15771
Amarsinghi	West Bengal	59641	
Kalmandasguri	West Bengal	39238	
Panahar	West Bengal	73990	78587
Tehang	Punjab	45515	65169

the high degree of variation across regions. The ratio of GVO per hectare in Rewasi, a Rajasthan village, and Alabujanahalli in the Cauvery belt of Mandya district in Karnataka was of the order of 1:7.[3]

If we compare the GVO per hectare of small farmers with that of larger farmers within a village (that is, within a given agro-ecological region), we find that it was lower in a majority of the villages.

Net incomes or returns to farmers depended, of course, on costs incurred. Table 5 shows Cost A2 per hectare for small farmers and large farmers in the study villages. Again, the costs varied widely across the villages, ranging from Rs 11,000 to 13,000 in Rewasi and Zhapur, both single-crop villages, to Rs 72,000 in Alabujanahalli and Rs 93,000 in Ananthavaram. In Ananthavaram, as explained later, an important component of small farmers' costs was rent payments. In Panahar the costs were high on account of fertilizer and pesticide use for the potato crop. At the same time, Cost A2 among small farmers was less than among large famers in all but three villages. One reason for this could be the fact that costs of family labour are not included in Cost A2, and this

---

[3] Our assumption of equivalence between 1 acre of irrigated land and 3 acres of unirrigated land may not be too high, on average.

is likely to underestimate the costs incurred by small farmers relative to large farmers (discussed later in this chapter).

Our main measure of profitability is farm business income or net income, that is, GVO minus Cost A2. Tables 6 and 7 pertain to this variable, profitability or returns per hectare of operational holding.

First, in terms of diversity, net incomes per hectare among small farmers ranged from around Rs 3,000 in Bukkacherla and Zhapur (rainfed villages) to over Rs 40,000 in Alabujanahalli, Nimshirgaon, Amarsinghi, and Tehang. Mean net incomes were negative in only one village – Ananthavaram. This was on account of high rent payments made by tenant farmers (see Chapter 6). To understand these variations, we propose a three-fold classification of villages

Table 6 *Average net incomes from crop production per hectare of operational holding by farmer category, study villages, at 2010–11 prices* in Rs per hectare

Village	State	Small farmers	Large farmers
		Net income/hectare	Net income/hectare
Ananthavaram	Andhra Pradesh	-1294	136275
Bukkacherla	Andhra Pradesh	3666	295
Kothapalle	Telangana	11244	7678
Mahatwar	Uttar Pradesh	8604	11155
Harevli	Uttar Pradesh	14317	33420
Nimshirgaon	Maharashtra	43208	32722
Warwat Khanderao	Maharashtra	24031	25399
Gharsondi	Madhya Pradesh	16462	18226
25F Gulabewala	Rajasthan	9671	25916
Siresandra	Karnataka	18402	55407
Zhapur	Karnataka	3123	7559
Alabujanahalli	Karnataka	40843	46660
Rewasi	Rajasthan	4254	13967
Amarsinghi	West Bengal	49064	–
Kalmandasguri	West Bengal	39527	–
Panahar	West Bengal	15323	29954
Tehang	Punjab	49332	65775

Table 7 *Categorisation of villages by level of net incomes per hectare from farming*

Category	Net income per hectare	Villages in ascending order of income
Low income	Rs 3000 to Rs 10,000	Zhapur, Bukkacherla, Rewasi, Mahatwar
Medium income	Rs 10,000 to Rs 40,000	Kothapalle, Harevli, Panahar, Gharsondi, Siresandra, Warwat Khanderao, Kalmandasguri
High income	Rs 40,000 to Rs 50,000	Alabujanahalli, Nimshirgaon, Amarsinghi, Tehang

on the basis of level of net incomes (excluding 25F Gulabewala in Ganganagar district, Rajasthan, as there were only two small farmers in this village).

The low-income villages were mainly the rainfed villages or villages that had experienced drought (Rewasi) or water shortage (Mahatwar) in the survey year. The high-income villages were those with assured irrigation, multiple cropping, and a crop mix that included some commercial crops (jute in Amarsinghi, sugarcane in Alabujanahalli, fruits and vegetables in Nimshirgaon). The medium-income category presented a mixed bag. Except Warwat Khanderao, all the villages had irrigation, but incomes were not as high as in the third category. Various factors accounted for this. For example, Panahar is a high cropping intensity, irrigated village, but in the survey year there were huge losses in potato cultivation that brought down incomes (see Chapter 6). Cultivators in Kalmandasguri were also affected by the crash in potato prices. Siresandra is a village with widespread mulberry cultivation, the returns from which were not included in crop incomes (they were counted in incomes from sericulture; see Chapter 5). The main *kharif* crop failed in Gharsondi in the survey year on account of a bad monsoon and pest attack. In other words, small farmers in most of the villages in the middle-income category may well have reported higher net incomes from farming in other years.

To illustrate this point, we recalculated gross and net incomes from crop production in Gharsondi village with certain assumptions based on district-level data for previous years. These assumptions relate to: (i) standard yields for soybean (monocropped and intercropped) and sesame; (ii) standard prices for soybean and sesame; and (iii) a cost escalation, assuming that harvest and post-harvest operations were carried out on the crops that actually failed. After making these adjustments, the gross value of output among small farmers in Gharsondi rose to Rs 69,294 per hectare (as compared to the actual estimate of Rs 46,799) and net income per hectare rose to Rs 44,511 (as compared to Rs 16,462 in Table 6). In other words, small farmers in Gharsondi were likely to get a net return of over Rs 40,000 in a good agricultural year and fall in the third category (see Table 7 above). Correspondingly, with adjusted incomes, the proportion of small farmers with income losses fell to 7 per cent (from 23 per cent in the survey year).

Two conclusions can be made from this data. On the one hand, it is clear that access to irrigation – and with it the cropping intensity and choice of crop-mix – makes a big difference to incomes. On the other hand, irrigation alone is not a guarantor of high incomes. Incomes can be lower than expected or lower than their potential on account of a variety of factors, including a poor monsoon, inadequate irrigation, crop failure, or market factors like price crash, to name but a few.

The riskiness of agriculture for small farmers is brought out further by the data presented in Tables 8 and 9 on the variability of incomes among small farmers, and the number and proportion of households reporting negative incomes or net losses (negative net incomes) from crop production in the survey year.

The data in Table 8 show that there was extreme variability in the returns that accrued to small farmers within a village. In every village there were some small farmers who made losses (minimum income was negative) in the survey year. Further, the median net return (Table 8) was lower than the mean net return (Table 6) in most villages. In fact in Rewasi, the mean income was positive though low (Rs 4,253), while the median income was negative (Rs 286). In the three villages of Andhra Pradesh, however, the mean net income was lower than the median income (in this case the large negative income pulled down the mean relative to the median). Lastly, the highest income among small farmers was a multiple of the mean income. In Amarsinghi, for example, the highest income per hectare among small farmers was 6.5 times the mean income per hectare.

Table 8 *Descriptive statistics on net income from crop production per hectare of operational holding for small farmers by village, 2010–11 prices* in Rs

Village	State	Median	Minimum	Maximum	CV
Ananthavaram	Andhra Pradesh	6920	−41800	439612	4.5
Bukkacherla	Andhra Pradesh	3376	−18723	31900	2.9
Kothapalle	Telangana	12896	−40604	52777	1.6
Harevli	Uttar Pradesh	13716	−72210	85269	1.7
Mahatwar	Uttar Pradesh	6730	−20899	60231	1.4
Nimshirgaon	Maharashtra	32708	−8612	111947	0.8
Warwat Khanderao	Maharashtra	23896	−20348	67391	0.6
Gharsondi	Madhya Pradesh	15500	−22776	133225	1.4
Siresandra	Karnataka	5167	−17271	246460	2.3
Zhapur	Karnataka	326	−30961	53854	3.4
Alabujanahalli	Karnataka	29416	−27709	253577	1.1
Rewasi	Rajasthan	−286	−164530	834010	5.4
Amarsinghi	West Bengal	3855	−15953	273644	2.6
Kalmandasguri	West Bengal	−1801	−14408	29856	3.8
Panahar	West Bengal	41390	−33684	180340	0.9
Tehang	Punjab	47538	−5767	104985	0.7

Note: (i) We have excluded 25F Gulabewala as there were only two small farmer households in that village.
(ii) CV = coefficient of variation.

Table 9 *Number and proportion of households having negative incomes from crop production in the survey year by farmer category, selected villages*

Village	State	Small farmers		Large farmers		Combined (%)
		Number	Proportion	Number	Proportion	
Ananthavaram	Andhra Pradesh	80	35	0	0	20
Bukkacherla	Andhra Pradesh	42	39	18	31	20
Kothapalle	Telangana	24	24	0	0	17
Harevli	Uttar Pradesh	6	13	1	4	7
Mahatwar	Uttar Pradesh	13	19	0	0	12
25F Gulabewala	Rajasthan	0	0	0	0	0
Rewasi	Rajasthan	56	52	2	4	24
Gharsondi	Madhya Pradesh	16	23	10	14	11
Nimshirgaon	Maharashtra	3	1	22	41	11
Warwat Khanderao	Maharashtra	3	3	1	6	4
Siresandra	Karnataka	11	21	0	0	11
Zhapur	Karnataka	13	50	4	29	19
Alabujanahalli	Karnataka	9	8	1	4	7
Amarsinghi	West Bengal	1	1	NA	NA	1
Kalmandasguri	West Bengal	6	8	NA	NA	8
Panahar	West Bengal	43	29	3	33	30
Tehang	Punjab	2	3	2	4	3

Note: The results for Nimshirgaon are based on a sample survey.

With the sole exception of 25F Gulabewala, in all the villages studied we found some cultivator households with negative incomes from crop production in the survey year (Table 9). At the village level, the proportion of loss-making households ranged from three in Tehang to 24 in Rewasi. As already noted, Rewasi suffered a drought and crop failure in the survey year. The proportion of cultivators with negative incomes was also high (20 per cent) in Ananthavaram and Bukkacherla villages of Andhra Pradesh.

In most of the study villages, the incidence of losses from farming was much higher among small farmers than large farmers. To illustrate, in Zhapur, a rainfed village in north Karnataka, 50 per cent of small farmers suffered losses in the survey year as compared to 29 per cent of large farmers. In Alabujanahalli, a village in the Cauvery belt of Karnataka, 8 per cent of small farmers and 4 per cent of large farmers suffered losses in the survey year. The only exception to this pattern was Nimshirgaon in Kolhapur district of Maharashtra (the results come from a sample-based multiplier). In four villages, there were no negative incomes among large farmer households in the survey year.

In terms of scale, 50 per cent of small farmer households made losses from crop production in the two dry villages – Zhapur of Kalaburagi district in Karnataka, and Rewasi of Sikar district in Rajasthan. As noted in Chapter 2, Sikar district was declared to be drought-affected in the reference year, 2009–10. In Bukkacherla, another rainfed village of Andhra Pradesh, 39 per cent of small farmers made losses. To put it differently, negative incomes were reported with a high frequency among small farmers in both rainfed and drought-prone villages. Animal rearing played an important role in the dry villages, and incomes from animal resources are not included in this calculation (see the next section of this chapter and Chapter 5).

The only irrigated village with a high proportion of loss-making cultivators was Ananthavaram in Guntur district of Andhra Pradesh. Almost 40 per cent of small farmers made losses in this village. This was on account of punishing rent payments (see Ramachandran, Rawal, and Swaminathan 2010). Income losses among small farmers were lowest in Nimshirgaon, a multi-crop village in Kolhapur district of Maharashtra, Warwat Khanderao, a cotton-growing village in the Vidarbha region of the same State, and Tehang in the Doaba region of Punjab.

The reasons for negative incomes from crop production varied across villages. In the rainfed village of Zhapur in Kalaburagi district and drought-affected Rewasi village in Sikar district, the losses were primarily on account of crop failure due to poor rainfall and other weather-related factors. The losses were likely to be lower in good rainfall years. In villages with better access to irrigation – such as Ananthavaram, Harevli, and Alabujanahalli – losses were due not so much to lower production (as our data on GVO showed) but to high costs (for example, of irrigation, machine labour, and rent payments). These factors were less related to natural conditions such as the monsoon, and more to markets and production relations. Chapter 6 examines some of the major costs of production in more detail.

While the risk to incomes came from diverse sources, only a handful of small farmers reported any kind of crop insurance.

CROP INSURANCE

Only a handful of small farmer households in the study villages had taken crop insurance (see table below). In nine villages, no household reported crop insurance; in all but one village, less than five households had taken crop insurance. Only in Warwat Khanderao, a cotton-growing village in Buldhana district, Maharashtra, 12 small farmer households reported buying crop insurance (for cotton and green gram).

Number and proportion of farmers reporting crop insurance by farmer category, PARI study villages

Village	State	Small farmers		Big farmers	
		No.	%	No.	%
Ananthavaram	Andhra Pradesh	1	2.1	1	3.6
Bukkacherla	Andhra Pradesh	1	2.8	3	15
Kothapalle	Telangana	0	0	0	0
Harevli	Uttar Pradesh	0	0	0	0
Mahatwar	Uttar Pradesh	0	0	1	20
Warwat Khanderao	Maharashtra	12	11.3	3	18.8
Nimishirgaon	Maharashtra	0	0	0	0
25F Gulabewala	Rajasthan	0	0	7	12.5
Rewasi	Rajasthan	1	0.9	1	2.2
Alabujanahalli	Karnataka	0	0	0	0
Siresandra	Karnataka	0	0	0	0
Zhapur	Karnataka	0	0	0	0
Gharsondi	Madhya Pradesh	3	4.2	5	7
Panahar	West Bengal	5	3.5	0	0
Amarsinghi	West Bengal	1	1.9	0	0
Kalmandasguri	West Bengal	3	4.4	0	0
Tehang	Punjab			0	0

Source: PARI survey data.

To sum up, there are large variations in the gross value of output (GVO) obtained by small farmers in different agro-ecological regions. Net incomes were calculated as gross incomes net of Cost A2 or paid-out costs. Net incomes or returns per hectare among small farmers showed big variations across villages, implying large regional differences in the returns obtained by small farmers. The major difference was as between rainfed and irrigated villages, with small farmers in the latter reporting net incomes that were ten times higher than that in the former.

Secondly, the three-fold categorisation of villages based on crop incomes per hectare showed both qualitative differences between villages with and without irrigation, and the fact that irrigation alone is not a determinant of returns from farming. One could argue that in the low-income group of villages, the major risk was due to natural and climatic factors, whereas in the middle-income group, the major risk was due to market factors. As shown in Chapter 6 of this volume, small farmers faced higher costs and risk from market factors (as, for example, due to a low output price or variable output price) than large farmers.

Thirdly, the data bring out high variability in net incomes among small farmers within a given agro-ecological context, a fact that is rarely discussed in the secondary data-based literature. In other words, small farmer households constitute a heterogeneous group even within a village.

Fourthly, the fact of negative incomes or losses among small farmer households in all the study villages (with the exception of 25F Gulabewala, which had only two small farmers) points to a serious crisis of incomes among small farmers. The burden of crop-related income risk was much higher among small farmers than large farmers. Although our data pertain to a single year and we were not able to track long-term incomes, nevertheless, the data showed that in a given year, a substantial proportion of small farmers in the study villages suffered losses.

Household-level Incomes from Crop Production

From return per unit of operational holding, we now turn to return or profit per household. The data presented in Table 10 show mean and median net incomes from crop cultivation per household by category of farmers.

First, the levels of income from crop production among small farmer households ranged from Rs 4,200 per annum to Rs 50,000 per annum at 2010–11 prices. Not surprisingly, incomes from farming were highest in Tehang village of the Doaba region of Punjab, followed by the two villages of Maharashtra (one a mixed cropping village and the other a cotton-growing village). Nevertheless, even in Nimshirgaon (Kolhapur district, Maharashtra), a village with sugarcane and grape cultivation, the mean annual income from farming was less than Rs 4,000 a month (at 2010–11 prices). Incomes were

Table 10 *Mean and median incomes from crop production per household, by farmer category, study villages, 2010–11 prices* in Rs per household

Village	Small farmers		Large farmers	
	Mean income	Median income	Mean income	Median income
Ananthavaram	4281	2932	301650	110021
Bukkacherla	11074	7528	38309	18380
Kothapalle	11465	11597	55681	13179
Harevli	18602	7585	186553	116746
Mahatwar	8181	2810	79100	73955
Nimshirgaon	44290	35986	192578	78758
Warwat Khanderao	46923	32496	269873	202320
Gharsondi	17371	13085	255635	61054
25F Gulabewala	23798	23798	471892	311651
Alabujanahalli	37280	29416	156622	165249
Siresandra	24163	6146	328855	331066
Zhapur	6401	326	166232	29824
Rewasi	6205	−491	65139	59119
Amarsinghi	23830	19350	−	−
Kalmandasguri	14935	13408	−	−
Panahar	7500	64347	156948	1993
Tehang	58797	52422	436024	292477

very low in villages with no or little irrigation. As already mentioned, net incomes were low in Ananthavaram, an irrigated paddy-growing village in Guntur district, Andhra Pradesh, on account of rent payments made by tenant small farmers.

Secondly, the mean or average net income from crop production was higher than the median among small farmers (as also large farmers) in all the study villages except Panahar in Bankura district, West Bengal.[4] In other words, there was unequal distribution of incomes among small farmers. For example, in Zhapur village (Kalaburagi district, Karnataka), the mean crop income among small farmers was Rs 6,401 a year while the median was only Rs 646 a year, at 2010–11 prices.

As may be expected, the mean crop incomes of small farmers were significantly lower than the corresponding averages for large farmers in *every* village except Kothapalle (Karimnagar district, Telangana) and Zhapur, where the number of large farmers was small (14 and 9, respectively). This observation

[4] In Panahar, there were many households with negative incomes, pulling the mean below the median.

of lower annual incomes from farming among small farmers is to be expected since the variable is income per household, and small farmers have less land than large farmers by definition. Nevertheless, what it shows is that in the current environment, small farmers are unable to overcome their disadvantage in terms of extent of holding by means of crop choice or intensive cultivation.

Lastly, and most crucially, the level of income generated from crop farming alone was low in relation to the requirements of subsistence. We have estimated subsistence requirements or minimum income on the basis of the official minimum wage (as discussed in Chapter 5). Based on the minimum wage criterion, the proportion of small farmer households with crop income less than the minimum requirement ranged from 45 per cent in Tehang (Jalandhar district, Punjab) to 97 per cent in Kalmandasguri (Koch Bihar district, West Bengal). The proportion of households with less than minimum income was higher among small farmers than large farmers. Further, in some villages, particularly rainfed villages, the proportion of large farmer households with less than minimum wage was also high. Crop incomes generated more than the minimum subsistence income for large farmers in the better irrigated villages of Harevli (Bijnor district, Uttar Pradesh), Alabujanahalli (Mandya district, Karnataka), and Tehang.

Table 11 *Proportion of farmer households with crop income less than minimum income, by farmer category, selected villages* in per cent

Village	Small farmers	Large farmers
Ananthavaram	91	20
Bukkacherla	83	62
Kothapalle	93	89
Harevli	80	8
Mahatwar	91	40
Nimshirgaon	42	41
Warwat Khanderao	44	19
Gharsondi	83	32
25F Gulabewala	0	0
Siresandra	86	0
Zhapur	92	57
Alabujanahalli	67	4
Rewasi	95	39
Amarsinghi	96	–
Kalmandasguri	97	–
Panahar	95	50
Tehang	45	7

Note: To calculate minimum wage, we have used the Minimum Wage for Agriculture by State in 2011, as reported by the Labour Bureau, and assumed 300 days of employment.

To sum up our discussion on absolute returns or profits from farming per household, the data showed large variations for small farmers across villages and among small farmers within a village. The main question examined in this section is the adequacy of income from crop production for subsistence (using the minimum wage as our criterion for subsistence). The answer is that in nine study villages, for 80 per cent or more of small farmers, returns from farming were less than the minimum wage. To put it differently, incomes from farming cannot be the mainstay for a large majority of small farmer households. After all, Rs 3,000 a hectare a year is low by any standard. Even in an irrigated village, income from crop production of around Rs 40,000 a hectare for an average small farmer is poor and insufficient.[5]

Household-level Incomes from Crop Production and Animal Resources

In many parts of India, particularly in rainfed regions, the livestock economy is an integral part of agriculture, and livelihoods depend on the crop–livestock cycle rather than on crop incomes alone. In Table 12, we have reported the mean income of small farmers from crop-cum-livestock farming. As column 3 of the table makes clear, incomes from crop and animal farming were higher than incomes from crop cultivation alone in all the villages. However, incomes from animal rearing were as important as incomes from crop cultivation (that is, double the mean from crop incomes) in several villages, notably the rainfed and drought-prone villages of Bukkacherla (Anantapur district, Andhra Pradesh), Rewasi (Sikar district, Rajasthan), and Zhapur (Kalaburagi district, Karnataka). The ratio was less than 1.5 in three irrigated villages – Harevli of Bijnor district, western Uttar Pradesh; Nimshirgaon of Kolhapur district, western Maharashtra; and Alabujanahalli of Mandya district, Karnataka. The case of Anathavaram (Guntur district, Andhra Pradesh) is special. Here, small farmers were tenants for whom a major benefit from paddy cultivation was the use of straw as fodder for livestock rearing (Ramachandran, Rawal, and Swaminathan 2010).

Let us take the case of Rewasi. In this village, the mean income of small famer households from crop production alone was Rs 4,254 (Table 6), but with incomes from animal resources included, the mean income rose five-fold to Rs 28,272. Correspondingly, the proportion of small farmer households with negative incomes fell from 52 to 16.

[5] If we assume that a small farmer typically operates 1 hectare, the numbers above are equivalent to the income per year for a household. The level and composition of household incomes among small farmers are discussed further in Chapter 5.

Table 12 *Average income from crop production and animal husbandry (CPAL) among small farmers, selected villages, at 2010–11 prices* in Rs per household

Village	State	Mean income from crop production and animal husbandry (CPAL)	Ratio of CPAL to mean income from crop farming*	Proportion of households with negative crop incomes plus animal rearing incomes
Ananthavaram	Andhra Pradesh	36235	8.5	15
Bukkacherla	Andhra Pradesh	21622	1.9	34
Kothapalle	Telangana	32977	2.9	12
Harevli	Uttar Pradesh	26957	1.4	11
Mahatwar	Uttar Pradesh	14409	1.8	9
Nimshirgaon	Maharashtra	64893	1.5	0
Warwat Khanderao	Maharashtra	57000	1.2	3
Siresandra	Karnataka	56715	4.6	5
Zhapur	Karnataka	11940	1.8	38
Alabujanahalli	Karnataka	48090	1.3	4
Rewasi	Rajasthan	28272	4.6	16
Amarsinghi	West Bengal	28668	1.2	0
Kalmandasguri	West Bengal	28419	1.5	6
Panahar	West Bengal	10030	1.3	22
Gharsondi	Madhya Pradesh	31759	1.8	13
Tehang	Punjab	97511	1.7	1

Note: * Mean crop incomes are as reported in Table 6.

Animal rearing was an integral part of the small farmer economy in most villages, and including incomes from animal rearing raised the income per household in all villages. In some villages, incomes from animal resources were more than double the total farm incomes.

Relative Returns of Small Farmers and Large Farmers

We now turn to the question of "efficiency" or relative performance of small farmers vis-à-vis large farmers (that is, farmers with more than 2 hectares of irrigated land or 6 hectares of unirrigated land). The net income or profit per hectare was invariably lower among small farmers than large farmers within the boundaries of a village. For example, in Harevli (Bijnor district, western Uttar Pradesh), the average net income per hectare was Rs 14,317 among small farmers and Rs 33,420 among large farmers. In Rewasi (Sikar district, Rajasthan) the gap was larger: Rs 21,485 for small farmers and Rs 695,746 for large farmers.

Table 13 *Test for difference in means (t-test) of net incomes from crop production per hectare of operational holding, by farmer category, study villages*

Village	State	Net income	
		T value	P value
Ananthavaram	Andhra Pradesh	6.1	0*
Bukkacherla	Andhra Pradesh	−0.653	0.515
Harevli	Uttar Pradesh	2.88	0.006*
Warwat Khanderao	Maharashtra	0.24	0.81
Zhapur	Karnataka	0.82	0.41
Alabujanahalli	Karnataka	0.91	0.36
Rewasi	Rajasthan	4.51	0*
Gharsondi	Madhya Pradesh	0.44	0.622
Tehang	Tehang	2.19	0.029*

Note: * statistically significant at 5 per cent level of confidence.

We tested for difference in means of net incomes per hectare by farmer category (Table 13). The comparison was not very relevant for the small farmer-dominated villages of Siresandra (Kolar district, Karnataka), Kothapalle (Karimnagar district, Andhra Pradesh), and Mahatwar (Ballia district Uttar Pradesh), so these were excluded from the analysis. We also excluded the three villages of West Bengal (which comprised predominantly small farmers) and 25F Gulabewala in Rajasthan (with only two small farmer households) from this analysis.

The difference in net incomes was statistically significant at a 5 per cent level of significance in four of the nine villages: Ananthavaram (Guntur district, Andhra Pradesh), Harevli, Rewasi, and Tehang (Jalandhar district, Punjab). Except for Rewasi, these were all irrigated villages cultivating multiple crops. In Ananthavaram, the costs of cultivation were lower for owner–cultivators as compared to tenant small farmers (Chapter 6). In Rewasi, larger farmers with access to tubewells were able to manage better during drought. Differences in net incomes across categories of cultivators in Harevli arose from differences in cropping pattern (more area under sugarcane among larger farmers relative to smaller farmers) and access to irrigation (there was substantial inequality in ownership of irrigation equipment and access to additional sources of water) (Rawal 2013). Similarly, in Rewasi, the big landowners had more irrigated land and, as a result, the extent of crop failure on their land was less than on the land of small farmers (*ibid.*)

Another comparison that is telling is between the average net income of small farmers and the highest net income reported in the same village (invariably that obtained by a big landlord or capitalist farmer). The gap between the

Table 14 *Net incomes from crop production per hectare of operational holding by farmer category, study villages, at 2010–11 prices* in Rs per hectare

Village	State	Small farmers	Large farmers	Landlords/ Capitalist farmers	Ratio of column 3/column 1
		Mean	Mean	Maximum	
Ananthavaram	Andhra Pradesh	–1294	159606	42676	NA
Bukkacherla	Andhra Pradesh	3666	11785	14583	4
Kothapalle	Telangana	11244	4381	26356	2
Harevli	Uttar Pradesh	14317	35047	57899	4
Mahatwar	Uttar Pradesh	8604	11409	17627	2
Warwat Khanderao	Maharashtra	24029	23890	39668	2
Nimshirgaon	Maharashtra	43208	27427	221379	5
25F Gulabewala	Rajasthan	9671	25609	46257	5
Gharsondi	Madhya Pradesh	16447	15038	83498	5
Zhapur	Karnataka	3122	6477	35780	11
Alabujanahalli	Karnataka	42381	50585	48769	1.1
Rewasi	Rajasthan	21585	69746	137557	6
Panahar	West Bengal	15491.14	–21723.2	90527	6

Note: We have dropped three villages without any landlord/capitalist farmer household, namely, Siresandra, Amarsinghi, and Kalmandasguri.

mean incomes of small farmers and the highest income per hectare obtained by a landlord or capitalist farmer within the same village was very large (Table 14). The highest net income was more than twice the mean income of a small farmer in all but one village, namely, Alabujanahalli in Mandya district of Karnataka. In Rewasi, the ratio of the highest income to the mean among small farmers was 5:1, and it was 10:1 in Zhapur (Kalaburagi district, Karnataka).[6] In other words, if the constraints faced by small farmers are addressed, there is scope for raising the return per hectare.

Relative Returns with Cost of Family Labour Included

Most small farmer households use family labour in the process of cultivation. The Cost A2 measure of costs does not include any imputed value for the use of family labour. In this section, we have attempted to place a value on family labour (FL) and computed Cost A2 + FL. The cost imputed to family labour is the prevailing daily wage rate in the village. The labour performed by women and children is valued at the prevailing daily wage rate for female

---

[6] We have not carried out statistical tests of difference as there were only one or two landlords/ capitalist farmers in a village.

labour in the village, and the labour input of men is valued at the prevailing male wage rate.

To see the difference in use of family labour between small farmers and large farmers, we computed two costs per unit of operational holding. Costs with family labour (A2 + FL) were distinctly higher than Cost A2 alone among small farmers in all villages, implying that small farmers did use family labour for crop cultivation (Table 15).[7] Family labour was intensively used in the villages of West Bengal and in Mahatwar of eastern Uttar Pradesh. In Kalmandasguri of Koch Bihar district, West Bengal, the costs of cultivation nearly tripled when the imputed value of family labour was added to paid-out costs.

As expected, net incomes reduced when the imputed value of family labour was included in costs of cultivation, that is Cost A2 + FL, and, of course, the reduction was more for small farmers as they deployed more family labour than large farmers. With costs of family labour included, net incomes were negative for small farmers in eight villages. In these eight villages, the returns from crop production were not even equivalent to the wage incomes that small farmers could have gained if the days of family labour had been days of labouring-out instead. This raises an important question, as to why small farmers continue to engage in crop production.

The above two sections establish that the profits or net returns per hectare for small farmers were not higher than for large farmers. In fact, on average, net returns for small farmers were lower than for large farmers in the same village, and this difference was statistically significant in four villages, of which three were high-productivity, irrigated villages. Secondly, the disadvantage of small farmers relative to large farmers was likely to widen if the cost of family labour was included, as small farmers inevitably used more family labour than large farmers. Thirdly, the gap between net returns obtained by a small farmer and a landlord/capitalist farmer in the same village was large.

Crop-wise Net Incomes per Hectare

We now briefly turn to incomes from production of selected crops – paddy, wheat, and sugarcane – among small farmers.

In paddy cultivation, it is interesting to note that across five villages located in Andhra Pradesh, Telangana, Karnataka, and Punjab, the GVO obtained by small farmers was around Rs 50,000 per hectare on average (Table 16). There were, however, major variations in the costs of cultivation, as a result of which net incomes varied from Rs 4,000 to Rs 33,000 per hectare. Net incomes were

---

[7] This in no way precluded the use of hired labour (see Chapter 3).

Table 15 *Average Cost A2 and Cost A2 + FL per hectare of operational holding for small farmers, study villages, at 2010–11 prices in Rs per hectare*

Village	State	Cost A2	Net income over Cost A2	Cost A2 + FL	Net income over Cost A2+FL	Ratio of Cost A2 + FL / Cost A2
Ananthavaram	Andhra Pradesh	93703	−1294	99920	−7511	1.1
Bukkacherla	Andhra Pradesh	19407	3666	21956	1108	1.1
Kothapalle	Telangana	32662	11244	41572	2334	1.3
Harevli	Uttar Pradesh	45331	14372	65929	−6225	1.4
Mahatwar	Uttar Pradesh	26967	8604	48492	−12833	1.8
Gharsondi	Madhya Pradesh	24338	16462	28048	12752	1.2
Warwat Khanderao	Maharashtra	17618	24031	20594	21054	1.2
Nimshirgaon	Maharashtra	42991	43208	50131	34913	1.2
Siresandra	Karnataka	35594	18402	42874	11122	1.2
Zhapur	Karnataka	13544	3123	17583	−584	1.3
Alabujanahalli	Karnataka	75658	46060	100946	48602	2.1
Rewasi	Rajasthan	11795	4373	20610	−4442	1.7
Amarsinghi	West Bengal	59641	49064	110948	−1801	1.9
Kalmandasguri	West Bengal	39238	39527	107384	−28299	2.7
Panahar	West Bengal	73990	15323	121297	−31579	1.6

Note: 25F Gulabewala and Tehang have been excluded because of too few observations.

Table 16 *Average GVO, Cost A2, and net income per hectare for paddy (summer), by farmer category, study villages, at 2010–11 prices* in Rs per hectare

Village	State	Small farmers			Large farmers		
		GVO	Cost A2	Net income	GVO	Cost A2	Net income
Ananthavaram	Andhra Pradesh	50304	46255	4049	48461	38996	9467
Bukkacherla	Andhra Pradesh	51377	44517	6860	49490	44856	4635
Kothapalle	Telangana	49792	37341	12453	43298	39345	3954
Alabujanahalli	Karnataka	53490	39796	13694	56970	38004	18966
Amarsinghi	West Bengal	44215	24629	19586			
Kalmandasguri	West Bengal	32532	12026	20506			
Panahar	West Bengal	46239	26936	19303	48972	29818	19153
Tehang	Punjab	57418	23961	33457	62311	31807	30503

Table 17 *Average GVO, Cost A2, and net income per hectare for paddy (winter), by farmer category, study villages, at 2010–11 prices* in Rs per hectare

Village	State	Small farmers			Large farmers		
		GVO	Cost A2	Net income	GVO	Cost A2	Net income
Amarsinghi	West Bengal	68829	41554	27274			
Panahar	West Bengal	51400	33851	17549	56146	30414	25731

highest in the village of Tehang (where cultivation was relatively mechanised) and lowest in Ananthavaram (due to high costs including rent). Further, net incomes of small farmers from paddy cultivation were not necessarily lower than that of large farmers.

In West Bengal, *rabi* or *boro* paddy showed higher GVO, but costs were also higher (Table 17).

The picture is a little different in the case of wheat (Table 18). First, GVO per hectare itself varied widely across the villages, being the lowest in Mahatwar, Ballia district, Uttar Pradesh (due to both poor yield and low prices), and highest, and more than twice as much as in Mahatwar, in Tehang, Jalandhar district, Punjab (due to high yield and minimum support prices). Variation was less in costs, but nevertheless, net incomes varied. Further, except in Tehang (with only seven small farmer households), the net income of small farmers was less than that of large farmers in respect of wheat cultivation in all villages.

What this implies is that when considering returns from farming for small farmers, the crop and crop environment (by which we mean not just inputs and technology, but also marketing and prices) make a big difference, as

Table 18 *Average GVO, Cost A2, and net income per hectare for wheat, by farmer category, study villages, at 2010–11 prices* in Rs per hectare

Village	State	Small farmers			Large farmers		
		GVO	Cost A2	Net income	GVO	Cost A2	Net income
Harevli	Uttar Pradesh	31887	28643	3243	36015	27612	8405
Mahatwar	Uttar Pradesh	24080	16447	7632	22041	12841	9200
Gharsondi	Madhya Pradesh	33126	18705	14421	38634	20001	18632
Rewasi	Rajasthan	49872	25361	24511	49826	20130	29697
Tehang	Punjab	62031	23863	38168	60128	28967	31025

all small farmers do not face similar conditions. Further, differences in net incomes across farmer categories were not large in the case of paddy but were large in the case of wheat. Differences in incomes as between small farmers and large farmers partly depended on crop choice.

To sum up, even at the crop level, there were notable differences across villages, deriving from differences in productive forces and production relations across villages.

CONCLUSIONS

This chapter has examined incomes from farming for small farmers in 17 villages across nine States. The two main questions addressed were: (i) the extent to which absolute incomes generated from farming were adequate for basic subsistence, and (ii) the relative returns of small farmers as compared to other strata in a village.

Crop-mix and crop choice
(i) First, the choice of crop-mix depended on multiple factors, and cropping patterns varied across agro-ecological regions. In dry or rainfed villages, intercropping was more prominent among small farmers than large farmers. This choice can be interpreted as a risk-reducing strategy in an uncertain agricultural environment.
(ii) Secondly, allocation of land to foodgrains was generally higher among small farmers within a given village context. Proportionately less area under non-foodgrains or purely commercial crops among small farmers may reflect a lack of resources (including irrigation). The preference for foodgrain cultivation, however, should not be conflated with subsistence production, as almost all small farmers in the study villages participated in the product market in some way, even for foodgrains.

Net incomes from farming per hectare

(iii) There was tremendous variation in net crop incomes of small farmers across the study villages and, correspondingly, across different agro-ecological regimes. In other words, small farmers in different agrarian regimes were very different from each other.

(iv) Incomes from farming alone were low for the average small farmer, particularly in rainfed and drought-prone villages. In irrigated villages, gross incomes were higher, but net incomes varied depending on cost and prices (see Chapter 6).

(v) In almost every village, a proportion of small farmers made losses in crop production in the survey year. This was particularly marked in rainfed and drought-affected villages.

Net incomes from farming per household

(vi) Net incomes from crop production per household fell below the minimum wage for a large majority of small farmers. In other words, incomes from crop production alone could not ensure subsistence income for the majority of small farmers.

(vii) Incomes from animal rearing made up for net losses from cropping in dry regions of the country such as Rajasthan. This does suggest that income from agriculture and allied activities should be examined together at the household level.

(viii) The extent of family labour used in crop production was higher for small farmers than large farmers. Correspondingly, including the imputed cost of family labour reduced net incomes more sharply for small farmers than for large farmers. The fact of negative net incomes shows that the return to family labour could have been less than the return to wage labour.

Relative returns: small farmers versus large farmers

(ix) While the gross value of output (GVO) per hectare did not vary much between small farmers and large famers, there were significant differences in the net incomes of small and large farmers, particularly in irrigated villages. Thus, our village survey data do not show any evidence of higher returns per hectare for small farmers relative to large farmers, that is, the inverse farm size profitability hypothesis.

(x) The differences in returns per hectare across farmer category were due to multiple factors, and some of these differences are explored in the following chapters on costs and prices (Chapter 6), fertilizer use (Chapter 7), and access to credit (Chapter 8).

(xi) Lastly, there was a big gap between the profit per hectare of an average small farmer and the farmer with the highest crop income (a landlord or capitalist farmer) in every village. This shows that small farmers operated under multiple constraints and were unable to obtain returns similar to that of big farmers.

To conclude, this chapter brings out the low incomes and vulnerability of a majority of small farmers in our study villages. It also brings out the income disadvantage of small farmers relative to farmers with larger landholdings. The observed differentiation among cultivators in a village is a warning against labelling entire agrarian and farming communities as being in crisis.

REFERENCES

Foundation for Agrarian Studies (FAS) (2015), "Calculation of Household Incomes: A Note on Methodology," available at http://fas.org.in/wp-content/themes/zakat/pdf/Survey-method-tool/Calculation%20of%20Household%20Incomes%20%20A%20Note%20on%20Methodology.pdf, viewed on 20 February 2017.

Ramachandran, V. K., Rawal, Vikas, and Swaminathan, Madhura (2010), *Socioeconomic Surveys of Three Villages of Andhra Pradesh: A Study of Agrarian Relations*, Tulika Books, New Delhi.

Rawal, Vikas (2013), "Cost of Cultivation and Farm Business Incomes in India," Institute of Economic Research, Hitotsubashi University, Working Paper Series No. 2012–15, March.

Rawal, Vikas, and Swaminathan, Madhura (2011), "Are There Benefits from the Cultivation of Bt Cotton? A Comment Based on Data from a Vidarbha Village," *Review of Agrarian Studies*, vol. 1, no. 1, available at http://ras.org.in/are_there_benefits_from_the_cultivation_of_bt_cotton_a_comment_based_on_data_from_a_vidarbha_village, viewed on 20 February 2017.

Rawal, Vikas, and Swaminathan, Madhura (2012), "Returns from Crop Cultivation and Scale of Production," paper presented at the Workshop on Policy Options and Investment Priorities for Accelerating Agricultural Productivity and Development in India, New Delhi.

Surjit, V. (2008), "Farm Business Income in India: A Study of Two Rice Growing Villages of Thanjavur Region, Tamil Nadu," unpublished PhD thesis, University of Calcutta, Kolkata.

Appendix Table 1 *Gross cropped area (GCA) by season, study villages* in acres and per cent

Village	Farmer category	Kharif		Rabi		Annual		Total	
		Extent	Per cent of GCA	Extent	Per cent of GCA	Extent	Per cent of GCA	Extent	Per cent of GCA
Ananthavaram	Large	150	43	152	44	47	13	349	100
	Small	179	51	171	48	4	1	353	100
Bukkacherla	Large	290	85	41	12	9	3	341	100
	Small	258	89	31	11	1	0	289	100
Kothapalle	Large	30	64	12	26	4	9	47	100
	Small	70	51	57	41	11	8	138	100
Harevli	Large	24	16	37	25	86	59	147	100
	Small	14	32	10	23	20	45	44	100
Mahatwar	Large	19	46	20	49	2	4	41	100
	Small	37	47	38	48	4	5	80	100
Nimshirgaon	Large	120	48	46	18	82	33	249	100
	Small	125	40	85	27	106	34	316	100
Warwat Khanderao	Large	157	92	13	8	0	0	171	100
	Small	208	97	7	3	0	0	215	100
Gharsondi	Large	557	50	564	50	3	0	1124	100
	Small	86	51	83	49	0	0	168	100
25F Gulabewala	Large	380	30	872	69	10	1	1263	100
	Small	1	34	1.7	65	0	0	2.7	100
Siresandra	Large	7	26	10	37	9	33	27	100
	Small	37	61	11	18	12	20	61	100

(continued)

Appendix Table 1 (continued)

Village	Farmer category	Kharif		Rabi		Annual		Total	
		Extent	Per cent of GCA	Extent	Per cent of GCA	Extent	Per cent of GCA	Extent	Per cent of GCA
Zhapur	Large	137	82	29	17	1	1	167	100
	Small	49	83	10	17	0	0	59	100
Alabujanahalli	Large	31	27	24	21	59	51	114	100
	Small	39	29	33	24	65	48	137	100
Rewasi	Large	151	49	158	51	0	0	309	100
	Small	150	70	64	30	0	0	214	100
Amarsinghi	Small	25	50	25	50	0	0	50	100
Kalmandasguri	Small	47	77	20	21	1	1	68	100
Panahar	Large	28	47	31	53	0	0	59	100
	Small	50	47	57	53	0	0	107	100
Tehang	Large	632	51	629	49	16	1	1314	100
	Small	48	49	49	51	0	0	97	100

APPENDIX 1

The components of Cost A1 are:

- Value of hired human labour
- Value of hired and owned bullock labour
- Value of owned machine labour
- Value of hired machine charges
- Value of seed (both farm-produced and purchased)
- Value of insecticides and pesticides
- Value of manure (owned and purchased)
- Value of fertilizer
- Irrigation charge
- Depreciation of implements and farm buildings
- Land revenue, cesses, and other taxes
- Interest on working capital
- Other miscellaneous expenses

Cost A2 = Cost A1 + Rent paid for leased-in land
Cost A2 + FL = Cost A1 plus value of imputed family labour

5

Incomes of Small Farmer Households

Aparajita Bakshi, with Tapas Singh Modak

INTRODUCTION

An important contribution of recent literature on rural incomes and livelihoods is the understanding that rural households are not solely dependent on agriculture, and that other sources of income and employment have a significant impact on rural well-being and rural poverty. This understanding, of course, emanated from the myriad processes of transformation that were experienced in developing economies in the 1980s and 1990s. Urbanisation, rapid penetration of markets, industrialisation – no matter how uneven or narrowly based, and the spread of education – no matter how thin, changed the development contexts in post-colonial nations, and this needed a new approach to understanding rural realities.

This chapter analyses the household incomes of small farmer households in India; the different kinds of activities within and outside agriculture that members of small farmer households engage in; and the incomes they receive from different sources. The chapter specifically tries to address three questions:

(i) What is the extent of income deprivation experienced by small farmer households?

(ii) How important is farming to the incomes and livelihoods of small farmer households? What are the other sources of income that small farmers depend on, and to what extent?

(iii) How does the access of small farmer households to different sources of income, particularly non-agricultural income, impact household income poverty?

In this analysis we have also tried to understand, though in a very limited way, the income choices of small farmer households in the larger context of development in India.

INCOMES AND LIVELIHOODS OF FARMER HOUSEHOLDS:
A BRIEF REVIEW OF THE LITERATURE

Locating the Small Farmer in the Process of Structural Change and Rural Transformation

The trajectory of structural change and economic development of current transition economies like India are vastly different from that experienced by developed economies in the past. Lewis (1952) described the process of development in a less developed economy as one characterised by the migration of surplus labour from a traditional subsistence sector to a modern industrial sector. The traditional sector was viewed largely as rural and agricultural, while the modern industrial sector was urban. Development, as perceived by Lewis, was necessarily external to agriculture, and any growth of agricultural productivity and wages could only decelerate the process of rapid industrialisation by increasing the prices of wage goods and industrial inputs, distorting the terms of trade between agriculture and industry, and reducing surplus accumulation in the industrial sector. In this scheme, the small farmer was traditional, unproductive, and would exit agriculture to join the industrial work force as development progressed. The process of development, which is ultimately the process of modernisation of the whole economy including agriculture, was thus also envisaged as a process of de-peasantisation and proletarianisation of the peasantry as modern industrial relations and technology replaced traditional peasant farming, much in the same way as it historically occurred in the developed countries.

The notion of the peasant as traditional and unproductive, and incapable of adopting new technologies, was challenged in the 1960s and 1970s. Schultz (1962) characterised the peasant as a rational farmer who takes informed decisions on agriculture and technology. The new characterisation of farmers as rational and efficient decision makers also challenged the old-school view of the drudgery of peasant farming mired in problems of small-scale agriculture, and emphasised the higher productivity of small farms[1] and the scale-neutrality of modern agricultural technology (Lipton and Longhurst 1989). This, they argued, is particularly so in the changing context of agriculture in developing countries, with introduction of the green revolution technology of hybrid seeds, chemical fertilizers, and irrigation-intensive agriculture.

The changing characterisation of small farms and small farmers also led to a new theorisation of structural change. The rural development school posited

---

[1] The debate on the inverse relationship between farm size and productivity is discussed in earlier chapters, and hence is not cited here.

that technological changes and productivity growth in agriculture could spur demand-led industrial growth (Mellor 1976). An increase in farmers' incomes and agricultural wages could increase demand for non-agricultural commodities through forward and backward demand linkages, and thus lead to growth of the 'rural non-farm sector' (as opposed to the urban industrial sector). Thus, the rural is transformed from a mere supplier of work force and raw materials to an active participant and agent of modern industrial growth. In fact, the rural becomes the very site of economic transformation, and farmers the architects of such transformation, according to the rural development theory.

The optimism that surrounded the notion of equitable economic growth led by highly productive small farms, advocated by populist rural development theorists and policy makers, cast aside old worries of income poverty and backwardness of small farmers. Nor was the role of large public investments in agricultural research, extension, infrastructure, and subsidised provision of inputs that sustained the small farmer advantage discussed in much detail, though Mellor (1976) did suggest that governments should invest in rural development and rural infrastructure to strengthen demand-led linkages for the growth of the rural non-farm sector.

The literature on rural livelihoods that emerged in the 1990s examined more critically the viability of small farms and the role of the rural non-farm sector.

Livelihoods and Livelihood Diversification of Farmer Households

The disenchantment with rural growth theories rested on two sets of empirical evidence. First, it was found that large sections of small farmers were poor (Fann and Chan-Kang 2005; Fann et al. 1999; Singh et al. 2002). The viability of small farms in a situation of growing fragmentation of landholdings, liberalised trade regimes, and unfair competition faced by small farmers in developing countries from rich countries with highly protected agricultural policies, was increasingly being questioned (Hazell 2005; Fann and Chan-Kang 2005).

Secondly, empirical evidence suggested that rural households in a large part of the developing (as well as developed) world, besides being actively engaged in agriculture, derived incomes from non-agricultural sources. Reardon et al. (2007) reviewed a number of income surveys in developing countries and concluded that:

Rural non-farm income (RNFY) constitutes roughly 35 per cent of rural household income in Africa and about 50 per cent in Asia and Latin America.

Suffice it to say that most of the studies reported showed from moderate to fast growth in the share of RNFY in total income over the past two decades. In China, for instance, in 1981 only 15 per cent of rural households worked off-farm, compared to 32 per cent in 1995 (De Brauw *et al.* 2002). In Bangladesh 42 per cent of rural income came from RNFY in 1987, but by 2000 the share was 54 per cent (Hossain 2004). Clearly integrated farm–nonfarm households are a common sight across the developing world, and the upward trend is steep. (*Ibid.*, p. 117)

Thus, this new stream of literature broke the myth of a self-sustaining small farmer household and emphasised the inadequacy of incomes from small farms. It also refuted the segregation of farm or agricultural households/workers and non-farm or non-agricultural households/ workers, and showed that households and workers engaged in both kinds of activities simultaneously. This phenomenon was termed as "livelihood diversification." Ellis (1998) defined livelihood diversification as "the process by which rural families construct a diverse portfolio of activities and social support capabilities in their struggle for survival and in order to improve their standards of living."

The major differences between the perspective of the livelihoods school and the earlier discourse arise from the fact that while the earlier literature understood transformation of the rural economy as part of a process of structural change and modernisation (or as a process of technological progress and development of capitalism) in developing economies, the livelihoods literature tried to explain livelihood and occupational diversification as individual and conscious choices of the rural people. Rural households and workers were rational agents making choices regarding their occupations and income portfolios to maximise their well-being, subject to the various labour, capital, assets, and other meso- and macro-level structural constraints they face. How these micro-, meso-, and macro-level constraints come into being, or how the larger economy takes shapes in the course of time was not a major concern. In the preoccupation with poverty and the well-being of households, the larger vision of development was lost. This approach emphasised household decision-making, the livelihood choices that households make in order to accumulate wealth and overcome poverty. However, the approach is unable to reconcile household decisions to larger processes of economic transformation in a country and region. It is ironic that while individuals gain some agency in this theory to make informed choices about their employment and occupations, they also lose the agency to collectively bring about changes to shape the world they live in.

By definition, escaping poverty is the prime motivation for livelihood diversification, and much of the theoretical and policy discussions on rural livelihoods focus on this issue.[2] However, it must be noted that the relationship between livelihood diversification and poverty is not linear.

> Diversification may occur both as a deliberate household strategy [Stark 1994] and as an involuntary response to crisis [Davies 1996]. It is found both to diminish [Adams 1994] and to accentuate [Evans and Ngau 1991] rural inequality. It can act both as a safety valve for the rural poor [Zoomers and Kleinpenning 1996] and as a means of accumulation for the rural rich [Hart 1994]. It can benefit farm investment and productivity [Carter 1997] or impoverish agriculture by withdrawing critical resources [Low 1986]. (Cited in Ellis 1998)

There is a large body of empirical literature that has emerged which analyses the role of non-farm incomes in poverty reduction in India (Ravallion and Dutt 1999; Farrington *et al.* 2006; Deb *et al.* 2002; Dev and Mahajan 2005; Lanjouw and Shariff 2004; Jayaraj 2004). The literature explores various household and individual characteristics (such as land, education, social stratification) that explain households' and workers' access to non-farm employment, and the diversification strategies (type of agricultural and non-agricultural occupations) adopted by households in order to mitigate poverty and accumulate wealth. However, it rarely engages with macro-level issues of economic growth or structural change.

The Future of Small-Scale Farming

Three broad trends in the discussion on incomes of farmers in general, and small farmers in particular, take three different positions on the future of small-scale farming as an economy develops. The classical literature predicts a gradual demise of small-scale farming with economic development. The rural development and neo-populist school of thought of the 1970s and 1980s foresees the survival and stability of efficient and productive small-scale farming. The livelihood literature generally finds the question of the future and survival of small farms redundant, as small farmers are no longer tied to agriculture, but also receive incomes from a vibrant, non-farm economy (Wiggins 2014).

---

[2] The livelihoods discourse gave rise to different sustainable livelihood frameworks to guide policy research and formulation, and these were adopted by international development agencies, such as the sustainable livelihoods framework of the Department for International Development (DfID), the United Nations Development Programme (UNDP) promoting sustainable livelihoods approach. The discourse had wide influence on development interventions by international agencies and poverty alleviation programmes implemented by governments of the developing nations.

However, it is also acknowledged that the survival and economic well-being of the small farmer, whether solely dependent on agriculture or with a diversified income portfolio, crucially depend on a large number of external factors, such as urbanisation, infrastructure, and connectivity; development of the rural non-farm sector; government support to agriculture and to the rural non-farm economy. According to Fann and Chan-Kang (2005), for small farmers to survive in the era of globalisation, land policies should be conducive to their better access to land through lease and purchase; targeted credit, input, and technology subsidies for them; development of high-value commodities; development of the rural non-farm sector; and protection of rights and social services at their place of work for farmers who migrate to urban areas seasonally. Hazell (2005) also outlined similar policies including agricultural research and extension related to technologies specifically appropriate for small farmers, and credit and insurance policies to reduce production and marketing risks. Investment in rural public goods – electricity and transport, road infrastructure, and education – are important for the development of the rural non-farm sector (Wiggins 2014).

ESTIMATION OF HOUSEHOLD INCOMES

Any analysis of the household incomes of small farmer households in particular, and households in informal economies in general, faces serious limitations of data. Collection of data and estimation of household incomes of small farmer households in a country such as India involves two problems. The first is at the conceptual and definitional level, and the second at the operational level. There are no uniform definitions of household incomes – that is, the different types of income sources or flows that are to be included as household incomes – in the way that there are uniform definitions of national incomes in the system of national accounts (Bakshi 2011). Thus, different studies using survey-based data of household incomes rely on varying definitions and methods of estimation of different components of income. Any comparison of household income estimates across different surveys or available macro-aggregates becomes impossible.

The problem of estimating household incomes at the operational level is complex in developing country contexts. First, employment and consequently income flows are highly variable within the year, which makes collection of accurate data through surveys difficult. Most surveys use the interview method, and respondents are required to recall past events and report accurately. Longer recall periods and high variability of the variables they are required to recall adversely affect the quality of data. Secondly, there are problems of valuation of goods in a non-monetised economy. Many inputs and outputs are not transacted in the market, or sometimes the transactions are not monetised. Inputs are home-produced, outputs are produced for home consumption, and part of the wages is paid in kind. Such exchanges are common in underdeveloped economies, which present economists with the tedious task of accounting for and valuing such goods and services.

The Foundation for Agrarian Studies (FAS) has developed a detailed methodology to gather data and arrive at reasonably good estimates of household incomes. The methodology can be obtained from the *Income Calculation Manual* prepared by FAS (Foundation for Agrarian Studies 2015). Here we flag a few positive aspects of the income calculation methodology.

(i) Income is a derived variable. Hence, detailed and disaggregated data on all sources of income are collected. The FAS–PARI (Project on Agrarian Relations in India) interview schedule uses separate modules for different income sources. About twelve such sources of income are reported for each household (see Appendix 1).

(ii) Household incomes include all cash and kind flows. Kind flows are valued at annualised local market prices or reported valuation by respondents. Price data from local markets and from reliable respondents are collected separately.

(iii) To the extent possible, an accounting framework is adopted to collect data on incomes from self-employment, such as crop production, livestock, and non-farm business. Income is defined as gross value of output less paid-out cost of production.

(iv) For estimating incomes from crop production, concepts and definitions followed by the Commission for Agricultural Costs and Prices (CACP) of the Government of India are broadly adhered to. Crop income is gross value to output (main and by-products) less Cost A2. Cost A2 excludes cost of family labour and rental value of owned land. It includes all paid-out costs, including imputed value of depreciation of own machinery and interest on working capital.

(v) Data are collected at various levels of disaggregation, to assist in more accurate reporting and better recall from respondents. For example, crop incomes and costs of production are reported and calculated for each crop and crop-mix during the agricultural year. Livestock incomes are calculated for each type of livestock. Wage incomes are calculated for each type of wage employment by every member of household.

(vi) The reference period is the agricultural year.

Having discussed at some length the methodology of household income estimation, let us now focus on some of the results.

INCOME DEPRIVATION

Household well-being would depend on total household income, and not on income per acre of land owned. Hence, irrespective of whether small farmers are more productive or not, the smallness of farm size will always remain a constraint on total household income (unless productivity gains in small farms are exponential compared to larger farms) and consequently on household well-being. In this section we try to understand the income disadvantage of small farmer households.

Low Levels of Income of Small Farmer Households

Table 1 shows the mean and median per capita incomes of small farmer households in the PARI study villages at 2010–11 prices. State-specific consumer price indices for rural labour were used to convert all income estimates at 2010–11 constant prices. The estimates at current prices are reported in Appendix Table 1. The last two columns of the table convert annual incomes to income per day.

Table 1 *Mean and median per capita incomes of small farmer households at 2010-11 prices in Rs*

Village	State	Annual		Per day	
		Mean	Median	Mean	Median
Ananthavaram	Andhra Pradesh	16,688	13,556	45.7	37.1
Bukkacherla	Andhra Pradesh	9,528	7,675	26.1	21.0
Kothapalle	Telangana	11,711	10,322	32.1	28.3
Harevli	Uttar Pradesh	9,181	6,105	25.2	16.7
Mahatwar	Uttar Pradesh	6,167	4,704	16.9	12.9
Nimshirgaon	Maharashtra	19,908	16,321	54.5	44.7
Warwat Khanderao	Maharashtra	15,999	13,303	43.8	36.4
25F Gulabewala*	Rajasthan	30,682	30,682	84.1	84.1
Rewasi	Rajasthan	22,851	18,225	62.6	52
Gharsondi	Madhya Pradesh	7,539	5,878	20.7	16.1
Alabujanahalli	Karnataka	17,932	14,282	49.1	39.1
Siresandra	Karnataka	31,005	20,417	84.9	55.9
Zhapur	Karnataka	8,373	7,001	22.9	19.2
Amarsinghi	West Bengal	15,959	11,463	43.7	31.4
Panahar	West Bengal	10,549	7,046	28.9	19.3
Kalmandasguri	West Bengal	13,676	12,111	37.5	33.2
Tehang	Punjab	55,929	44,955	153.2	123.6

Note: * There were only two small farmer households in 25F Gulabewala.

Among the villages with high small farmer incomes were irrigated villages characterised by highly developed agriculture, such as Tehang (Jalandhar district, Punjab), 25F Gulabewala (Ganganagar district, Rajasthan), and Nimshirgaon (Kolhapur district, Makarashtra) – but also moderately irrigated villages where small farmers had income opportunities other than crop production, such as Siresandra (Kolar district, Karnataka) where sericulture was practised, and Rewasi (Sikar district, Rajasthan) where non-agricultural incomes were high. Among the villages with low small farmer incomes were villages such as Panahar (Bankura district, West Bengal), Gharsondi (Gwalior district, Madhya Pradesh), and Mahatwar (Ballia district, Uttar Pradesh), where farmers faced adverse agricultural conditions during the survey year; low irrigation villages such as Zhapur (Kalaburagi district, Karnataka) and Bukkacherla (Anantapur district, Andhra Pradesh); and villages where the average size of holdings was very small, such as Amarsinghi (Malda district, West Bengal) and Kalmandasguri (Koch Bihar district, West Bengal). The levels of income of small farmer households were thus dependent on a number of factors related to crop incomes and opportunities for receiving incomes outside crop production.

The very low levels of household income received by small farmer households in post-liberalisation India has been frequently noted in the literature. Bhalla (2006) examined the NSS Situation Assessment Survey (2003) data and noted:

> In 14 out of 18 states, the income for the farmer households with landholding up to 2 hectares was insufficient to meet the consumption expenditure, not to talk about meeting the expenditure on capital assets. (Bhalla 2006, p. 74)

Earlier studies based on FAS–PARI data also pointed to the same symptoms. As Ramachandran and Rawal (2010) stated:

> Our survey data from Andhra Pradesh, Uttar Pradesh, and Maharashtra indicate the near-impossibility, in the present circumstances, of peasant households with 2 hectares of operational holdings or less earning an income sufficient for family survival. (*Ibid.*, p. 74)

In the FAS–PARI surveys, household consumption expenditures are not estimated. Hence it is not possible to assess how household incomes compare with consumption estimates, or if household incomes are sufficient to cover consumption needs. To understand the extent of income deprivation in the study villages, we need a benchmark against which to compare the incomes. One of the benchmarks we can use is the minimum wages in agriculture specified under the Minimum Wage Act in respective States of India. The minimum wage in India should meet the calorie requirements and basic necessities such as clothing, housing, fuel, and lighting for a working-class family of three consumption units (two adults and two adolescents) (Labour Bureau 2015). Minimum wages are fixed by central or State governments for a list of scheduled employments. Table 2 makes a comparison of mean and median household incomes at 2010–11 prices with the minimum wages for agriculture in the respective States in the year 2011.

In our calculations we have used total household incomes instead of per capita incomes, since the minimum wage is also considered as family wage in India, for a family of four. However, we have not adjusted total household incomes for family size in our analysis. The ratios of mean and median household incomes per day to minimum wage in agriculture are close to 1 in most of the villages. This indicates that small farmers on average received incomes that were only adequate to cover the bare necessities. The median income did not exceed two times the minimum wage in any village except 25F Gulabewala (Sri Ganganagar district, Rajasthan), where there were only two small farmer households, and Tehang (Jalandhar district, Punjab), two villages with the highest income levels and advanced irrigated agriculture. In half the villages, the ratio was less than 1, indicating that the average earnings of small farmer households were below minimum wages. It must be kept in mind that we considered all types of incomes received by small farmer households, including non-agricultural incomes, transfers, and remittances. If incomes from agriculture exclusively are considered, incomes received by small farmer households would fall short of minimum wages.

Thus, the observation made earlier that incomes from crop production are insufficient to meet the basic consumption requirements of small farmer households is the stark reality of small-scale farming in contemporary India.

Comparison of Incomes of Small Farmers with Other Sections of the Peasantry

Income deprivation of small farmer households in India is a well-known fact and cannot be emphasised enough. However, do low levels of income affect only small farmers, or is it widespread among other sections of the peasantry as well? Table 3 shows the ratio of mean and median per capita household incomes of landlord and capitalist farmers, and large farmers to small farmers.[3]

It is clear that the average household income received by landlord and capitalist farmer households was significantly higher than that received by small farmer households. The ratio was particularly high, more than ten-fold,

---

[3] See Appendix Table 1 for the mean and median per capita incomes for each class. We have excluded Tehang from our analysis, as classification of households by occupational groups was incomplete in this village.

Table 2 *Comparison of mean and median household incomes (per day) of small farmer households with minimum wages in agriculture, at 2010–11 prices* in Rs

Village	State	Mean household income	Median household income	Minimum wage (MW) in agriculture (1)	Ratio Mean:MW	Ratio Median:MW
Ananthavaram	Andhra Pradesh	158.1	133	112	1.4	1.2
Bukkacherla	Andhra Pradesh	103.5	81.1	112	0.9	0.7
Kothapalle	Telangana	124.6	113.1	157.3	0.8	0.7
Harevli	Uttar Pradesh	131.5	106.8	100	1.3	1.1
Mahatwar	Uttar Pradesh	116.2	78.8	100	1.2	0.8
Nimshirgaon	Maharashtra	271.4	189.2	100	2.7	1.9
Warwat Khanderao	Maharashtra	204.2	157.4	100	2	1.6
25F Gulabewala	Rajasthan	431.0	431	135	3.2	3.2
Rewasi	Rajasthan	307.9	241.8	135	2.3	1.8
Gharsondi	Madhya Pradesh	129.1	112.5	124	1	0.9
Alabujanahalli	Karnataka	230.2	190	157.3	1.5	1.2
Siresandra	Karnataka	416.5	297.8	157.3	2.6	1.9
Zhapur	Karnataka	146.5	123.8	157.3	0.9	0.8
Amarsinghi	West Bengal	161.4	113.3	167	1	0.7
Panahar	West Bengal	141.2	75.3	167	0.8	0.5
Kalmandasguri	West Bengal	184.8	154.8	167	1.1	0.9
Tehang	Punjab	710.4	530.3	160.5	4.4	3.3

Note: State-specific minimum wages for the year 2011 are used.
Source: The sources for the data are *Indian Labour Statistics*, 2011 and 2012, Labour Bureau, Government of India (2015).

Table 3 *Ratio of per capita household incomes of landlord and capitalist farmer households, and large farmer households to small farmer households*

Village	State	Landlord and capitalist farmers to small farmers		Large farmers to small farmers	
		Mean	Median	Mean	Median
Ananthavaram	Andhra Pradesh	5.6	6.6	5	2.9
Bukkacherla	Andhra Pradesh	4.9	5.4	1.9	1.8
Kothapalle	Telangana	11.9	1.9	1.4	1.5
Harevli	Uttar Pradesh	12.8	18.5	5	6
Mahatwar	Uttar Pradesh	4.1	3.9	0.9	1.1
Nimshirgaon	Maharashtra	12	16.2	1.5	0.8
Warwat Khanderao	Maharashtra	7.2	4.8	1.8	2
25F Gulabewala[3]	Rajasthan	7.4	3	1.9	1.7
Rewasi	Rajasthan	3.6	1.7	1.1	1
Gharsondi	Madhya Pradesh	34	13.8	2.5	2.3
Alabujanahalli	Karnataka	-	-	3.8	2.5
Siresandra	Karnataka	-	-	2	2.8
Zhapur	Karnataka	11.5	7.2	2	1.5
Amarsinghi[1]	West Bengal	-	-	-	-
Panahar[2]	West Bengal	8.3	8.9	-0.5	-0.8
Kalmandasguri[1]	West Bengal	-	-	-	-
Tehang	Punjab	-	-	2.8	2.3

Notes: [1] There were no capitalist farmer or large farmer households in Amarsinghi and Kalmandasguri.
[2] There was only one large farmer household in Panahar.
[3] There were only two small farmer households in 25F Gulabewala.

in Gharsondi (Gwalior district, Madhya Pradesh), Harevli (Bijnor district, Uttar Pradesh), and Nimshirgaon (Kolhapur district, Maharashtra) – all three irrigated villages. The differences in less irrigated villages such as Rewasi (Sikar district, Rajasthan), Mahatwar (Ballia district, Uttar Pradesh), and Bukkacherla (Anantapur district, Andhra Pradesh) were relatively small. However, irrigation alone did not explain the differences, as there are exceptions to the rule. For example, Zhapur (Kalaburagi district, Karnataka) was an unirrigated village, but landlords and big farmers here received 11.5 times higher household incomes than small farmer households. On the other hand, the ratio was 5.6 in the irrigated village of Ananthavaram (Guntur district, Andhra Pradesh). However, in spite of variations in the ratios between villages, the fact remains that landlord and large farmer households received much higher incomes than small farmers. In fact, the classes of landlords and large farmers were quite distinct from the rest of the peasantry in terms of incomes received.

The income differences between small and large farmers (largely middle peasants) were narrower. In a few villages such as Nimshirgaon, Kothapalle (Karimnagar district, Telangana), Mahatwar, and Rewasi, the ratio was close to 1, while in other villages it was above 1 but lower than 2.5. Only in three villages – Harevli in Bijnor, Ananthavaram in Guntur, and Alabujanahalli in Mandya – did the ratio exceed 2.5. Thus, in most villages the middle peasantry was not better off than the small farmers.

Poverty in Small Farmer Households

An issue that is frequently discussed in the literature on livelihoods is the incidence of poverty in small farmer households and the possible ways out of this poverty. Income poverty is difficult to assess in the Indian context as there is no official income poverty line. In India, unlike in many other countries, there is no official source of data on household incomes, and hence poverty is measured using a consumption poverty line. It would be theoretically erroneous to use the consumption poverty line to identify income-poor households. In many countries it has been found that survey-based estimates of household incomes are lower than estimates of household incomes derived from macro-aggregates in national accounts statistics (Anand 1983; Deaton 1997; Smeeding and Weinberg 2001). It has also been found for many countries, including India, that a sizeable section of the bottom income deciles dis-save – that is, their consumption exceeds their income (Deaton 1997; Anand and Harris 1994). This may happen due to various reasons, including year-to-year variations in the incomes of poor households, underestimation of certain types of income flows, and exclusion of borrowings and dis-savings from income estimates. Hence, in the assessment of poverty in our analysis, we have refrained from using the official consumption poverty line. There are further problems with using the official poverty line: the methodology for calculating the official poverty line in India was mired in controversy during the years the PARI village surveys were conducted, that is, between 2005 and 2012. In this period, two separate expert committees were set up by the Government of India – the Tendulkar Committee in 2005 and the Rangarajan Committee in 2012 – to devise an acceptable methodology for determining the poverty line, amidst wide criticism of official estimates of poverty that showed a rapid decline in the head count ratios though most direct indicators of consumption (and nutrition) and standard of living indicated otherwise.

Hence, in our analysis below, we use a range of poverty lines, instead of a single poverty line, to understand the extent of poverty in the study villages. First, we use a range of poverty lines based on minimum wages in agriculture. Of course, the calculation of minimum wage in India is tied to estimates of

Table 4 *Percentage of small farmer households in poverty, using minimum wage-based poverty lines*

Village	State	P1 = Minimum wage x 300 days	P2 = Minimum wage x 350 days
Ananthavaram	Andhra Pradesh	37	40
Bukkacherla	Andhra Pradesh	55	61
Kothapalle	Telangana	32	40
Harevli	Uttar Pradesh	33	44
Mahatwar	Uttar Pradesh	52	59
Nimshirgaon	Maharashtra	4	10
Warwat Khanderao	Maharashtra	21	27
25F Gulabewala	Rajasthan	0	0
Rewasi	Rajasthan	27	30
Gharsondi	Madhya Pradesh	41	55
Alabujanahalli	Karnataka	29	35
Siresandra	Karnataka	18	18
Zhapur	Karnataka	50	54
Amarsinghi	West Bengal	63	70
Panahar	West Bengal	75	82
Kalmandasguri	West Bengal	43	54
Tehang	Punjab	2	2

consumption expenditure and hence it is quite close to the official poverty line (see Appendix Table 2). Secondly, we use a relative poverty line based on the median income levels in each village.

To understand the level of absolute poverty in the villages, we use two poverty lines based on minimum wages in agriculture. Minimum wages are specified for a working day, and it is unclear how many days of rest should be accounted for such that the family would still be able to achieve the minimum necessities of life. According to the Minimum Wages Act, 1948, workers should be provided one paid day of rest every week. The reality in casual wage employment is different, however, since workers do not receive payment for days they do not work. Taking a minimalist approach, we use two poverty lines based on the assumption that 300 to 350 days of employment at minimum wages should enable an agricultural worker to provide for basic family needs. Thus, a household is in poverty if total household income is less than the earnings of one worker for 300 to 350 days of employment at minimum wage. The proportion of small farmer households in poverty, based on the two cut-offs of days of employment, is reported in Table 4.

The levels of absolute poverty indicate that the proportion of poor small farmer households in the study villages ranged from 2 per cent in Tehang,

4 per cent to 10 per cent in Nimshirgaon, to 75 per cent to 82 per cent in Panahar. Among the villages with a high incidence of poverty (above 50 per cent by the P2 measure) were the three Bengal villages where average land size was the smallest among all the villages, the two dry villages of Bukkacherla and Zhapur, and Mahatwar.

To understand how small farmers fared in their respective villages vis-à-vis other types of households, we examine small farmer households that are in relative poverty. We use the median level of per capita income in each village as the relative poverty line, and measure the proportion of households (within each occupational class) with incomes below this relative poverty line. In every village, the incidence of poverty was highest among manual worker households. Small farmer households were poorer than all other occupational groups except manual workers in most of the villages. Among farmer households, small farmer households were the poorest.

In our previous analysis we found that the average income levels of small farmer households were not very different from that of large farmer households, especially in Mahatwar (Ballia district, Uttar Pradesh), Rewasi (Sikar district, Rajasthan), Nimshirgaon (Kolhapur district, Maharashtra), and Kothapalle (Karimnagar district, Telangana). However, when we compare the relative poverty figures of these two groups, we find that a higher proportion of small farmers were relatively poor in these villages. The proportion of relatively poor small farmer households ranged from 37 per cent in Kalmandasguri (Koch Bihar district, West Bengal) to 77 per cent in Zhapur (Kalaburagi district, Karnataka). In contrast, there were no poor landlord households in many of the villages. There were a few landlord households among the relatively poor – one household each in Rewasi, Bukkacherla (Anantapur district, Andhra Pradesh), Panahar (Bankura district, West Bengal), Gharsondi (Gwalior district, Madhya Pradesh), and Zhapur. The proportions reported in Table 5 are high because of the small number of landlord households in the villages. In each of these five cases, income poverty was transitory in nature and caused by large crop losses in the survey year.

Among the relatively poor small farmers, a higher incidence of poverty is observed among Dalit, Adivasi, and Muslim farmers as compared to other caste/religious groups in most villages (Table 6). Thus, while small farmers received lower incomes than large farmers and a higher proportion of small farmer households were among the relatively poor, there was differentiation within the small farmers. Dalit farmers were relatively poorer than small farmers belonging to social groups that were higher in the caste hierarchy of the villages.

Table 5 *Proportion of households in relative poverty* in per cent

Village	State	Small farmers	Large farmers	Landlord and capitalist farmers	Manual workers	Non-farmers
Ananthavaram	Andhra Pradesh	44	0	0	84	33
Bukkacherla	Andhra Pradesh	53	38	30	57	50
Kothapalle	Telangana	44	0	0	62	39
Harevli	Uttar Pradesh	58	5	0	71	58
Mahatwar	Uttar Pradesh	57	0	0	62	35
Nimshirgaon	Maharashtra	45	43	0	72	17
Warwat Khanderao	Maharashtra	40	23	0	82	35
25F Gulabewala	Rajasthan	0	3	0	81	28
Rewasi	Rajasthan	46	39	13	76	54
Gharsondi	Madhya Pradesh	61	31	8	76	31
Alabujanahalli	Karnataka	53	24		63	28
Siresandra	Karnataka	50	0		69	33
Zhapur	Karnataka	77	60	25	39	39
Amarsinghi	West Bengal	41			62	43
Panahar*	West Bengal	54	100	14	54	30
Kalmandasguri	West Bengal	37			67	46
Tehang	Punjab	6	7	-	-	59

Notes: * There was only one large farmer household in Panahar.
Relatively poor households are households that received incomes below the village median. The proportion of poor households is calculated within each occupational group.

Table 6 *Relatively poor small farmer households by social groups, as percentage of households within each social group*

Village	State	Scheduled Castes (SCs)	Scheduled Tribes (STs)	Muslims	All other caste and religious groups
Ananthavaram	Andhra Pradesh	43	100	0	42
Bukkacherla	Andhra Pradesh	71		0	50
Kothapalle	Telangana	33			51
Harevli	Uttar Pradesh	67		43	57
Mahatwar	Uttar Pradesh	67			47
Nimshirgaon	Maharashtra	100		100	32
Warwat Khanderao	Maharashtra	67		71	31
25F Gulabewala	Rajasthan	0			0
Rewasi	Rajasthan	57	33		47
Gharsondi	Madhya Pradesh	75	100	100	57
Alabujanahalli	Karnataka	75			51
Siresandra	Karnataka	80			39
Zhapur	Karnataka	69	100		80
Amarsinghi	West Bengal	33			43
Panahar	West Bengal	64	38	0	47
Kalmandasguri	West Bengal	29	33	56	31
Tehang	Punjab	25	0		9

Note: There was only one Scheduled Tribe household in Tehang.

COMPOSITION OF HOUSEHOLD INCOMES

Small farmer households did not depend on cultivation alone for incomes and employment; rather, they participated in other agricultural and non-agricultural occupations. This was a phenomenon not restricted to small farmer households. In fact, secondary data on employment as well as primary data-based studies have shown increasing trends in non-agricultural employment in rural India since the late 1980s (Unni 1991; Vaidyanathan 1986; Unni 1998). Several state-level reports of FAS–PARI surveys also indicate that a significant share of rural incomes is from non-agricultural sources (Ramachandran, Rawal, and Swaminathan, eds. 2010; Swaminathan and Rawal, eds. 2014; Swaminathan and Das, eds. 2017).

Though data are collected on twelve sources in the FAS–PARI surveys (see Appendix 1), we have used a six-fold classification of income sources in this chapter. These are:

(i) Self-employment in agriculture, which includes all incomes from crop cultivation including tree crops (for fruits or timber), rental incomes from leasing out agricultural land, and income from livestock. Incomes from sericulture in Karnataka and pisciculture in West Bengal are also included.

(ii) Agricultural wage incomes, which include all incomes received from casual and long-term wage employment in agriculture.

(iii) Self-employment in non-agriculture, which includes incomes from business and trade; incomes from owned non-agricultural establishments; rental incomes from machinery, buildings, etc.

(iv) Non-agricultural wage incomes, which include all incomes from manual wage employment in non-agriculture.

(v) Salaries, which include incomes from regular skilled wage employment in non-agriculture.

(vi) All other sources, which include incomes from pensions, scholarships and other cash transfers, remittances, interest, financial income, and any other income received by households.

The first two categories comprise agricultural incomes and the remaining four categories comprise non-agricultural incomes. It is difficult to classify non-agricultural incomes under secondary and tertiary sector incomes, as the FAS–PARI surveys use an occupational classification of incomes rather than an industrial classification.

Table 7 shows that though there are variations across villages, self-employment in agriculture remains the single most important source of household incomes for small farmer households. The share of total household income from self-employment in agriculture ranged from 22.3 per cent and 24.4 per cent, respectively, in Panahar (Bankura district, West Bengal) and Zhapur (Kalaburagi district, Karnataka), to 75 per cent and above in Alabujanahalli (Mandya district, Karnataka), Warwat Khanderao (Buldhana district, Maharashtra), and Kothapalle (Karimnagar district, Telangana). Apart from cultivation and livestock, wages and self-employment in non-agriculture were important sources of income, and small farmer households in most of the villages received almost equal shares of income from these sources. In a few villages such as Rewasi (Sikar district, Rajasthan), Kalmandasguri (Koch Bihar district, West Bengal), Panahar, Mahatwar (Ballia district, Uttar Pradesh), and Tehang (Jalandhar district, Punjab), significant incomes accrued from other sources, particularly remittances. Thus small farmers were primarily dependent on cultivation and livestock, though they also diversified to manual labour and petty business and trade to supplement their incomes.

It is difficult to discern any pattern to identify individual factors that might impact the composition of incomes of small farmer households. Rather, the

Table 7 *Composition of household incomes of small farmer households by source, as percentage of total household income*

Village	State	Self-employment in agriculture	Agricultural wages	Self-employment in non-agriculture	Non-agricultural wages	Salaries	All other sources	Total income
Ananthavaram	Andhra Pradesh	63	20.5	0.8	8.4	5.2	2.1	100
Bukkacherla	Andhra Pradesh	67.1	13.6	11.1	5.1	0	3	100
Kothapalle	Telangana	78	4.1	3.7	5.3	8.2	0.7	100
Harevli	Uttar Pradesh	60	24.2	1.3	4.9	3.4	6.2	100
Mahatwar	Uttar Pradesh	34.1	1.9	21.4	24.9	3.3	14.5	100
Nimshirgaon	Maharashtra	65.6	2.2	3.2	11.9	4.6	12.6	100
Warwat Khanderao	Maharashtra	77.6	10	4.7	2.4	1.4	4	100
25F Gulabewala	Rajasthan	45.3	5	8.4	0	0	41.3	100
Rewasi	Rajasthan	30.1	1.6	16.8	4.7	1.6	45.2	100
Gharsondi	Madhya Pradesh	67.4	5.5	2.2	12.9	5.3	6.6	100
Alabujanahalli	Karnataka	75	4.6	4.7	0.8	2.8	12.1	100
Siresandra	Karnataka	68.1	8.5	5.2	6.6	6.5	5.1	100
Zhapur	Karnataka	24.4	17.8	0.4	27.9	24.5	4.9	100
Amarsinghi	West Bengal	52.2	2.9	17.5	7.7	10	9.7	100
Panahar	West Bengal	22.8	17.9	17.4	11.2	15.6	15.3	100
Kalmandasguri	West Bengal	44.4	8.3	15.8	11.9	0.5	19	100
Tehang	Punjab	43.9	0	25.8	5.6	4.8	20	100
All villages		54.5	7.1	10.6	7.6	5.1	15.1	100

composition of incomes depended on a large set of village-specific conditions that interacted to result in the given outcomes. Some of these were agro-climatic conditions, pattern of land ownership, irrigation and agricultural mechanisation, availability of employment outside agriculture, the history of outmigration from the village, and so on. For example, among the low and moderately irrigated villages, namely, Kothapalle, Warwat Khanderao, Siresandra (Kolar district, Karnataka), Bukkacherla (Anantapur district, Andhra Pradesh), Rewasi, and Zhapur, the first four had high shares of incomes from farming while the latter two had low shares. The low share of incomes from farming in Rewasi was due to a long history of outmigration from this village, and from the entire Shekhawati region in the arid deserts of Rajasthan where the village is situated. Thus, other non-agricultural incomes, particularly remittance incomes, were high in this village. Zhapur is situated among stone quarries in Gulbarga, and the residents of this village found wage employment in these quarries. Such migration networks or wage employment opportunities were not present in the other four moderate to low irrigation villages, and hence, in spite of adverse agricultural conditions, small farmers could not diversify to other sources of income, resulting in higher shares of farm incomes. There was considerable diversification within agriculture in Siresandra, as many farmer households cultivated mulberry and reared silk worms. Additional incomes and employment from sericulture resulted in a high share of farm incomes in this village.

We find the same contrast between the highly irrigated village of Tehang, as opposed to Ananthavaram (Guntur district, Andhra Pradesh) and Harevli (Bijnor district, Uttar Pradesh). Tehang, a village that was resettled after the India–Pakistan partition, witnessed steady outmigration of its inhabitants to countries outside India and, consequently, high remittance incomes. Though Ananthavaram too had a sizeable section of upper-caste households whose members migrated to other countries or cities within India, this migration was largely restricted to rich, large landowning households rather than small farmer households. Similarly, in Harevli there were instances of family members of rich, large farmer households staying outside the village but this was not so common among small farmer households.

Two types of differentiation processes are underway in the study villages, among the small peasantry. One is a process of proletarianisation and depeasantisation wherein small peasants exit agriculture. The method of studying villages at a single point of time cannot adequately address this process of occupational change and proletarianisation of the peasantry, for the simple reason that it is unable to capture the peasant households who migrate to towns. However, remittance flows indirectly indicate depeasantisation and proletarianisation of individual members and parts of extended households. The second process is one of proletarianisation without depeasantisation, wherein small farmers complement their incomes from farming with other types of wage income and incomes from non-agricultural petty production. The process of "proletarianisation without depeasantisation" was emphasised by Harris (1986, 1989), and was observed in other village studies in the 1980s (Ramachandran 1990). In the PARI study villages, the significant share of wage and other types of income in the income portfolio of small farmer households also indicates the process of proletarianisation without depeasantisation.

FEATURES OF INCOME DIVERSIFICATION

In this section we discuss some features of the income diversification of small farmer households in the study villages, and we proceed to examine the association between income diversification and poverty in subsequent sections.

Pluriactivity of Small Farmer Households

In the literature on livelihoods, simultaneously engaging in multiple activities is termed "pluriactivity." Pluriactivity can occur at the household level (where households engage in multiple activities) or at the individual level (where individual workers engage in multiple activities over a fixed time-period). Household-level pluriactivity need not imply individual-level pluriactivity, as individual members of households may specialise in particular occupations. In fact, according to Ellis (2000), individual-level diversification is often indicative of income stress and poverty, while household-level diversification (with individual specialisation) may indicate an accumulative strategy of households. In our analysis here, we confine ourselves to household-level pluriactivity. There are very few studies that deal with both types of pluriactivity, though multiple activities at the household level may be more common than at the individual level (Reardon *et al.* 2007).

The PARI village survey data show that a majority of small farmer households were pluriactive – that is, they received incomes from more than one source (Table 8). In a large majority of the villages more than 80 per cent of the households received incomes from more than one source. Only in Alabujanahalli (Mandya district, Karnataka), Bukkacherla (Anantapur district, Andhra Pradesh), and Kothapalle (Karimnagar district, Telangana) was the proportion lower. Most small farmer households received incomes from two to three sources in each study village. The average number of sources of income per small farmer household varied from 1.9 in Bukkacherla and Kothapalle, to 3.5 in Kalmandasguri (Koch Bihar district, West Bengal).

Table 8 *Distribution of small farmer households by number of sources of income as percentage of all households, and mean number of sources per household*

Village	State	Number of sources of income per household						Mean no. of sources	No. of households
		1	2	3	4	> 4	All households		
Ananthavaram	Andhra Pradesh	3.5	38.3	54.6	3.5	0	100	2.6	227
Bukkacherla	Andhra Pradesh	41.7	30.6	27.8	0	0	100	1.9	108
Kothapalle	Telangana	41.8	33.7	20.4	4.1	0	100	1.9	98
Harevli	Uttar Pradesh	20	31.1	35.6	13.3	0	100	2.4	45
Mahatwar	Uttar Pradesh	4.3	26.1	47.8	17.4	4.3	100	2.9	69
Nimshirgaon	Maharashtra	18.1	60.1	21.8	0	0	100	2	238
Warwat Khanderao	Maharashtra	14.2	51.9	23.6	8.5	1.9	100	2.3	106
25F Gulabewala	Rajasthan	0	50	0	50	0	100	3	2
Rewasi	Rajasthan	11.2	45.8	25.2	15	2.8	100	2.5	107
Gharsondi	Madhya Pradesh	12.7	31	36.6	16.9	2.8	100	2.7	71
Alabujanahalli	Karnataka	25.7	45.1	15.9	11.5	1.8	100	2.2	113
Siresandra	Karnataka	17.9	33.9	19.6	25	3.6	100	2.6	56
Zhapur	Karnataka	3.8	26.9	38.5	19.2	11.5	100	3.1	26
Amarsinghi	West Bengal	5.6	27.8	44.4	14.8	7.4	100	2.9	54
Panahar	West Bengal	4.2	18.1	51.4	23.6	2.8	100	3	144
Kalmandasguri	West Bengal	1.5	13.2	35.3	36.8	13.2	100	3.5	68
Tehang	Punjab	28.6	42.9	20.4	8.2	0	100	2.1	49
All villages		15.7	38.4	33.1	10.6	2.1	100	2.5	1579

Proletarianisation and Diversification to Manual Wage Incomes

A large number of small farmer households participated in manual wage employment, in both agriculture and non-agriculture. Results from the study villages show that 21 to 85.9 per cent of small farmer households participated in wage employment during the survey year. Further, their participation was not limited to agricultural wage employment; participation in non-agricultural wage employment was also high. In many villages, such as Rewasi (Sikar district, Rajasthan), Mahatwar (Ballia district, Uttar Pradesh), Alabujanahalli (Mandya district, Karnataka), and the three West Bengal villages, a larger number of small farmer households received incomes from non-agricultural wages than agricultural wages. In Tehang (Jalandhar district, Punjab), there was no agricultural labouring-out among small farmers. In this village most agricultural operations were mechanised and only paddy transplantation was done manually – by migrant workers from Bihar.

The participation of small farmers in wage work without necessarily exiting agriculture is a reflection of their gradual proletarianisation. Ramachandran (2011) observed that in most villages it is difficult to distinguish between the poor peasantry and manual workers. Based on his analysis of data on

Table 9 *Proportion of small farmer households receiving incomes from wage employment*

Village	State	Agricultural wages	Non-agricultural wages	All or any kind of wages	Total no. of small farmer households
Ananthavaram	Andhra Pradesh	85.9	30	85.9	227
Bukkacherla	Andhra Pradesh	49.5	16.8	52.8	106
Kothapalle	Telangana	36.7	20.4	48.5	99
Harevli	Uttar Pradesh	60	24.4	62.2	45
Mahatwar	Uttar Pradesh	26.1	62.3	63.8	69
Nimshirgaon	Maharashtra	31.5	31.9	50.4	238
Warwat Khanderao	Maharashtra	67.9	22.6	72.6	106
25F Gulabewala	Rajasthan	50	0	50	2
Rewasi	Rajasthan	16.8	43	45.8	107
Gharsondi	Madhya Pradesh	47.9	43.7	64.8	71
Alabujanahalli	Karnataka	18.6	28.3	31.9	113
Siresandra	Karnataka	46.4	41.1	55.4	56
Zhapur	Karnataka	73.1	42.3	76.9	26
Amarsinghi	West Bengal	29.6	70.4	75.9	54
Panahar	West Bengal	66.7	74.3	77.8	144
Kalmandasguri	West Bengal	63.2	73.5	80.9	68
Tehang	Punjab	0	21	21	49

labour-days hired in and hired out by residents of Ananthavaram village in coastal Andhra, he concluded that poor peasants were "substantially and characteristically semi-proletarians" as they laboured out heavily, while they were also heavy employers of labour (*ibid.*, p. 69).

In spite of such a large number of small farmer households labouring out on wage employment, the contribution of wage earnings to total household income was less than 30 per cent in all the PARI study villages except Zhapur (Kalaburagi district, Karnataka) (Table 7). The share of wage earnings in Zhapur was higher at 45.7 per cent (due to non-agricultural wages from stone mining, as discussed in the previous section). This difference between the proportion of households participating in manual wage employment and the share of wage earnings in total incomes indicates low average incomes from wages for the households vis-à-vis other sources of incomes.

The low wage earnings of small farmer households can be explained by two factors: first, the number of days of wage employment they obtained, and second, the average daily wages (see Appendix Tables 2A and 2B). The total number of days of wage employment (in agricultural and non-agricultural activities) obtained by small farmer households in the study villages was less than 180 days – that is, six months for one member of the household – in most villages. The exceptions were Siresandra (Kolar district, Karnataka) where workers were engaged in sericulture, Mahatwar (Ballia district, Uttar Pradesh) and Zhapur where workers laboured out in non-agriculture (in semi-skilled jobs of well-sinking in Mahatwar and quarry work in Zhapur), and in Harevli (Bijnor district, Uttar Pradesh) where small farmers laboured out heavily on the lands and orchards of large farmers. It is also important to note that small farmers in most of the villages participated almost equally in agriculture and non-agriculture.

The average daily wage rates received by small farmer households are reported in Appendix Table 2B. The most important features of these wage rates are as follows. First, in all the villages, wages received by women workers were much lower than that received by male workers. Secondly, non-agricultural wages in most villages were much lower than agricultural wages (the exceptions were, predictably, Mahatwar, Rewasi, and Zhapur) – for reasons discussed above. Thirdly, in half the villages, the agricultural wages for male workers fell short of the minimum wages in agriculture declared by the respective State governments in 2010–11 (see Appendix Table 2C).

To summarise, though a large number of small farmer households did participate in manual wage employment activities, wage employment did not provide them with adequate incomes. This was reflected in the low share of incomes from wage employment in the household income portfolio of small farmers in most of the study villages. Low wage rates in general, and particularly for women and in the non-agricultural sector, were a major contributory factor to such outcomes.

Participation in Non-Agricultural Activities

More than 60 per cent of small farmer households in most villages received incomes from non-agriculture. Non-agricultural incomes comprised incomes from non-agricultural self-employment, manual wage employment, salaries, and other non-agricultural incomes. Only in three villages – Bukkacherla (Anantapur district, Andhra Pradesh), Kothapalle (Karimnagar district, Telangana), and Warwat Khanderao (Buldhana district, Maharashtra) – was the share of non-agricultural incomes lower than 60 per cent.

Given that such a large number of small farmers participated in non-agricultural activities, an obvious question arises as to why they remained in agriculture at all. Why did small farmer households not exit agriculture and join the non-agricultural work force? The answer, perhaps, lies in the limited nature of structural change in India, and the low growth of employment in the secondary and tertiary sectors. The non-agricultural sector is unable to provide regular and sustained income to rural households. This is made clear in Table 11

Table 10 *Small farmer households receiving incomes from non-agricultural sources*

Village	State	No. of households	As percentage of all small farmer households
Ananthavaram	Andhra Pradesh	151	66.8
Bukkacherla	Andhra Pradesh	36	33.6
Kothapalle	Telangana	41	41.8
Harevli	Uttar Pradesh	29	64.4
Mahatwar	Uttar Pradesh	66	95.7
Nimshirgaon	Maharashtra	151	63.4
Warwat Khanderao	Maharashtra	52	49.1
25F Gulabewala	Rajasthan	2	100
Rewasi	Rajasthan	94	87.9
Gharsondi	Madhya Pradesh	56	78.9
Alabujanahalli	Karnataka	83	73.5
Siresandra	Karnataka	42	75
Zhapur	Karnataka	21	80.8
Amarsinghi	West Bengal	49	90.7
Panahar	West Bengal	133	92.4
Kalmandasguri	West Bengal	67	98.5
Tehang	Punjab	36	73.5
All villages		1104	69.9

Table 11 *Small farmer households receiving incomes from different non-agricultural sources, as percentage of all small farmer households*

Village	State	Self-employed in non-agriculture	Non-agricultural wages	Salaries	All other non-agricultural sources
Ananthavaram	Andhra Pradesh	14	30	14	14
Bukkacherla	Andhra Pradesh	11	17	0	8
Kothapalle	Telangana	4	20	20	6
Harevli	Uttar Pradesh	13	24	4	40
Mahatwar	Uttar Pradesh	22	62	10	71
Nimshirgaon	Maharashtra	11	32	10	21
Warwat Khanderao	Maharashtra	15	23	6	21
25F Gulabewala	Rajasthan	50	0	0	100
Rewasi	Rajasthan	19	43	7	67
Gharsondi	Madhya Pradesh	13	44	10	52
Alabujanahalli	Karnataka	14	28	8	50
Siresandra	Karnataka	18	41	18	39
Zhapur	Karnataka	8	42	31	54
Amarsinghi	West Bengal	30	70	19	43
Panahar	West Bengal	18	74	10	35
Kalmandasguri	West Bengal	44	74	7	59
Tehang	Punjab	33	20	10	47

where we find that only a small section of small farmer households received salary incomes, while the remaining participated in wage employment and non-farm enterprises.

The small farmer households in the study villages were caught in a low income and low employment trap, with neither agriculture nor non-agriculture providing them adequate days of employment or income. The non-agricultural sector at best provided them with limited wages and self-employment opportunities to partially mitigate the income shortages. Whether such low levels of non-agricultural income and employment can improve the livelihood and standard of living of small farmers is questionable.

Non-Agricultural Incomes

Average non-agricultural incomes received by small farmer households in most villages were not very high, and in most cases were lower than agricultural incomes. The only exceptions were Mahatwar (Ballia district, Uttar Pradesh), Panahar (Bankura district, West Bengal), Rewasi (Sikar district, Rajasthan), Zhapur (Kalaburagi district, Karnataka) and Tehang (Jalandhar district,

Punjab).[4] The results in these five villages were due to exceptional prevailing circumstances. In Mahatwar, a number of households participated in well-digging, a specialised semi-skilled wage employment activity, which fetched them higher incomes than small-scale agriculture and agricultural wages. High non-agricultural incomes in Rewasi were due to high remittance incomes and incomes from non-agricultural self-employment. This village as well as the region in which it is situated were characterised by out-migration to different parts of India and outside the country for various kinds of jobs, particularly jobs in the armed forces, in specialised marble and stone constructions, and business and trade activities. Households residing in the village also ran small businesses, particularly in transportation, in nearby towns. Households in Zhapur, which was a dry village with low agricultural incomes, received non-agricultural incomes from casual wage employment in nearby stone quarries, and salaries and remittances from Gulbarga town. Tehang was a prosperous, early green revolution village in Punjab. Its relative prosperity stimulated non-agricultural demand and investment. About 30 per cent of small farmer households received incomes from own businesses located within the village and in the nearby town of Phillaur. Besides business incomes, small farmer households also received a significant proportion of their total income from remittances, as many family members had migrated and settled in other countries. In villages where such exceptional opportunities were not present, non-agricultural incomes were lower than and supplemented agricultural incomes in the household economy.

Income Poverty and Sources of Income

The relationship between income diversification and poverty is not unidirectional, as diversification may occur due to push factors related to income stress, or pull factors when households participate in various activities to accumulate higher incomes. Thus both rich and poor households diversify. The impact of distress diversification on household incomes and income poverty is also difficult to assess because when households are pushed out of cultivation diversification increases their incomes compared to the status quo even though it may not necessarily lift households out of poverty. Thorough research on income diversification and its implications for poverty requires long-term panel data on household incomes. Therefore, we do not attempt such analysis in this section. The question we pose here instead is: are there differences in the composition of incomes between relatively rich and poor

---

[4] One must be cautious in interpreting the results in Panahar and Mahatwar, as these villages had lower than normal crop incomes in the survey year.

Table 12 *Average agricultural and non-agricultural incomes received by small farmer households*

Village	State	Agricultural incomes			Non-agricultural incomes		
		No. of households	Mean	Median	No. of households	Mean	Median
Ananthavaram	Andhra Pradesh	227	48,213	33,660	151	14,247	11,592
Bukkacherla	Andhra Pradesh	107	30,514	21,002	36	21,791	8,292
Kothapalle	Telangana	98	37,328	35,950	41	19,429	11,592
Harevli	Uttar Pradesh	45	40,436	32,651	29	11,749	4,908
Mahatwar	Uttar Pradesh	69	15,257	8,163	66	28,379	18,573
Nimshirgaon	Maharashtra	238	67,130	62,603	151	50,233	28,688
Warwat Khanderao	Maharashtra	106	65,267	51,791	52	18,909	9,486
25F Gulabewala	Rajasthan	2	79,215	79,215	2	78,110	78,110
Rewasi	Rajasthan	107	35,689	25,205	94	87,309	59,953
Gharsondi	Madhya Pradesh	71	34,357	30,743	56	16,171	3,180
Alabujanahalli	Karnataka	113	66,874	62,196	83	23,334	11,135
Siresandra	Karnataka	56	116,555	88,395	42	47,313	30,960
Zhapur	Karnataka	26	22,584	16,591	21	38,258	25,542
Amarsinghi	West Bengal	54	32,433	29,234	49	29,187	15,984
Kalmandasguri	West Bengal	68	35,598	33,245	67	32,320	22,977
Panahar	West Bengal	144	20,932	15,498	133	32,836	9,812
Tehang	Punjab	49	113,731	992,38	36	198,123	121,300

Note: Average incomes are calculated for households that received incomes from the source.

households?[5] By comparing the income portfolios of the two groups we will be able to identify the specific income sources to which the relatively better-off have access. Tables 13a and 13b show the income composition of relatively rich and poor small farmer households. A few patterns emerge from these data.

First, in most of the study villages the share of agricultural incomes among poor small farmer households was similar to or higher than that of rich small farmer households. The notable exceptions were Ananthavaram (Guntur district, Andhra Pradesh) and Bukkacherla (Anantapur district, Andhra Pradesh).

Secondly, the shares of agricultural and non-agricultural wages were higher in the income portfolios of poorer households. Agricultural and non-agricultural wage employment constituted 17.1 per cent and 13.7 per cent in the aggregate, respectively, of household incomes of relatively poor small farmer households, and only 4.7 per cent and 6.2 per cent of household incomes of the relatively rich. The pattern held true for all villages in the case of agricultural wages, but in some villages (Amarsinghi in Malda district, West Bengal; Gharsondi in Gwalior district, Madhya Pradesh; Kothapalle in Karimnagar district, Telangana; Warwat Khanderao in Buldhana district, Maharashtra; and Zhapur in Kalaburagi district, Karnataka), non-agricultural wage incomes were higher for richer households. Thus we can say that participation in manual wage employment in general, and agricultural labour in particular, was the clearest indicator of income poverty. Insufficient income from cultivation was supplemented by wage labour, particularly in agriculture. Wage employment was the occupation of last resort. At the same time, participation in wage labour did not provide such levels of income that may make a serious dent to income poverty.

Thirdly, the share of salaries and non-agricultural self-employment in the income portfolios of small farmer households was relatively low in the aggregate. In almost all the villages richer households received a marginally higher or similar share of income from salaries.[6] Though the aggregate share of non-agricultural self-employment activities was higher for richer households compared to relatively poor households, there were wide variations across the

---

[5] Relatively rich and poor households are defined in the same way as in the previous section on income poverty. The median level of per capita income for all households in each village is used as the relative poverty line, and households with median per capita incomes below the relative poverty line are identified as relatively poor, while the remaining are identified as relatively rich.

[6] The only exception was Ananthavaram, where the share of salary income of the poorer households was higher (13 per cent) than that of richer households (3.6 per cent). However, the high share of salary incomes of poorer households in Ananthavaram was due to such incomes received by only two households. Since salaries are relatively much higher than other income sources, the percentage share was high; but we have to read this share carefully as only two households were involved.

villages. In eight out of 17 villages, non-agricultural self-employment incomes were somewhat higher for richer households; in four villages they were lower; and in four villages the shares were similar. The literature on livelihoods has often noted the heterogeneous nature of self-employment activities (Lanjouw and Shariff 2007). Self-employment activities may include petty trade such as vending daily consumption items, or larger trade and production activities. Depending on the capital invested and the scale of such activities, the income flows vary over a large range. Self-employment activities that poor households are able to engage in generally yield low and uncertain incomes. The impact of such activities on poverty reduction should be carefully assessed. The levels of income received from such activities and the associated income uncertainties often do not make non-agricultural self-employment a better livelihood option than farming for rural households.

Finally, the share of "other" incomes was much higher for relatively rich households as compared to poor households. The most important sources of other incomes were remittances and pensions. Among the PARI study villages there were five villages with high remittances – Rewasi (Sikar district, Rajasthan), Kalmandasguri (Koch Bihar district, West Bengal), Alabujanahalli (Mandya district, Karnataka), Amarsinghi, and Tehang (Jalandhar district, Punjab). In each of these five villages, the share of other incomes in the income portfolios of relatively rich households was significantly higher than for poorer households. Thus, remittances by members of households who lived and worked in urban areas had a very important role to play in poverty reduction. The migration of household members also signified the classic kind of proletarianisation of the peasantry, and the results indicate the importance of larger structural changes in the economy in poverty reduction.

WHY DO SMALL FARMERS CONTINUE FARMING?

The stability and persistence of small-scale farming have perplexed researchers for long. We do not intend to deliberate on the theoretical positions on this issue. Rather, in this section we try to gain some insights into this issue from the PARI village surveys.

We need to emphasise at the very outset that our data do not support the proposition that small farmers continue farming because it is a remunerative occupation. Small-scale farming does not give small farmers high incomes, even when they receive incomes from other sources to supplement their farm incomes. This is the rural reality in India. The question then arises, why don't small farmers exit from agriculture altogether? Why do they choose pluriactivity instead? Our analysis in the previous sections seems to suggest

Table 13a *Income composition of relatively poor households*

Village	State	Self-employed in agriculture	Agricultural wages	Total agriculture	Self-employed in non-agriculture	Non-agricultural wages	Salaries	All other sources	Total
Ananthavaram	Andhra Pradesh	12.2	49.9	62.1	1.4	21.1	13	2.4	100
Bukkacherla	Andhra Pradesh	20.9	44.2	65.1	19.9	13.6	0	1.3	100
Kothapalle	Telangana	81.9	13.3	95.1	0	1.3	3.6	0	100
Harevli	Uttar Pradesh	37	47.6	84.6	1.5	9.2	0	4.7	100
Mahatwar	Uttar Pradesh	29.6	4.7	34.2	8.5	38	4.1	15.1	100
Nimshirgaon	Maharashtra	69.5	9	78.5	0.3	19.7	0	1.5	100
Warwat Khanderao	Maharashtra	71	19.7	90.7	3.4	1.6	1.7	2.6	100
25F Gulabewala	Rajasthan								
Rewasi	Rajasthan	44.8	6.1	50.9	14.2	10.6	2.2	22.2	100
Gharsondi	Madhya Pradesh	68.7	10.4	79.2	2.5	9.3	0.9	8.1	100
Alabujanahalli	Karnataka	78.9	4.4	83.3	4.3	1.6	1.9	8.8	100
Siresandra	Karnataka	55.2	18.5	73.7	6.4	13.1	5.3	1.4	100
Zhapur	Karnataka	28.2	16.8	45	0.7	22.7	25.6	6	100
Amarsinghi	West Bengal	74.2	5.2	79.4	7.8	6.7	1.7	4.4	100
Panahar	West Bengal	21.9	37.6	59.5	4.5	24.3	2.1	9.7	100
Kalmandasguri	West Bengal	45.5	10.6	56.1	17.7	15.7	0.1	10.5	100
Tehang	Punjab	70	0	70	6.5	19.5	4	0	100
Total		53.8	17.1	70.8	5.3	13.7	3.5	6.6	100

Table 13b *Income composition of relatively rich households*

Village	State	Self-employed in agriculture	Agricultural wages	Total agriculture	Self-employed in non-agriculture	Non-agricultural wages	Salaries	All other sources	Total
Ananthavaram	Andhra Pradesh	73.2	14.6	87.8	0.7	5.8	3.6	2	100
Bukkacherla	Andhra Pradesh	75.3	8.2	83.5	9.6	3.6	0	3.3	100
Kothapalle	Telangana	77.3	2.6	79.9	4.4	6	8.9	0.8	100
Harevli	Uttar Pradesh	70.4	13.7	84.1	1.3	3	4.9	6.8	100
Mahatwar	Uttar Pradesh	36	0.7	36.7	26.7	19.4	2.9	14.3	100
Nimshirgaon	Maharashtra	64.4	0.1	64.5	4	9.5	6	16	100
Warwat Khanderao	Maharashtra	79	7.8	86.9	5	2.5	1.3	4.3	100
25F Gulabewala	Rajasthan	45.3	5	50.4	8.4	0	0	41.3	100
Rewasi	Rajasthan	27.2	0.7	27.9	17.3	3.5	1.4	49.9	100
Gharsondi	Madhya Pradesh	66.7	2.8	69.5	2.1	15	7.7	5.8	100
Alabujanahalli	Karnataka	73.3	4.7	78	4.9	0.5	3.1	13.5	100
Siresandra	Karnataka	73.1	4.7	77.8	4.7	4.1	7	6.5	100
Zhapur	Karnataka	18.3	19.4	37.7	0	36.4	22.6	3.3	100
Amarsinghi	West Bengal	46.5	2.3	48.8	20	8	12.2	11.1	100
Panahar	West Bengal	23	13	36	20.5	8	18.9	16.7	100
Kalmandasguri	West Bengal	44.2	7.8	52	15.4	11	0.6	21	100
Tehang	Punjab	42.9	0	42.9	26.5	5.1	4.8	20.7	100
Total		54.7	4.7	59.4	11.9	6.2	5.5	17.1	100

that small farmers continue in agriculture because other occupations do not provide them with certain and adequate days of employment and incomes. We have seen that this is true of wage employment, where wage rates are low. This is also true of non-agricultural incomes on the whole, as such incomes are on average lower than agricultural incomes, and farmers do not have access to regular non-agricultural wage employment (salaries) that could provide them with sustained employment and fairly high incomes.[7]

Nevertheless, where the opportunities and circumstances were conducive to acquiring sustained non-farm incomes, small farmers made significant gains. In some cases it eventually led to the exit of small farmer households from agriculture, though in some cases it did not. To elaborate on this point, we describe below four case studies of contrasting situations. The first two case studies – in Tehang (Jalandhar district, Punjab) and Zhapur (Kalaburagi district, Karnataka) – show the propensity of small farmer households to exit from agriculture. In Tehang, the process was accelerated by high agricultural incomes that led to investments in non-agricultural opportunities in nearby towns and to migration outside India. In Zhapur, the process of exit was mediated by low agricultural production and availability of non-agricultural wage employment nearby. The latter two case studies illustrate instances where small farmers did not exit agriculture. Nimshirgaon (Kolhapur district, Maharashtra) had high agricultural incomes and non-agricultural opportunities in nearby towns; nevertheless, small farmer households in this village remained in agriculture even while engaging in non-farm employment. Rewasi (Sikar district, Rajasthan) was a unique case study where remittances supported investments in agriculture and family incomes.

Tehang: remittances and urbanisation
KS was a small farmer in Tehang village operating 2.25 acres of land – 2 acres owned and 0.25 acre leased in on annual contract. His family consisted of five members, all adults and working. They cultivated paddy, wheat, fodder, and *berseem* (lucerne grass) in 2011. Their crop income was supplemented by income from livestock. However, only 19 per cent of the total income was received from agricultural activity. Out of the total household income of Rs 634,916, Rs 516,000 came from non-agricultural activities.

---

[7] The situation is similar to the formulation of the Hariss–Todaro model on rural–urban migration. Farmers' expected incomes (as reflected in the mean incomes in our analysis) from non-agricultural sources are low, owing to high uncertainty of obtaining employment, low number of days of employment per year available in the non-agricultural sector, and not very high rates of returns from non-agricultural income sources.

KS engaged in real estate dealings and also owned a small transport business. His real estate business was in Phillaur, a town 5 kilometres from Tehang. He owned a mini-truck that he rented out for transportation in Phillaur, Tehang, and surrounding areas. KS made a net income of about Rs 120,000 from his businesses during 2010–11.

KS had three sons. The eldest had migrated to Dubai where he worked as a car driver. He remitted Rs 25,000 to his family every month during the survey year. The remittance was thus a major source of income for KS and his family. The second son was a carpenter and worked at Ludhiana, to where he commuted from Tehang every day. He received piece-rated wages for his carpentry work, and was employed round the year. His annual income was Rs 96,000. The youngest son was studying in Phillaur at the time of the survey, and he helped his family in activities related to cultivation and animal husbandry.

While KS and his older two sons were engaged in non-agricultural activities, KS's wife supervised the agricultural work. She took all the major decisions on cultivation, while the sons participated in labour operations when required.

Connectivity with nearby towns, easy availability of non-agricultural employment opportunities all the year round, and international migration were features of Tehang village. A large proportion of small farmer households in this village participated in and received high incomes from non-agricultural employment. The fact that agriculture too was remunerative in this green revolution village provided households with adequate initial investment to set up businesses or to migrate abroad. The half-a-century-long history and steady flow of migration to countries abroad by residents of the village made for suitable social and kinship networks. These factors propelled a cycle of wealth creation and accumulation for many peasant households in the village, which eventually led to an exit from agriculture for several of them. Such tendencies were apparent from the high number of tenancies in the village, as out-migrants leased out their land to residents of the village. KS's household was on its way out of agriculture and may not necessarily be classified as a small peasant household any longer.[8] But the household exemplified a process of transition that was common in Tehang.

---

[8] In Tehang, the households were not classified in occupational categories (such as small peasants, other peasants, landlords and capitalist farmers, manual labour households) using detailed income and labour criteria, similar to all the other study villages, at the time of writing this book. Hence, identification of small farmers in this village was based only on area of land operated. KS and his household may well be classified as a non-agricultural household if we adopt the rigorous income and labour criteria used for other villages.

Zhapur: stone quarrying

Zhapur was an unirrigated village in Kalaburagi district of Karnataka, with a dry and rocky landscape. The rocky soil was a major constraint for crop production and irrigation, and most farmers cultivated only in the *kharif* (monsoon) season. A variety of pulses and millets were grown in the village. There were a few stone quarries near the village, which mined stone chips for the construction industry. These stone quarries provided employment to male workers in the village. Children were also employed in these quarries at times, working at a fraction of the wages given to adults. The work was gruelling, but available all the year round. The wages were much higher than what could be earned in agriculture.

RS owned 4 acres of unirrigated cropland in Zhapur, on which he cultivated pigeon pea, green gram, sorghum, sunflower, and a small extent of paddy in the *kharif* season. His was a large household of 13 members – comprising his wife, six sons, one daughter, two daughters-in-law, and two grandchildren. RS, along with his eldest son and the adult female members of the household, worked on their 4-acre plot. However, agricultural incomes were uncertain and during the survey year 2008–09, the household incurred losses. RS's eldest son also laboured out on others' fields during the agricultural season.

Three of RS's sons worked at the quarry sites near Zhapur. AR worked as quarry labour for 300 days in the survey year, at a weekly wage of Rs 600. AS also worked at a quarry site. His job was to supervise the loading of stone chips. He received wages on a monthly wage of Rs 2,000, and he was employed for 10 months in the survey year. SS worked as a machine operator and he received Rs 5,000 per month for 12 months in the survey year. The combined incomes of these three sons constituted 98 per cent of the total household income. The household can barely be characterised as a peasant household, as adverse agricultural conditions together with low-skill non-agricultural opportunities nearby accelerated the pace of proletarianisation and depeasantisation.

Nimshirgaon: industrialisation and regular wage employment

Nimshirgaon village is in the sugarcane-growing region of Maharashtra and there are several sugar mills in nearby towns. Kolhapur district, where Nimshirgaon is located, has witnessed industrialisation and urbanisation since the late colonial period (late 1800s and early 1900s), when cotton textile mills were set up in Ichhalkaranji and adjoining locations. Thus there were many industries and small factories in small and large towns such as Jaisinghpur, Shirol, and Ichhalkaranji near Nimshirgaon. Semi-skilled regular wage employment and skilled salaried employment were available in these factories.

DC and his family of five members owned 2 acres of crop land in the village. They cultivated sugarcane, maize, *jowar* (sorghum), and vegetables. Sugarcane and tomato were marketed, while the rest of the produce was for consumption by the household. The income from crop production in 2006–07 was Rs 59,220, with an additional income of Rs 6,083 from animal husbandry.

DC worked as a machine operator in a cooperative sugar mill in Shirol, about 17 kilometres from Nimshirgaon. During the lean season he worked in the sugar mill as a full-time employee for a period of four months, at a salary of Rs 5,000 per month. During the eight months of peak agricultural production, he worked as a part-time labourer and received Rs 1,250 per month. The share of non-agricultural income in total household income was 31 per cent.

Commercial crop production, well-developed marketing facilities, and availability of non-agricultural employment during the lean season meant that DC could optimise his income portfolio by remaining in agriculture.

Rewasi: outmigration and remittances

Rewasi is situated in the midst of arid desert land in the State of Rajasthan. The soil in the village is sandy, the climate extremely dry, and water scarce. Cultivation is carried out with the help of sprinkler irrigation for only one season during the year. Hence people from this village and the region, particularly males, have been migrating out for generations in search of employment and income. They were and remain traders, soldiers, and specialised workers in marble and stone construction. Women, children, and elderly family members of migrant workers stayed back in the village and continued the seasonal cultivation. Livestock, particularly goats that feed on the thorny *loong* trees of the desert, were extremely important for the subsistence farm economy, providing much needed sources of nourishment and cash income. Women formed the major part of the work force in this village.

The large joint family of 60-year-old NJ and his sons consisted of 20 members. NJ had six sons, four of whom worked outside the village and two in the village. The wives and young children (eight in all) of the sons who worked outside stayed in the village.

The household owned 5.69 acres of unirrigated land, which yielded a net income of only Rs 645 during the survey year 2009–10 due to poor rainfall. The household owned livestock as well, from which the net income was Rs 22,890. It was mainly the women of the household who worked on the land and reared animals. Remittances sent by the four sons were the major source of income for this small farmer household.

NJ's first son was a factory worker in Jindal Aluminium in Bangalore. His monthly salary was Rs 7,000 and he sent home Rs 4,000 every month.

The second and third sons worked as masons in Abu Dhabi and Riyadh, and remitted Rs 12,000 and Rs 15,000 respectively every month. The fourth son had recently found employment in a cloth company in Faridabad, and had sent home Rs 2,000 per month for eight months during the survey year. The fifth son aspired to find employment outside India, like his two brothers. Unfortunately, NJ's youngest son was disabled by polio. He received a disability pension of Rs 1,000 each month.

An interesting feature of Rewasi is that in spite of low agricultural incomes and out-migration, most households did not exit agriculture or sever ties with the village. Women and children stayed back in the village, older men returned after spending years working away from their families, and a new stream of impatient young men were setting out in search of jobs in places far and near.

SUMMARY OF MAJOR FINDINGS

In this chapter we have examined household incomes and income poverty of small farmer households in the PARI study villages. The major points that emerge from our analysis are summarised below.

The levels of income received by small farmer households were low, in both absolute and relative terms. The average incomes received by small farmers were not much higher than the minimum wages in agriculture stipulated by State governments. Minimum wage in India is defined as subsistence wage; hence incomes received by small farmer households were inadequate to meet investments or any requirements other than daily consumption needs.

Small farmer households were among the poorest in the villages. Using a relative poverty line defined by the median village income, it was found that the proportion of poor households among small farmer households were lower only in comparison to manual labour households. All other occupation groups had a lower proportion of poor households.

Comparing the average income levels of small farmers with that of landlord and capitalist farmers, and large farmers, it was found that the income levels of landlords and capitalist farmers were much higher than that of small farmers, while the income difference between large farmers and small farmers was narrower. This finding is in consonance with our understanding of the contemporary rural reality in India within a market-oriented policy regime, wherein sections of the richer landowning classes and the larger peasantry have made income gains, while the small and medium peasantry have borne the brunt of contractionary agricultural policies. The schism between the rich and the poor within the peasantry may have intensified (Ramachandran 2011; Ramakumar 2014).

Small farmer households were not dependent on production of crops and rearing of animals alone for their livelihood; they received incomes from multiple sources. Though crop production and livestock rearing were the most important source of income for small farmer households, agricultural wages and incomes from non-agricultural sources also contributed significantly to their household incomes. While there were variations in the composition of incomes of small farmer households across villages, the share of self-employment in agriculture in total household incomes ranged from one-fourth to three-fourths in the PARI villages.

A majority of small farmer households were pluriactive; in other words, they received incomes from two to three sources. More than 30 per cent of small farmer households in the villages were engaged in manual wage work, indicating a gradual proletarianisation of the class. A significant section of small farmer households was also engaged in petty trade or other activities in the non-agricultural sector.

There were differences in the composition of incomes between the relatively rich and poor small farmer households. While poorer small farmer households depended on manual wage incomes to supplement incomes from farming, richer households had a higher share of "other incomes," which mainly comprised transfers and remittances. Remittances from family members working outside the village – those who had migrated to urban areas and joined the urban proletariat, or were receiving government pensions – were an important source of supplementary income for small farmers to avoid poverty and compensate for low farm incomes.

As noted earlier, the reasons why small farmer households do not exit farming are many. The average incomes received by small farmer households from non-agriculture were lower than the average incomes from farming. Though a large number of small farmer households participated in manual labour, the days of employment and wage rates were low. This does not, however, mean that farming was a lucrative venture for small farmers. Incomes from farming were low, but limited opportunities in the non-farm sector, low levels of non-agricultural incomes, inadequate wage employment opportunities, and low wage rates were factors that restricted the pace of rural transformation in the Indian countryside.

The question of small farmers exiting farming is a complex one, and it depends on a number of village-specific factors. There are no straightforward, unidirectional patterns – a fact that we illustrated using contrasting case studies. Agricultural productivity, availability of non-agricultural employment, urbanisation, industrialisation, and migration networks are all factors that influence farmers' decisions to continue with or quit farming.

REFERENCES

Anand, Sudhir (1983), *Inequality and Poverty in Malaysia*, Oxford University Press, New York.

Anand, Sudhir, and Hariss, Christopher J. (1994), "Choosing a Welfare Indicator," *American Economic Review*, vol. 84, no. 2.

Bakshi, Aparajita (2011), "Rural Household Income," PhD thesis submitted to University of Calcutta, Kolkata.

Bhalla, G. S. (2006), *Condition of the Indian Peasantry*, National Book Trust, New Delhi.

Deaton, Angus (1997), *The Analysis of Household Surveys: A Micro Econometric Approach to Development Policy*, World Bank, Washington D. C.

Deb, Uttam Kumar, Rao, G. D. Nageswara, Rao, Y. Mohan, and Slater, Rachel (2002), "Diversification and Livelihood Options: A Study of Two Villages in Andhra Pradesh, India, 1975–2001," Working Paper No. 178, Overseas Development Institute (ODI), London.

Dev, Mahendra S., and Mahajan, Vijay (2005), "Transforming Rural Economy in Andhra Pradesh," in Rohini Nayyar and Alakh N. Sharma (eds.), *Rural Transformation in India: The Role of Non-farm Sector*, Institute for Human Development, New Delhi.

Ellis, Frank (1998), "Household strategies and rural livelihood diversification," *Journal of Development Studies*, vol. 35 (1), pp. 1–38.

Ellis, Frank (2000), *Rural Livelihood and Diversity in Developing Countries*, Oxford University Press, New York.

Ellis, Frank, and Biggs, Stephen (2001), "Evolving Themes in Rural Development 1950s–2000s," *Development Policy Review*, vol. 19 (4).

Fann, Shengenn, and Chan-Kang, Connie (2005), "Is Small Beautiful? Farm Size, Productivity and Poverty in Asian Agriculture," *Agricultural Economics*, vol. 32 (1), January, pp. 135–46.

Fann, Shengenn, Hazell, P. B. R, and Thorat, S. (1999), *Linkages between Government Spending, Growth and Poverty*, IFPRI Research Report 110, available at http://ebrary. ifpri.org/cdm/ref/collection/p15738coll2/id/125916, viewed on 18 January 2017.

Farrington, John, Deshingkar, Priya, Johnson, Craig, and Start, Daniel (eds.) (2006), *Policy Windows and Livelihood Futures: Prospects of Poverty Reduction in Rural India*, Oxford University Press, New Delhi.

Foundation for Agrarian Studies (2015), "Calculation of Household Incomes: A Note on Methodology," available at http://fas.org.in/wp-content/themes/zakat/pdf/ Survey-method-tool/Calculation%20of%20Household%20Incomes%20-%20A%20 Note%20on%20Methodology.pdf, viewed on 18 January 2017.

Harriss, John (1985), *What Happened to the Green Revolution in South India? Economic Trends, Household Mobility and the Politics of an "Awkward Class"*, School of Development Studies, University of East Anglia.

Harriss, John (1989), "Knowing About Rural Economic Change: Problems Arising from a Comparison of the Result of 'Macro' and 'Micro' Research in Tamil Nadu," in P. Bardhan (ed.), *Conversations Between Economists and Anthropologists*, Oxford University Press, Oxford.

Hazell, Peter B. R. (2005), "Is There a Future for Small Farms?" *Agricultural Economics*, vol. 32 (1), January, pp. 93–101.

Jayaraj, D. (2004), "Social Institutions and Structural Transformation of the Non-farm Economy," in Barbara Harriss-White and S. Janakarajan (eds.), *Rural India Facing the 21st Century*, Anthem Press, London.

Labour Bureau (2015), *Report on the Working of the Minimum Wages Act, 1948 for the year 2013*, Ministry of Labour and Employment, Government of India.

Lanjouw, Peter, and Shariff, Abusaleh (2004), "Rural Non-farm Employment in India: Access, Incomes and Poverty Impact," *Economic and Political Weekly*, vol. 39, no. 40.

Lewis, Arthur W. (1954), "Economic Development with Unlimited Supply of Labour," *Manchester School of Economic and Social Studies*, vol. 22, no. 2.

Lipton, M., and Longhurst, R. (1989), *New Seeds and Poor People*, Unwin Hyman, London.

Mellor, John W. (1976), *The New Economics of Growth: A Strategy for India and the Developing World*, Cornell University Press, Ithaca/New York.

Ramachandran, V. K (1990), *Wage Labour and Unfreedom in Agriculture: An Indian Case Study*, Clarendon Press, Oxford.

Ramachandran, V. K. (2011), "State of Agrarian Relations in India Today," *Marxist*, January.

Ramachandran, V. K., and Rawal, Vikas (2010), "The Impact of Liberalisation and Globalisation on India's Agrarian Economy," *Global Labour Journal*, vol. 1, issue 1, pp. 56–91.

Ramachandran, V. K., Rawal, Vikas, and Swaminathan, Madhura (2010), *Socio-economic Surveys of Three Villages in Andhra Pradesh: A Study of Agrarian Relations*, Tulika Books, New Delhi.

Ramakumar, R. (2014), "Economic Reforms and Agricultural Policy in India," paper presented at the Foundation for Agrarian Studies Tenth Anniversary Conference, Kochi.

Ravallion, Martin, and Dutt, Gaurav (1999), "When is Growth Pro-Poor? Evidence from Diverse Experiences in Indian States," World Bank Policy Research Working Paper No. 2263, Washington, D. C.

Reardon, T., Berdegue, J., Barrett, C. B., and Stamoulis, K. (2007), "Household Income Diversification into Rural Non-farm Activities," in S. Haggblade, P. Hazell, and T. Reardon (eds.), *Transforming the Rural Nonfarm Economy: Opportunities and Threats in the Developing World*, International Food and Policy Research Institute and The Johns Hopkins University Press, Baltimore.

Schultz, T. W. (1964), *Transforming Traditional Agriculture*, Yale University Press, New Haven, Connecticut.

Singh, R. B., Kumar, P., and Woodhead, T. (2002), *Smallholder Farmers in India: Food Security and Agricultural Policy*, FAO Regional Office for Asia and Pacific, Bangkok, available at ftp://ftp.fao.org/docrep/fao/005/ac484e/ac484e00.pdf, viewed on 18 January 2017.

Smeeding, Timothy M., and Weinberg, Daniel H. (2001), "Toward a Uniform Definition of Household Income," *The Review of Income and Wealth*, vol. 47, no. 1.

Start, Daniel (2001), "The Rise and Fall of the Rural Non-farm Economy: Poverty Impacts and Policy Options," *Development Policy Review*, vol. 19, no. 4, pp. 491–505.

Swaminathan, Madhura, and Das, Arindam (eds.) (2017), *Socio-economic Surveys of Three Villages in Karnataka: A Study of Agrarian Relations*, Tulika Books, New Delhi.

Swaminathan, Madhura, and Rawal, Vikas (eds.) (2015), *Socio-economic Surveys of Two Villages in Rajasthan: A Study of Agrarian Relations*, Tulika Books, New Delhi.

Unni, Jeemol (1991), "Regional Variations in Rural Non-Agricultural Employment: An Exploratory Analysis," *Economic and Political Weekly*, January 19.

Unni, Jeemol (1998), "Non-agricultural Employment and Poverty in India: A Review of Evidence," *Economic and Political Weekly*, March 28.

Vaidyanathan, A. (1986), "Labour Use in Rural India: A Study of Spatial and Temporal Variations," *Economic and Political Weekly*, vol. 21, no. 52.

Wiggins, Steve (2014), "Rural Nonfarm Economy: Current Understandings, Policy Options and Future Possibilities," in Peter Hazell and Atiqur Rahman (eds.), *New Directions for Smallholder Agriculture*, International Fund for Agricultural Development (IFAD) and Oxford University Press, Oxford.

APPENDIX 1
SOURCES OF INCOME IN THE *FAS–PARI* SURVEYS

Incomes of households in the FAS–PARI villages are estimated separately for the following sources.[9]

1. Crop production
2. Animal resources (including rental income from animals)
3. Wage labour
    (a) Agricultural labour (casual)
    (b) Agricultural labour (long-term)
    (c) Non-agricultural labour (casual)
    (d) Non-agricultural labour (monthly/long-term)
4. Salaried jobs
    (e) Government jobs
    (f) Other jobs
5. Business and trade
6. Moneylending
7. Income from savings in financial institutions and equity
8. Pensions and scholarships
9. Remittances and gifts
10. Rental income
    (g) From agricultural land
    (h) From machinery
    (i) From other assets
11. Artisanal work and work at traditional caste calling
12. Any other sources

[9] The following section is drawn from the Manual of Income Calculation, Foundation for Agrarian Studies, available at http://fas.org.in/wp-content/themes/zakat/pdf/Survey-method-tool/ Calculation%20of%20Household%20Incomes%20-%20A%20Note%20on%20Methodology. pdf, accessed on 15 September 2016.

Appendix Table 1 *Mean and median per capita household incomes of small farmer households, at current prices*

Village (Year of survey)	State	Landlord and capitalist farmers		Large farmers		Small farmers		Manual workers		Non-farmers	
		Mean	Median	Mean	Median	Mean	Median	Mean	Median	Mean	Median
Ananthavaram (2006)	Andhra Pradesh	58,458	55,844	51,609	24,724	10,365	8,420	5,152	4,175	20,886	11,600
Bukkacherla (2006)	Andhra Pradesh	29,168	25,623	11,493	8,471	5,918	4,767	6,795	5,099	9,607	5,550
Kothapalle (2006)	Telangana	86,912	12,357	9,844	9,844	7,274	6,411	6,949	4,700	8,570	7,157
Harevli (2006)	Uttar Pradesh	77,765	74,609	30,551	24,424	6,080	4,043	4,509	3,981	6,587	4,987
Mahatwar (2006)	Uttar Pradesh	16,636	12,036	3,532	3,532	4,084	3,115	2,953	2,618	7,059	4,186
Nimshirgaon (2007)	Maharashtra	1,56,297	1,72,982	20,021	8,786	13,012	10,667	6,830	6,033	21,181	17,143
Warwat Khanderao (2007)	Maharashtra	75,091	41,505	18,608	17,760	10,457	8,695	5,010	3,915	12,552	9,202
25F Gulabewala (2007)	Rajasthan	1,54,650	62,769	40,699	35,750	21,015	21,015	5,388	4,834	18,812	13,853
Rewasi (2010)	Rajasthan	77,173	30,904	23,877	18,309	21,558	17,913	13,964	11,676	32,236	14,102
Gharsondi (2008)	Madhya Pradesh	1,84,452	58,238	13,648	9,784	5,424	4,230	4,416	3,415	12,080	7,768
Alabujanahalli (2009)	Karnataka			53,039	27,586	13,901	11,071	11,590	10,952	22,343	17,008
Siresandra (2009)	Karnataka			49,141	45,008	24,035	15,827	11,389	10,310	17,272	16,657
Zhapur (2009)	Karnataka	74,634	38,936	13,148	8,147	6,491	5,427	10,111	9,364	13,742	10,130
Amarsinghi (2010)	West Bengal					12,371	8,886	7,440	6,436	11,046	9,212
Panahar (2010)	West Bengal	78,442	56,376	-5,025	-5,025	9,504	6,348	6,722	6,133	16,861	9,658
Kalmandasguri (2010)	West Bengal					12,321	10,911	9,015	8,020	16,631	10,866
Tehang	Punjab			157948	103945	55929	44955			27705	15476

Appendix Table 2A *Average number of days of wage employment obtained by small farmer households during the survey year, PARI villages* in 8-hour working day

Village	State	Agriculture	Non-Agriculture	Total
Ananthavaram	Andhra Pradesh	87	99	186
Bukkacherla	Andhra Pradesh	90	74	164
Kothapalle	Telangana	31	39	70
Harevli	Uttar Pradesh	313	26	339
Mahatwar	Uttar Pradesh	22	218	240
Nimshirgaon	Maharashtra	60	55	114
Warwat Khanderao	Maharashtra	106	29	135
Rewasi	Rajasthan	20	83	103
Gharsondi	Madhya Pradesh	45	50	96
Alabujanahalli	Karnataka	68	21	89
Siresandra	Karnataka	200	0	200
Zhapur	Karnataka	135	128	263
Amarsinghi	West Bengal	3	57	60
Panahar	West Bengal	84	74	159
Kalmandasguri	West Bengal	72	75	125

Appendix Table 2B *Average wages received by small farmer households during the survey year, PARI villages, at 2010–11 prices* in Rs per day

Village	Agricultural wages		Non-agricultural wages	
	Females	Males	Females	Males
Ananthavaram	92	174	37	137
Bukkacherla	87	166	0	225
Kothapalle	68	208	64	135
Harevli	51	74	20	93
Mahatwar	53	53	12	92
Nimshirgaon	78	203	0	145
Warwat Khanderao	84	132	42	93
25F Gulabewala			118	73
Rewasi	165	174	91	166
Gharsondi	106	115	60	114
Alabujanahalli	97	240	101	104
Siresandra	93	173	99	147
Zhapur	61	132	55	152
Amarsinghi	107	125	89	100
Panahar	110	114	99	108
Kalmandasguri	100	142	112	142

Appendix Table 2C *Mean agricultural wage received by male workers of small farmer households at 2010–11 prices, and minimum wages in agriculture, 2011*

Village	Mean agricultural wage received by male workers	Minimum wage in agriculture[1]
Ananthavaram	174	112
Bukkacherla	166	112
Kothapalle	208	157.34
Harevli	74	100
Mahatwar	53	100
Nimshirgaon	203	100
Warwat Khanderao	132	100
25F Gulabewala		135
Rewasi	174	135
Gharsondi	115	124
Alabujanahalli	240	157.34
Siresandra	173	157.34
Zhapur	132	157.34
Amarsinghi	125	167
Kalmandasguri	142	167
Panahar	114	167

Note: [1] State-specific minimum wages for the year 2011 are used.
Source: Indian Labour Statistics, 2011 and 2012, Labour Bureau, Government of India (2015)

6

Costs and Prices

*Arindam Das, Tapas S. Modak, Biplab Sarkar, Madhura Swaminathan,
with Vijay Kumar and Ritam Dutta*

INTRODUCTION

Multiple factors affect the levels of income from crop production of small
farmer households in different agro-ecological regions. Costs of cultivation
and prices of agricultural output are two important factors, as income from
crop production or crop income is the gross value of output less cost of
cultivation. This chapter examines output prices and component parts of the
costs of cultivation based on data from 15 village surveys conducted by the
Foundation for Agrarian Studies (FAS) under its Project on Agrarian Relations
in India (PARI).

Data on cost of cultivation here refer to Cost A2 as defined by the
Commission for Agricultural Costs and Prices (CACP). Cost A2 includes the
costs of home-produced and purchased seeds, the value of home-produced and
purchased manure, the value of chemical fertilizer, plant protection, irrigation
charges, hired labour, the costs of owned and hired animal labour, the costs of
owned and hired machinery, rent paid for leased-in land, marketing expenses,
land revenue, interest on working capital, depreciation of own machinery,
and crop insurance expenses. Data on agricultural output prices here refer to
farm harvest prices (FHP), defined as the output prices at which farmers sold
their produce to traders at the village site or a nearby market during a specified
marketing period after the beginning of the harvest season.

The second section of this chapter discusses the component parts of the
cost of cultivation and its variation across small farmers in the PARI study
villages located in different agro-ecological regions of India. In particular,
labour absorption and cost of labour, cost of irrigation, machines, seed, and
rent on leased-in land are dealt with in detail. Variations in farm harvest prices
and their deviation from minimum support prices announced by the central
government are dealt with in the third section. The final section records key
findings from the analysis.

The objective of this chapter is to identify specific factors that contribute to low returns from farming as reported by small farmers, as well as identify how and why specific costs differ as between small and large farmers.

PATTERN OF INPUT USE IN THE STUDY VILLAGES

Chapter 4 of this volume showed that the cost of cultivation per hectare varied hugely among small farmer households across the PARI study villages. Per hectare cost of cultivation was relatively low in villages that had a low level of cropping intensity, and where cereals, millets, and pulses were cultivated. Take, for example, Rewasi in Sikar district of Rajasthan or Zhapur in Kalaburagi district of Karnataka, where the cost of cultivation per hectare was Rs 11,000 and Rs 13,000, respectively. On the other hand, in Alabujanahalli, an irrigated sugarcane-growing village in Mandya district of Karnataka, or Panahar, an irrigated potato-growing village in Bankura district of West Bengal, the cost of cultivation was around Rs 70,000 per hectare. Cropping intensity as well as crop-mix had an important influence on the cost of cultivation.

The total cost of cultivation masks significant differences in the distribution of costs of various inputs. Appendix Tables 1.1 to 1.3 of this chapter show the total cost of cultivation (Cost A2) by components in 16 villages across seven States for small farmer households. Costs of hired human labour, hired machinery, fertilizer, irrigation, seed, and rent on leased-in land constituted around two-thirds of the total cost of cultivation for small farmer households in most of the study villages. The share of each cost item in the total cost, however, varied across villages because of differences in production conditions and relations of production. We shall investigate some of these cost items separately, with particular emphasis on the pattern of input use in different production systems.[1] Small farmers' access to technical information was an important influence on farmers' practices, including use of inputs. We discuss this in Appendix 2.

Labour Absorption and Cost of Labour

The labour used in agricultural production can be broadly classified into three categories: family labour, hired labour, and exchange labour. Family labour includes labour performed by any member of the household on its own land for agricultural production. Hired labour includes casual workers who are hired on a daily or piece-rate basis for agricultural operations, and long-term workers who are hired, most commonly, for a season or for a year. Apart

---

[1] Fertilizer use and fertilizer costs have been dealt with separately in chapter 7 of this volume.

from these, in some cases, labour exchange among farmer households was also practised.[2]

As shown in chapter 3 of this volume, use of family labour was predominant on small farms in eight out of 15 villages studied. In Harevli (Bijnor district, Uttar Pradesh), Mahatwar (Ballia district, Uttar Pradesh), Gharsondi (Gwalior district, Madhya Pradesh), and Rewasi (Sikar district, Rajasthan), 70 per cent or more of the total labour absorbed by small farms was family labour. The ratio was higher than 50 per cent in Nimshirgaon (Kolhapur district, Maharashtra), Alabujanahalli (Mandya district, Karnataka), Siresandra (Kolar district, Karnataka), Zhapur (Kalaburagi district, Karnataka), and Kalmandasguri (Koch Bihar district, West Bengal). However it is important to mention here that hired labour was employed by small farmers in all the villages, and, in some villages, the use of hired labour on small farms was greater than the use of family labour. The contribution of exchange labour to small farmers' overall labour input was low, at less than 10 per cent.

We now turn to the variations in costs of hired labour among small farmer households in the study villages, and assess the role played by crop-mix and family labour use. There are three major findings, as shown in Tables 1 and 2 below.

First, the expense on hired labour was a major component of total paid-out cost (Cost A2) of small farmer households in the study villages. It was more than one-fifth of the total paid-out cost of small farmers in nine villages (Table 1). Among all the study villages, the share of hired labour cost in total paid-out cost was the highest in Alabujanahalli and Amarsinghi (Malda district, West Bengal) – more than 40 per cent. This was mainly because of intensive cultivation and a high share of labour-intensive crops such as sugarcane and jute in the gross cropped area of small farmer households.[3] However, expenses on hired labour as a share of total paid-out cost were less than 10 per cent in two of the study villages: Mahatwar and Gharsondi.

Secondly, per hectare hired labour cost varied hugely among small farmer households across the study villages. For instance, on average, per hectare hired labour cost of small farmer households in Alabujanahalli and Amarsinghi

---

[2] Exchange labour is defined as follows: "members of a household join members of another household in working without payment at certain tasks on the land of the second household; in exchange of this, members of the second household join members of the first household in labouring on the latter's land" (Ramachandran 1990).

[3] Given the present state of technological development in Indian agriculture, the scope for mechanisation is limited in sugarcane and jute cultivation.

Table 1 *Hired labour cost per hectare of operational holding, small farmers, study villages, at 2010–11 prices*

Village	State	Hired labour cost (in Rupees per hectare)	Share in total Cost A2 (per cent)
Ananthavaram	Andhra Pradesh	15,545	17
Bukkacherla	Andhra Pradesh	5,053	26
Kothapalle	Telangana	7,560	23
Harevli	Uttar Pradesh	4,828	11
Mahatwar	Uttar Pradesh	1,528	6
Nimshirgaon	Maharashtra	7,958	19
Warwat Khanderao	Maharashtra	5,679	32
Gharsondi	Madhya Pradesh	1,933	8
Alabujanahalli	Karnataka	32,079	44
Siresandra	Karnataka	7,093	22
Zhapur	Karnataka	3,266	24
Rewasi	Rajasthan	6,225	16
Amarsinghi	West Bengal	25,190	41
Kalmandasguri	West Bengal	11,779	30
Panahar	West Bengal	111,532	16

Source: PARI survey data.

was around Rs 32,000 and Rs 25,000, respectively (Table 1). On the other hand, in Mahatwar and Gharsondi, per hectare hired labour cost was around Rs 1,500 or almost 20 times lower than the hired labour cost incurred by small farmer households in Alabujanahalli.

Thirdly, in most of the PARI villages where expenses on hired labour use in crop production were low for small farms, exploitation of family labour was evident. For example, in Mahatwar village of eastern Uttar Pradesh, the average cost incurred by small farmer households on hired labour was the lowest, at Rs 1,528 per hectare, and labour on crop production was mainly provided by family members.

The data suggest that variations in costs of hired labour as well as total paid-out costs among small farms across different parts of India largely depend on forms of agricultural production, specifically the use of unpaid family labour versus wage labour. It should also be noted that in estimates of total paid-out cost (Cost A2), the cost of family labour was not imputed. Therefore, other things remaining the same, total paid-out costs will be lower and estimated crop incomes net of paid-out costs will be higher for small farmer households located in family labour-based production systems than is truly warranted.

To correct for this observation, we imputed the cost of family labour (FL) of small farmer households in nine villages.[4] As expected, Cost A2 (paid-out cost) was lower than Cost A2 + FL for all villages (Table 2). The difference between Cost A2 and Cost A2 + FL among the study villages, however, varied. In absolute terms, this difference ranged from Rs 774 in Zhapur to Rs 21,570 in Mahatwar. In percentage terms, the increase in cost of cultivation after imputing family labour was around 6 per cent in Zhapur and 80 per cent in Mahatwar. Taking a cost definition inclusive of family labour is likely to lower variations in cost across different production systems.

Cost of Irrigation

There was a significant variation in average irrigation costs among small farmer households across the study villages (Appendix Tables 1.1 to 1.3). In villages located in scarce rainfall regions, small farmer households primarily practised dry land agriculture and, consequently, the cost of irrigation was negligible. Bukkacherla in the dry, drought-prone region of Rayalaseema in Andhra Pradesh, Warwat Khanderao in the unirrigated cotton-growing tracts of Vidarbha in Maharashtra, and Zhapur in the semi-arid Deccan Plateau region of north Karnataka belong to this category; the exception was Rewasi in Sikar district of Rajasthan. In Zhapur, for example, irrigation was not used by small farmer households and therefore the cost of irrigation was zero. However, large farmers in this village had irrigated 15 per cent of their farm land by means of tubewells during the year 2008–09. In contrast, small farmer households in irrigated regions had access to irrigation, but the costs varied hugely, ranging from Rs 1,000 per hectare in Alabujanahalli, a canal-irrigated village, to Rs 8,265 per hectare in Amarsinghi, a groundwater-irrigated village.

In general, differences in costs of irrigation were related to the type of irrigation infrastructure (public investment in canal or groundwater irrigation versus private investment in groundwater irrigation), distribution of ownership of irrigation equipment (particularly diesel or electric pumps), and State policies on access to irrigation. These are discussed in greater detail below.

Canal irrigation

The cost of irrigation in canal-irrigated villages such as Alubujanahalli was substantially low. This village receives irrigation water from a network of

[4] The cost of family labour (FL) was calculated by imputing a wage, since family labour is unpaid. The cost was imputed on the basis of the prevailing daily wage rate in the village. Since men, women, and children participated in cultivation, we valued the days of labour performed by women and children at the prevailing daily wage rate for female labour in the village, and the days of labour performed by men at the prevailing male wage rate (for daily-paid casual labour tasks).

Table 2 *Cost of cultivation after imputation of family labour, small farmers, study villages*

Village	State	Imputed family labour cost (at 2010–11 prices, in Rs per hectare)	Cost A2 + FL (at 2010–11 prices, in Rs per hectare)	Difference between Cost A2 + FL and Cost A2 (at 2010–11 prices)	Percentage increase in cost after imputation of family labour
Ananthavaram	Andhra Pradesh	6,217	99,921	6,217	7
Bukkacherla	Andhra Pradesh	2,594	22,002	2,594	13
Kothapalle	Telangana	9,180	41,843	9,180	28
Harevli	Uttar Pradesh	20,598	65,931	20,598	45
Mahatwar	Uttar Pradesh	21,570	48,538	21,570	80
Nimshirgaon	Maharashtra	8,163	50,155	8,163	19
Warwat Khanderao	Maharashtra	2,977	20,595	2,977	17
Zhapur	Karnataka	774	14,651	774	6
Rewasi	Rajasthan	18,098	57,541	18,098	46

Note: FL = family labour.
Source: PARI survey data.

system tanks that is fed by canals from the Krishnarajasagar dam on the Cauvery river. Two-thirds of the total crop land was irrigated solely by tanks, while the remaining was irrigated by a combination of tanks and tubewells (Kumar 2017). The average cost of irrigation from system tanks was Rs 290 per hectare in the agricultural year 2008–09. The fact of secure and low cost irrigation led small farmer households to cultivate water-intensive crops such as paddy and sugarcane.

Combination of canal and groundwater irrigation
Ananthavaram village in the command area of the Varalapuram canal on the Krishna river, Harevli village in the command area of the Eastern Ganga canal, and Gharsondi village in the command area of the Harsi dam received irrigation water from a combination of canals and tubewells. Due to insufficient supply of canal water, tubewell irrigation emerged as an alternative source, particularly for cultivation during the *rabi* season. In the case of groundwater, the pattern of ownership of tubewells and irrigation equipment made a difference to costs. Table 3 shows the inequality in ownership of private irrigation equipment in the above-mentioned three villages. The proportion of households that owned irrigation equipment was much lower among small farmers than among large farmers.

In Ananthavaram, per hectare irrigation cost for small farmers was more than twice that for large farmers. Small farmers incurred an expense of Rs 4,878 per hectare for irrigation, as compared to Rs 2,044 incurred by large farmer households. This difference was on account of the cost of irrigation for crops cultivated in the *rabi* season. An examination of cropping pattern and access to water shows that irrigation costs were almost the same for *kharif* paddy cultivation across different classes of farmers in the village. This was because adequate canal water from the Krishna delta irrigation system was available for cultivation in the *kharif* season. Canal charges were similar for all cultivators in the village. Cultivation of *rabi* crops such as maize and sorghum, however, was mostly dependent on tubewells. Tubewells were under the control of rich farmers in the village, and small farmers purchased water from these rich farmers, which increased their irrigation cost for *rabi* crops.

In Gharsondi, farmers faced a severe shortage of canal water in the years preceding 2007–08 (Rawal 2010). This continued shortage of canal water created demand for an alternative source of irrigation, and rich farmers in the village invested in tubewells and electric pumps. Small farmer households had very little access to tubewells. Table 4 shows that 88 per cent of the gross irrigated area of small farmer households was irrigated by canal water and only

Table 3 *Proportion of households that owned irrigation equipment and average value of irrigation equipment by farmer category, study villages, at 2010–11 prices*

Village	State	Small farmers		Large farmers	
		Proportion of households that owned irrigation equipment (%)	Value of irrigation equipment per household (Rs)	Proportion of households that owned irrigation equipment (%)	Value of irrigation equipment per household (Rs)
Ananthavaram	Andhra Pradesh	12	3,217	60	31,147
Harevli	Uttar Pradesh	27	7,245	75	49,009
Gharsondi	Madhya Pradesh	18	3,137	55	43,999

Source: PARI survey data.

Table 4 *Proportion of gross irrigated area by source of irrigation and farmer category,
in Gharsondi (2008) and Harevli (2006)*

Sources	Gharsondi		Harevli	
	Small farmers	Large farmers	Small farmers	Large farmers
Canal	88	82	47	45
Tubewell irrigation with diesel pump	7	2	51	23
Tubewell irrigation with electric pump	21	68	35	74

Source: PARI survey data.

28 per cent by tubewells. Purchase of water from rich farmers increased the
cost of cultivation for small farmers in the village.

In Harevli, *kharif* crops were irrigated by the Eastern Ganga canal project.
Tubewells fitted with diesel or electric pumps were used for *rabi* cultivation,
and as supplementary irrigation for *kharif* crops. Most of the tubewells run by
electric pumps in the village belonged to large landowning Tyagi households.
Small farmers mainly owned diesel pumps. Table 4 shows that small farmers
irrigated 51 per cent of gross irrigated area by means of diesel pumps (mostly
self-owned) and 35 per cent by electric pumps. The ownership pattern of
irrigation equipment, particularly diesel and electric pumps, and access to
tubewell irrigation through the water market increased the cost of irrigation
for small farmer households in Harevli.

To sum up, the above discussion of irrigation in three study villages shows
that canal irrigation in Anathavaram, Gharsondi, and Harevli was available
only for *kharif* crop cultivation. The most common sources of irrigation in
the *rabi* season were private tubewells run by electric or diesel pumps. Small
farmer households hardly owned any tubewells with electric pumps. Lack of
access to electricity connections or the required investment for submersible
tubewells implied that small farmers had to depend on diesel pumps or the
water market.

Groundwater irrigation

The cost of irrigation was higher for villages dependent on groundwater
irrigation as compared to canal-irrigated villages. In Amarsinghi (Malda district,
West Bengal), Panahar (Bankura district, West Bengal), Mahatwar (Ballia
district, Uttar Pradesh), Nimshirgaon (Kolhapur district , Maharashtra), and
Tehang (Jalandhar district , Punjab) – all groundwater-irrigated villages – per
hectare irrigation costs for small farmer households were Rs 8,265, Rs 6,333,

Rs 6,181, Rs 4,195, and Rs 7,909, respectively.[5] Small farmer households in these villages obtained groundwater for irrigation through informal water markets. In contrast, in Punjab, 77 per cent of small farmers owned tubewells and electric pumps. The State government provided free electricity for irrigation in Punjab. However, due to erratic electricity supply, farmers used diesel-powered generators, which increased irrigation costs among small farmer households.

Dry villages with groundwater irrigation

Cultivation in Rewasi (in the western dry region of Rajasthan) and Siresandra (in the semi-arid south-eastern dry region of Karnataka) was mainly rainfed, supplemented by some irrigation by means of tubewells. There was no public source of irrigation in these villages, and irrigation was entirely dependent on private investment.

In Rewasi, the groundwater level is low. To extract groundwater, tubewells need to be powered by electric pumps. Therefore, the installation of tubewells depends on the capacity of landowners to make the requisite investment for electric pumps and electricity connections. As only irrigated land could be cultivated in the *rabi* season, small farmers had very little land under *rabi* cultivation in Rewasi (Swaminathan and Rawal 2015). In cases where small farmers had access to irrigation in the *rabi* season, the cost structure shows that they had to incur much higher costs than large farmers, particularly in wheat and rapeseed cultivation.

In Siresandra, only 35 per cent of total operated land was irrigated by means of tubewells during 2008–09 (Sarkar 2017). To reduce wastage of water, drip irrigation technology was used for vegetable cultivation. However, the spread of irrigation in Siresandra was limited to rich farmers because of the high cost involved in extending the depth of tubewells. In other words, continuous depletion of groundwater led to higher costs of installation of tubewells, which restricted the spread of groundwater irrigation among small farmers, even though electricity for irrigation was free in the State of Karnataka.

To sum up, the skewed distribution of irrigation equipment, the nature of private investment in groundwater irrigation systems, and State policies significantly affected the cost of irrigation of small farmers. Ownership of

---

[5] In Amarsinghi the costs of irrigation were high despite some reduction between 2005 and 2015. Sarkar (2015, 2017), and Modak and Bakshi (2016) have shown that per hectare average irrigation cost in Amarsinghi village fell between 2005 and 2015 because of installation of a panchayat deep tubewell managed by farmers' beneficiary groups, and a shift from diesel pumps to electric pump-operated private tubewells in the subsequent period. Per hectare average cost of irrigation for *boro* paddy at 2014–15 prices fell from Rs 26,103 in 2005 to Rs 11,452 in 2010, and Rs 8,987 in 2015.

irrigation equipment, particularly electric-powered tubewells, was uncommon among small farmer households. They mostly owned diesel pumps, which were associated with a high cost of irrigation.

Secure canal irrigation has reduced the cost of irrigation in the total paid-out cost of small farmer households in Alabujanahalli (Mandya district, Karnataka). Unreliable canal irrigation pushed farmers to invest in private groundwater irrigation systems in Ananthavaram (Guntur district, Andhra Pradesh), Gharsondi (Gwalior district, Madhya Pradesh), and Harevli (Bijnor district, Uttar Pradesh). Since these investments were made by rich farmer households, small farmers were either excluded from the service or had to access it through high payment to tubewell owners.

The fragmented nature of landholdings and high installation cost of tubewells created an informal, private water market in groundwater-irrigated villages. Most small farmers accessed irrigation water through this private water market, which, as we have shown, increased the costs of irrigation. By contrast, public intervention in groundwater irrigation in terms of free electricity (such as in Tehang, Jalandhar district, Punjab) or installation of panchayat-owned deep tubewells (as in Amarsinghi, Malda district, West Bengal) reduced the costs of irrigation for small farmer households. In dry villages, small farmer households were mostly excluded from irrigated cultivation.

Expenses on Machine Labour

The cost incurred for machine labour was not high, other than in the highly mechanised villages of Harevli (Bijnor district) in Uttar Pradesh, Gharsondi (Gwalior district) in Madhya Pradesh, and Tehang (Jalandhar district) in Punjab (Appendix Tables 1.1 to 1.3). The use of machines for different operations in crop production was determined by the type of crop grown and the cost of machines. Small farmer households in the PARI study villages reported the use of tractors, power tillers, diesel and electric pumps, sprayers, seed drillers, combine harvesters, and threshers for crop operations. Machines were used mainly for land preparation, irrigation, plant protection, and threshing of the harvested crop. Seed drill machines were used for sowing of wheat in Gharsondi and Tehang. Combine harvesters were used for harvesting of paddy and wheat in Tehang, and for wheat in Gharsondi and Harevli.

In these three villages, a high proportion of agricultural operations undertaken in wheat cultivation by small farmer households used machine labour. In Gharsondi, small farmer households ploughed 97 per cent of their land with tractors, and operations like sowing and threshing were fully mechanised. Machine rental markets in the villages played an important role

Table 5 *Spread of mechanisation in different operations of wheat cultivation, small farmer households, selected villages* in per cent

Village	Proportion of land ploughed with tractor	Proportion of land sown with seed drill	Proportion of land harvested with combine harvester	Proportion of output threshed with thresher
Harevli	70	0	0	100
Gharsondi	97	100	37	100
Tehang	100	82	95	100

Source: PARI survey data.

Table 6 *Share of work done by rented machines in total machine labour used in different operations of wheat cultivation, small farmer households, selected villages* in per cent

Village	Ploughing	Sowing	Threshing
Harevli	84		100
Gharsondi	80	92	97
Tehang	69	76	96

Source: PARI survey data.

in providing access to machines for small farmer households who could not afford to own them (Table 6).

At the same time, there was a significant difference in the costs of machine labour across farmer categories. Small farmer households paid a higher price for machine labour than large farmers. This difference was mainly because of the unequal pattern of ownership of machines across farmer categories (Table 7). In 2010, there were 15 tractors, 10 sprayers, four threshers, and one seed-drill machine in Harevli village. The majority of these machines were owned by the class of landlords and capitalist farmers, or large farmers. Three landlord households owned one tractor each. Eight tractors were owned by large farmer households. Small farmers owned only three out of 15 tractors in the village.

Similarly, in Gharsondi, we observed high inequality in the distribution of agricultural machines across classes of farmers (Table 8). Small farmer households owned a handful of modern agricultural machines: just three tractors out of a total of 51 in the village, and only one thresher out of a total of 20.

To sum up, ownership of agricultural machinery was concentrated in the hands of farmers with relatively large operational holdings. Small farmer households rarely owned a tractor or power tiller or thresher. They mainly owned traditional wooden equipments, hand sprayers, and diesel pumps. The absence of control over machinery resulted in a higher cost of machine labour for small farmer households relative to other households.

Table 7 *Distribution of ownership of agricultural machines by farmer category, Harevli, 2006*

Farmer category	Number of households	Tractors	Sprayers	Seed drills	Threshers
Landlords and capitalist farmers	3	3	2	1	1
Large farmers	21	8	5	0	2
Small farmers	45	3	2	0	0
Total	69	15	10	1	4

Note: Agricultural machines owned by non-agricultural households in the village are included in the total number of particular machines.
Source: PARI survey data.

Table 8 *Distribution of ownership of agricultural machines by farmer category, Gharsondi, 2008*

Farmer category	Number of households	Tractors	Sprayers	Seed drills	Threshers
Landlords and capitalist farmers	12	15	26	12	7
Large farmers	59	31	31	19	11
Small farmers	71	3	6	3	1
Total	142	51	70	34	20

Note: Agricultural machines owned by non-agricultural households in the village are included in the total number of particular machines.
Source: PARI survey data.

Seed Costs

Seed is the most vital input for crop production. Other things remaining the same, one can increase productivity by planting quality seeds. The cost of seed is determined by the crop grown, and relative shares of farm-saved seed and purchased seed.

Table 9 shows the share of seed cost in total paid-out costs in the study villages. The share of seed cost in total cost of cultivation of small farmer households ranged from about 4 per cent in Alabujanahalli (Mandya district, Karnataka) to 23 per cent in Gharsondi (Gwalior district, Madhya Pradesh). In Alabujanahalli, the share of seed cost was low because of a special form of sugarcane cultivation known as "ratooning," where farmers, after the first sugarcane harvest, leave the roots and lower parts of the plant in the soil. Sarkar (2017) shows how farmers saved on the costs of preparing the field and planting by this process. In Bukkacherla (Anantapur district, Andhra Pradesh), Warwat Khanderao (Buldhana district, Maharashtra), Gharsondi,

Table 9 *Share of seed cost in total cost (Cost A2) of cultivation, study villages* in Rs per hectare and per cent

Village	State	Seed cost	Share in total cost
Ananthavaram	Andhra Pradesh	6,083	6
Bukkacherla	Andhra Pradesh	3,774	19
Kothapalle	Telangana	1,947	6
Harevli	Uttar Pradesh	2,458	5
Mahatwar	Uttar Pradesh	3,855	14
Nimshirgaon	Maharashtra	3,427	8
Warwat Khanderao	Maharashtra	2,688	15
Gharsondi	Madhya Pradesh	5,506	23
Alabujanahalli	Karnataka	2,700	4
Siresandra	Karnataka	3,486	10
Zhapur	Karnataka	690	5
Rewasi	Rajasthan	4,934	13
Amarsinghi	West Bengal	6,414	11
Kalmandasguri	West Bengal	5,867	15
Panahar	West Bengal	14,882	20

Source: PARI survey data.

Table 10 *Average seed rate used for paddy cultivation, by farmer category, study villages* in kg per hectare

Village	State	Small farmers	Large farmers
Harevli	Uttar Pradesh	77	15
Mahatwar	Uttar Pradesh	89	74
Gharsondi	Madhya Pradesh	59	30
Alabujanahalli	Karnataka	121	106
Amarsinghi	West Bengal	99	NA
Panahar	West Bengal	77	67
Kalmandasguri	West Bengal	40	NA

Source: PARI survey data.

Kalmandasguri (Koch Bihar district, West Bengal), and Panahar (Bankura district, West Bengal), the share of seed cost in total cost of cultivation was substantially high. In Kalmandasguri and Panahar, the share of seed cost was high on account of potato cultivation (for which seeds are usually purchased and prices are high). In Warwat Khanderao and Gharsondi, the seed cost was high on account of cotton and soybean cultivation, respectively (both are entirely commercial crops where seeds are purchased annually).

To investigate the seed rate (or intensity of seed use), and relative share of farm-saved and purchased seed in total seed use by small farmer households, we have chosen two major crops – paddy and wheat – in the study villages. In all

Table 11 *Share of homegrown seed in total seed used for paddy cultivation, small farmers, study villages* in per cent

Village	State	Share
Harevli	Uttar Pradesh	81
Mahatwar	Uttar Pradesh	49
Gharsondi	Madhya Pradesh	60
Alabujanahalli	Karnataka	45
Amarsinghi	West Bengal	83
Panahar	West Bengal	23
Kalmandasguri	West Bengal	60

Source: PARI survey data.

Table 12 *Average seed rate used for wheat cultivation, by farmer category, study villages* in kg per hectare

Village	State	Small farmers	Large farmers
Gharsondi	Madhya Pradesh	219	240
Harevli	Uttar Pradesh	129	131
Mahatwar	Uttar Pradesh	133	178
Nimshirgaon	Maharashtra	86	93
Rewasi	Rajasthan	126	124

Source: PARI survey data.

Table 13 *Share of homegrown seed in total seed used for wheat cultivation, small farmers, study villages* in per cent

Village	State	Share
Gharsondi	Madhya Pradesh	0
Harevli	Uttar Pradesh	0
Mahatwar	Uttar Pradesh	17
Rewasi	Rajasthan	3

Source: PARI survey data.

paddy-growing villages, it was observed that small farmers used higher seed rate per hectare than large farmer households (Table 10). This may be because small farmers anticipate high mortality of seedlings in case of water scarcity. Higher seed rate ensured that enough seedlings survived (Surjit 2006). It was also observed that almost half the seed used for paddy cultivation was farm-saved seed among small farmer households, except in Panahar village of West Bengal (Table 11). However, in the case of wheat, per hectare use of seed was higher for large farmer households than small farmer households (Table 12). Small farmers preferred purchased seed for wheat cultivation in the study villages (Table 13).

Cost of Rent for Leased-in Land

In many parts of India, the problems of small farmers are accentuated by the fact that they operate land that is not owned by them. The conditions of tenants and terms of tenancy contracts are subjects that require detailed and independent study. Rents on leased-in land constituted a substantial component of paid-out costs (Cost A2) of small farmer households in some of the study villages. In Harevli (Bijnor district, Uttar Pradesh) and Ananthavaram (Guntur district, Andhra Pradesh), the average rent paid on land leased-in by small farmer households constituted about 35 and 32 per cent, respectively, of Cost A2 (Appendix Table 1.1).

An important aspect of paddy cultivation in Harevli was that a large part of it was cultivated on the basis of seasonal tenancy contracts. Most commonly, large cultivators in Harevli leased out small plots of land on sharecropping contracts (in between two sugarcane crops) to their long-term workers for the cultivation of paddy (Rawal 2009). Land was leased out by landlord or rich peasant Tyagi households on seasonal share contracts to Dalit households for paddy cultivation. The average value of gross rent on land leased in for paddy cultivation was Rs 8,914 per acre for one season. This was about 61 per cent of the average gross value of output per acre. In comparison, the net income of the tenant was 14.8 per cent of the gross value of output. Data reported by landowners suggest that the average net rental income per acre of land leased out was Rs 6,294, or about 46 per cent of the average gross value of output from paddy on leased-in land (Rawal 2014).

A detailed account of tenancy in Ananthavaram village was first provided by P. Sundarayya (1977) in his article based on a survey of the village in 1974. At that time most tenant farmers leased in land for the cultivation of paddy, maize, and black gram. They paid a fixed rent in cash for the paddy crop and a fixed rent in kind for other crops. Based on the PARI resurvey of Ananthavaram in 2005 and 2006, Ramachandran, Rawal, and Swaminathan (2010, pp. 62–69) note:

> In Ananthavaram, landowners made no contribution to the costs of cultivation. In 2005–06, a tenant who leased land on a rent-in-kind tenancy contract produced about 18.9 quintals from an acre of land. Of this, 16.2 quintals were given away as rent . . . the tenant was allowed to keep all the straw from the paddy crop. This enabled tenants to maintain a much larger stock of animals than other households. Tenants cultivated maize in the rabi season . . . the net income from paddy and maize put together was Rs 984 per acre. In other words, a tenant who leased in land and paid rent in kind just managed to break even after the maize harvest.

To sum up, in both Harevli and Ananthavaram, landless households typically leased in land for foodgrain (paddy) cultivation and not for high-value commercial crops. The high cost of cultivation and low or negative return from paddy cultivation was on account of a specific form of tenancy relations characterised by punishing rents.

AGRICULTURAL OUTPUT PRICES

The analysis in this section deals with two sets of questions. First, we investigate how government interventions in support prices and procurement, and the physical location of agricultural commodity markets affected agricultural output prices in the study villages. Secondly, we examine the impact of the market price crash of a commercial crop on the profitability of small farmer households. We have chosen two major crops, paddy and wheat, in 15 villages to answer the first question. The second question has been investigated by a case study of potato cultivation in Panahar village in West Bengal, where small farmers incurred huge losses because of a price crash in the survey year.

Paddy

We estimated the weighted average farm harvest price (FHP) of paddy for 12 villages and calculated the deviation from the minimum support price (MSP) in the respective years.

In seven out of 12 villages the average FHP was lower than the official MSP in the relevant year. So MSP did not act as a floor price for paddy in all the villages. Average FHP was lower than MSP in villages belonging to the major rice-producing States of India.[6] This is not a one-off finding, as official FHP statistics for these States show that in most years, the State-weighted average FHP was lower than the official MSP.[7]

When we compared the MSP of paddy with the actual FHP for each small farmer household in Ananthavaram (Guntur district, Andhra Pradesh), the unit price obtained was below MSP in 36 per cent of small farmer households. The corresponding proportion was 40 per cent in Bukkacherla (in the dry zone of Rayalaseema, Anantapur district, Andhra Pradesh), 81 per cent in Kothapalle (Karimnagar district, north Telangana, Andhra Pradesh), and 30 per cent in Harevli (Bijnor district, western Uttar Pradesh).

---

[6] West Bengal is the largest producer of paddy, followed by Andhra Pradesh.
[7] Available at the website of the Directorate of Economics and Statistics, Ministry of Agriculture – http://eands.dacnet.nic.in/FHP%28District%29.htm – from 1998–99 to 2010–11.

Table 14 *Minimum support price (MSP) and farm harvest price (FHP) obtained by small farmers for paddy* in Rs per 100 kg

Village	State	Year	Minimum support price (MSP)	Farm harvest price (FHP)	FHP–MSP
Ananthavaram	Andhra Pradesh	2005–06	570	580	10
Bukkacherla	Andhra Pradesh	2005–06	570	625	55
Kothapalle	Telangana	2005–06	570	520	–50
Harevli	Uttar Pradesh	2005–06	570	600	30
Alabujanahalli	Karnataka	2008–09	900	860	–40
Panahar	West Bengal	2009–10	1050	870	–180
Amarsinghi	West Bengal	2009–10	1050	900	–150
Kalmandasguri	West Bengal	2009–10	1050	875	–175
Tehang	Punjab	2010–11	1000	1030	30

Source: PARI survey data.

In Alabujanahalli (Mandya district, Karnataka), surveyed in 2008–09, for 75 per cent of small farmer households in the pre-*kharif* season, and 50 per cent of small farmer households in the *kharif* and *rabi* seasons, MSP did not act as a floor price. The picture was more alarming in the West Bengal villages, where procurement seemed non-operational. By contrast, small farmer households in Tehang village in Jalandhar district of Punjab received prices either equal to or higher than the announced MSP. Paddy was cultivated as a commercial crop in Tehang, and almost 95 per cent of the produce was marketed with a strong government procurement system in place.

Wheat

We estimated the weighted average FHP of wheat for seven villages and calculated the deviation from MSP for the respective years.

The picture for wheat is very different from that for paddy. On average, the FHP of wheat in all the villages was higher than the announced MSP for that year. MSP did indeed act as a floor price for wheat. In other words, with active procurement, MSP can work as a floor price.

The output prices for wheat and paddy in our study villages showed that MSP was more effective for wheat than for paddy. The fact that small farmer households in Tehang (Jalandhar district, Punjab) obtained an FHP that was higher than MSP for both paddy and wheat reflects the success of procurement in Punjab.

Table 15 *Minimum support price (MSP) and farm harvest price (FHP) obtained by small farmers for wheat* in Rs per 100 kg

Village	State	Year*	Minimum support price (MSP)	Farm harvest price (FHP)	FHP–MSP
Harevli	Uttar Pradesh	2005–06	640	800	160
Mahatwar	Uttar Pradesh	2005–06	640	750	110
Gharsondi	Madhya Pradesh	2007–08	850	1100	250
Rewasi	Rajasthan	2009–10	1080	1300	220
Tehang	Punjab	2010–11	1100	1120	20

Note: *Indicates the agricultural year for which survey was done.
Source: PARI survey data.

Potato Cultivation in Panahar, Bankura District, West Bengal[8]

Potato is an important *rabi* crop in West Bengal. It offers significant returns for farmers, but at higher risk. Potato entails a higher cost of cultivation than paddy, with significant returns in normal years. In 2013–14, for example, the return from potato cultivation over paid-out cost was around Rs 47,000 per hectare, as compared to Rs 25,000 from *kharif* paddy in West Bengal.[9] However, the profitability of potato cultivation depends on access to markets and marketing arrangements, particularly for small farmers.

In 2010, farmers in Panahar village incurred, on average, a paid-out cost of Rs 271 per quintal of potato production, but obtained only Rs 181 per quintal (at 2014–15 prices). In 2015, the per quintal paid-out cost was Rs 260 and the average market price was Rs 200. As a result, more than 80 per cent of farmers in Panahar incurred losses from potato cultivation in 2010 and 2015.

A few farmer households, however, did not incur a loss, on account of the variety sown as well as the channels of marketing. Of the three potato varieties cultivated in Panahar – Jyoti, Pokhraj, and Atlantic – the Atlantic variety was cultivated by landlords and capitalist farmers, whereas Jyoti and Pokhraj were cultivated by all categories of farmers. There were two channels of potato marketing in the village: contract farming and open market sale. The Atlantic variety was grown by capitalist farmers in contract farming arrangements with Frito Lays, a subsidiary of PepsiCo Company Private Limited. These contracts offered a stable price of around Rs 600 per quintal at 2014–15 prices. The Jyoti and Pokhraj varieties were grown by all farmers for open market sale.

---

[8] This case study is based on Sarkar (2015, 2017).
[9] Extracted from the website of the Directorate of Economics and Statistics, Ministry of Agriculture and Farmers Welfare, Government of India: http://eands.dacnet.nic.in/Cost_of_Cultivation.htm

In the absence of a government support price, the price of potato was market-determined. Small farmers, however, did not sell directly at the market place or *mandi*, but through village traders who acted as commission agents of big traders. In some cases, such sales were based on arrangements with traders in exchange for inputs and credit, while in other cases, sales were conducted at the farm gate due to the high cost of transport and fluctuations in prices in the local market. Unlike small farmers who had no choice but to sell their produce immediately after the harvest, the landlords/capitalist farmers could use the facility of cold storages, and sell their produce later and at more favourable prices.

CONCLUSIONS

This chapter has investigated some components of the costs of cultivation and agricultural output prices in order to understand the composition of farm incomes of small farmer households. The chapter uses data from 16 villages surveyed under PARI by the Foundation for Agrarian Studies, in different agro-ecological regions. The following are the main findings from an analysis of Cost A2 or paid-out costs borne by small farmers.

First, and in consonance with the findings of chapter 3, the cost of hired labour was a significant component of paid-out costs in most of the study villages, the proportion going up to 40 per cent in some villages with labour-intensive crops such as jute and sugarcane. Family labour, of course, was more heavily used by small farmers as compared to large farmers.

Secondly, irrigation was a big cost for small farmers in many of the villages. The cost of irrigation varied with the type of irrigation, being the lowest for canal irrigation and the highest for private tubewell-based irrigation. Invariably, ownership of irrigation equipment was less among small farmers than large farmers, leading small farmers to dependence on water markets (and higher costs).

Thirdly, machine labour costs were higher for small farmers than large farmers, as ownership of machines was concentrated in the hands of the latter. Small farmers in the study villages rarely owned tractors or power tillers or threshers.

Fourthly, seed costs varied a lot across villages both in absolute terms and as a share of total paid-out costs. Expenditure on seeds was higher among small farmers in paddy-growing villages, relative to large farmers in the same villages. Differences across farmer category were not visible in the wheat-growing villages.

Fifthly, in villages with a high degree of tenancy, the rent on leased-in land was extremely high for small tenant farmers.

Sixthly, we undertook a quick and limited analysis on the price front, and the data showed that MSP acted as a floor in all villages for wheat but not for paddy. Not surprisingly, small farmers in Tehang village of Punjab received prices higher than the MSP for both wheat and paddy. For crops and regions without an effective MSP, market structures and marketing channels played a central role. The example of a potato-growing village in West Bengal brought out the risks faced by small farmers in selling to local traders.

Finally, we found that the limited reach of agricultural extension systems among small farmers (Appendix 2) is a matter of serious concern both for technological development, and control over the quality and cost of production.

In short, small farmers face multiple constraints in access to inputs and in markets for sale of output. Public support is required to address these constraints and improve the returns from crop production.

REFERENCES

Foundation for Agrarian Studies (FAS) (2015), "Calculation of Household Incomes: A Note on Methodology," available at http://fas.org.in/wp-content/themes/zakat/pdf/Survey-method-tool/Calculation%20of%20Household%20Incomes%20%20A%20Note%20on%20Methodology.pdf, viewed on 20 February 2017.

Government of India (GoI) (2009–2012), *Reports of the Commission for Agricultural Costs and Prices*, Commission for Agricultural Costs and Prices, Department of Agriculture and Cooperation, Ministry of Agriculture, Government of India, New Delhi.

Kumar, Deepak (2017), "Landholdings and Irrigation in the Study Villages," in Madhura Swaminathan, and Arindam Das (eds.), *Socio-Economic Surveys of Three Villages in Karnataka: A Study of Agrarian Relations*, Tulika Books, New Delhi.

Modak, Tapas S., and Bakshi, Aparajita (2016), "Tracing Changes in the Groundwater Market: Case Study of Amarsinghi Village, 2005 to 2015," paper presented at 76th Annual Conference of the Indian Society of Agricultural Economics, Assam Agricultural University, Jorhat.

Ramachandran, V. K., Rawal, Vikas, and Swaminathan, Madhura (eds.) (2010), *Socio-Economic Surveys of Three Villages of Andhra Pradesh: A Study of Agrarian Relations*, Tulika Books, New Delhi.

Rawal, Vikas (2010), "Cost of Cultivation and Farm Business Incomes in India," Working Paper Series No. 2012–15, Institute of Economic Research, Hitotsubashi University, March.

Rawal, Vikas (2013), "Economic Policies, Tenancy Relations and Household Incomes: Insights from Three Selected Villages in India," available at http://www.agrarianstudies.org/UserFiles/File/Rawal_Economic_Policies.pdf, viewed on 4 September 2017.

Rawal, Vikas, and Swaminathan, Madhura (2011), "Are There Benefits from the Cultivation of Bt Cotton? A Comment Based on Data from a Vidarbha Village," *Review of Agrarian Studies*, vol. 1, no. 1, available at http://ras.org.in/are_there_benefits_from_the_cultivation_of_bt_cotton_a_comment_based_on_data_from_a_vidarbha_village, viewed on 20 February 2017.

Sarkar, B., Ramachandran, V. K., and Swaminathan, Madhura (2014), "Aspects of Political Economy of Crop Incomes in India," *World Review of Political Economy*, 5 (3), pp. 392–413.

Sarkar, Biplab (2015), "Irrigation Development and Its Impact on Farm Business Incomes: Evidence from a Village Surveyed in West Bengal, 2005 to 2015," paper presented at the 75th Annual Conference of the Indian Society of Agricultural Economics, Punjab Agricultural University, Ludhiana, 19 to 21 November.

Sarkar, Biplab (2017), "The Economics of Household Farming: A Study with Special Reference to West Bengal," unpublished PhD thesis, University of North Bengal, West Bengal.

Sarkar, Biplab (2017), "Cropping Pattern, Yields, and Crop Income: Findings from Three Villages Surveyed in Karnataka," in Madhura Swaminathan, and Arindam Das (eds.), *Socio-Economic Surveys of Three Villages in Karnataka: A Study of Agrarian Relations*, Tulika Books, New Delhi.

Sen, A., and Bhatia, M. S. (2004), *Cost of Cultivation and Farm Income*, Vol. 14 of *State of the Indian Farmer: A Millennium Study*, Academic Foundation in association with Department of Agriculture and Cooperation, Ministry of Agriculture, Government of India, New Delhi.

Surjit, V. (2008), "Farm Business Income in India: A Study of Two Rice Growing Villages of Thanjavur Region, Tamil Nadu," unpublished PhD thesis, University of Calcutta, Kolkata.

Swaminathan, Madhura, and Das, Arindam (eds.) (2017), *Socio-Economic Surveys of Three Villages in Karnataka: A Study of Agrarian Relations*, Tulika Books, New Delhi.

Swaminathan, Madhura, and Rawal, Vikas (eds.) (2015), *Socio-Economic Surveys of Two Villages in Rajasthan: A Study of Agrarian Relations*, Tulika Books, New Delhi.

APPENDIX 1

Appendix Table 1.1 *Item-wise cost of cultivation on operational holdings of small farmer households, study villages, at 2010–11 prices* in Rs per ha and per cent

Cost items	Ananthavaram (Andhra Pradesh)		Bukkacherla (Andhra Pradesh)		Kothapalle (Telangana)		Harevli (Uttar Pradesh)		Mahatwar (Uttar Pradesh)	
	Cost (Rs/ha)	Share (%)	Cost (Rs/ha)	Share (%)	Cost (Rs/ha)	Share (%)	Cost (Rs/ha)	Share (%)	Cost (Rs/ha)	Share (%)
Seed	6,083	6	3,774	19	1,947	6	2,458	5	3,855	14
Manure	3,834	4	2,211	11	4,498	14	2,685	6	368	1
Fertilizer	16,711	18	1,275	7	6,547	20	2,836	6	4,027	15
Plant protection	5,937	6	619	3	852	3	183	0	205	1
Irrigation	4,878	5	30	0	409	1	2,856	6	6,181	23
Hired labour	15,545	17	5,053	26	7,560	23	4,828	11	1,528	6
Machine labour	5,783	6	1,512	8	2,203	7	4,421	10	4,675	17
Animal labour	775	1	1,122	6	1,553	5	5,653	12	35	0
Rent on leased-in land	30,368	32	1,554	8	3,632	11	15,833	35	3,854	14
Other costs	3,790	4	2,257	12	3,461	11	3,580	8	2,239	8
Cost A2 (paid-out cost)	93,704	100	19,408	100	32,663	100	45,333	100	26,968	100

Source: PARI survey data.

Appendix Table 1.2 *Item-wise cost of cultivation on operational holdings of small farmer households, study villages, at 2010–11 prices in Rs per ha and per cent*

Cost items	Nimshirgaon (Maharashtra)		Warwat Khanderao (Maharashtra)		Gharsondi (Madhya Pradesh)		Alabujanahalli (Karnataka)		Siresandra (Karnataka)	
	Cost (Rs/ha)	Share (%)	Cost (Rs/ha)	Share (%)	Cost (Rs/ha)	Share (%)	Cost (Rs/ha)	Share (%)	Cost (Rs/ha)	Share (%)
Seed	3,427	8	2,688	15	5,506	23	2,700	4	3,486	10
Manure	4,164	10	909	5	449	2	4,781	7	5,239	15
Fertilizer	6,540	16	1,981	11	2,970	12	14,339	20	4,897	14
Plant protection	1,754	4	1,593	9	1,312	5	1,132	2	1,856	5
Irrigation	4,195	10	50	0	1,345	6	1,001	1	1,664	5
Hired labour	7,958	19	5,679	32	1,933	8	32,079	44	7,093	20
Machine labour	5,214	12	896	5	7,911	33	3,266	4	2,058	6
Animal labour	840	2	2,268	13	74	0	4,027	6	3,008	8
Rent on leased-in land	2,529	6	261	1	1,287	5	3,078	4	69	0
Other costs	5,372	13	1,294	7	1,552	6	6,294	9	6,141	17
Cost A2 (paid-out cost)	41,992	100	17,618	100	24,338	100	72,697	100	35,510	100

Source: PARI survey data.

Apendix Table 1.3 *Item-wise cost of cultivation on operational holdings of small farmer households, study villages, at 2010–11 prices in Rs per ha and per cent*

Cost items	Zhapur (Karnataka)		Rewasi (Rajasthan)		Amarsinghi (West Bengal)		Kalmandasguri (West Bengal)		Panahar (West Bengal)		Tehang (Punjab)	
	Cost (Rs/ha)	Share (%)	Cost (Rs/ha)	Share (%)	Cost (Rs/ha)	Share (%)	Cost (Rs/ha)	Share (%)	Cost (Rs/ha)	Share (%)	Cost (Rs/ha)	Share (%)
Seed	690	5	4,934	13	6,414	11	5,867	15	14,882	20	2273	5
Manure	688	5	1,717	4	2,156	4	6,753	17	2,147	3	1790	3
Fertilizer	1,107	8	1,481	4	4,990	8	3,169	8	9,265	12	4709	12
Plant protection	2,282	16	243	1	2,064	3	1,991	5	5,135	7	1453	4
Irrigation	0	0	5,723	15	8,265	14	1,274	3	6,333	9	7909	11
Hired labour	3,266	24	6,225	16	25,190	41	11,779	30	11,532	16	10150	19
Machine labour	912	7	9,402	24	1,750	3	3,660	9	6,839	9	8286	21
Animal labour	3,454	25	351	1	3,358	6	964	2	968	1	0	0
Rent on leased-in land	375	3	4,952	13	3,648	6	1,129	3	13,543	18	7003	19
Other cost items	1,102	8	4,416	11	3,167	5	2,829	7	3,678	5	1942	5
Cost A2 (paid-out cost)	13,877	100	39,443	100	61,003	100	39,416	100	74,322	100	45515	100

Source: PARI survey data.

APPENDIX 2
AGRICULTURAL EXTENSION AND SMALL FARMERS' ACCESS TO INFORMATION

Madhura Swaminathan,
with Subhajit Patra, A. Bheemeshwar Reddy, and Shamsher Singh

In a recent review of agricultural extension in India, Sajesh and Suresh (2016) concluded that

> Indian agriculture is confronting serious issues such as a huge yield gap, a multitude of smallholders, imbalances with respect to input use and declining natural-resource productivity. Extension systems in India, which have an important role to play in addressing these concerns, are constrained by financial, infrastructural and human resource limitations.

They further noted that

> The inclusiveness of extension services remains a major concern. . . . The growth of smallholder agriculture will be determined by the extent to which institutions of research and extension are attuned to their priorities.

What does the agricultural extension system look like from the perspective of small farmers? A bleak picture emerges from a quick analysis of unit-level data from the recent National Sample Survey (NSS) survey on Situation of Households in India 2013.[10] At the all-India level, 40 per cent of small farmers and 63 per cent of large farmers reported gaining some technical advice during July–Decemeber 2012. Further, among small farmers, more than 50 per cent reported access to technical advice in seven states, and more than 65 per cent of small cultivators obtained some technical advice in only three States – Andhra Pradesh, Karnataka, and Kerala. The situation was similar during the second visit (in the *rabi* season), with more than 50 per cent of small cultivators reporting some access to information in only eight out of 18 States.

The sources of information on agriculture were diverse. The two most important sources of information for small farmers were "progressive farmers," and "radio/TV/newspaper/Internet" or media sources. Public sources of information including extension agents, Krishi Vigyan Kendras (KVKs), and agricultural universities played a minor role in providing technical information on agriculture. A similar situation prevailed during the second visit (in the *rabi*

---

[10] The unit-level data were analysed by Bheemeshwar Reddy using our farmer categorisation. Statistical tables based on NSS data were prepared by Bheemeshwar Reddy, and those based on PARI data were prepared by Subhajit Patra.

Appendix Table 2.1 *Number and proportion of cultivating households that accessed technical advice from any source, study villages, 2005–12*

Village	State	Large farmers		Small farmers	
		No.	%	No.	%
Ananthavaram	Andhra Pradesh	33	73	151	67
Bukkacherla	Andhra Pradesh	40	70	68	64
Kothapalle	Telangana	9	100	55	56
Harevli	Uttar Pradesh	21	58	28	57
Mahatwar	Uttar Pradesh	1	11	27	33
Nimshirgaon	Maharashtra	39	61	185	78
Warwat Khanderao	Maharashtra	12	75	81	76
Rewasi	Rajasthan	19	25	30	18
Gharsondi	Madhya Pradesh	32	45	39	55
Siresandra	Karnataka	0	0	0	0
Zhapur	Karnataka	0	0	0	0
Alabujanahalli	Karnataka	0	0	9	8
Amarsinghi	West Bengal	0	0	24	44
Kalmandasguri	West Bengal	0	0	32	47
Panahar	West Bengal	6	75	50	35
Tehang	Punjab	35	46.7	14	28.6

Source: PARI survey data.

season). As each farmer can report more than one source of information, it is not possible to aggregate across sources of information.

For small farmers, Kerala and Andhra Pradesh stand out in terms of better access to information from public agencies. In Kerala, 21 per cent of small farmers reported getting advice from KVKs and 14 per cent from extension agents. In Andhra Pradesh, 28 per cent of small farmers gained information from extension agents, though private commercial agents provided information to 36 per cent of small farmers. In short, the reach of agricultural extension, particularly public extension, appears very limited in most parts of India among small farmers.

What do the PARI village surveys show? First, there is huge variation across villages (Table 1). While over 75 per cent of small farmers in the two study villages of Maharashtra reported access to some technical advice, the figure was zero for the two villages of Karnataka.

For the seven villages in which more than 50 per cent of small farmers reported access to technical advice, we further examined the sources of information (Appendix Table 2.2). Small farmers reported getting information from progressive farmers, radio, TV, and other sources of media, private commercial agents, and agricultural extension and institutional sources. The

Appendix Table 2.2 *Number and proportion of small farmer households that accessed technical advice from each source, study villages, 2005–12*

Village	State	Public institutions		Private commercial agents		Radio/TV/ newspapers/ books/internet		Other progressive farmers	
		No.	%	No.	%	No.	%	No.	%
Ananthavaram	Andhra Pradesh	32	15	59	26	43	19	123	54
Bukkacherla	Andhra Pradesh	42	45	21	20	42	39	21	20
Kothapalle	Telangana	53	54	21	21	26	27	38	39
Harevli	Uttar Pradesh	9	18	8	16	16	33	20	41
Mahatwar	Uttar Pradesh	7	8	6	7	18	22	11	14
Gharsondi	Madhya Pradesh	19	27	12	17	23	32	28	39
Nimshirgaon	Maharashtra	94	39	125	53	67	28	47	20
Warwat Khanderao	Maharashtra	41	39	48	45	41	39	24	23
Amarsinghi	West Bengal	7	13	6	11	19	35	17	31
Kalmandasguri	West Bengal	6	8	20	29	2	3	10	15
Panahar	West Bengal	10	7	31	22	6	4	23	16
Tehang	Punjab	28	38	2	3	17	23	25	34

Source: PARI survey data.

importance of these sources varied across villages (and as each household could report more than one source, the proportions do not add up to 100). Nevertheless, it can be seen that there was clearly no single source that was the most important source of technical information for small farmers. It can also been seen that public institutions (be they universities or extension agents) were absent as a source of information in many villages, and, in any case, not more than one-half of small farmers had access to public institutions in any village.

Modern means of communication (such as the internet and mobile telephones) make it easier to provide access to information. Our village data provide an insight here. Appendix Table 2.3, prepared by Shamsher Singh, shows the ownership of TV, radio, mobile phones, and computers among small farmers in the study villages. Across the villages, less than 30 per cent of households owned radios or transistors. Ownership of televisions was more widespread, in the range of 40 to 70 per cent (with a few exceptions). Ownership of mobile phones was in the range of 40 to 60 per cent in the villages surveyed in more recent years (that is, Karnataka in 2009 and West Bengal in 2010) as compared to the villages surveyed earlier. There has clearly been a huge expansion in access to mobile phones in recent years. Except for one household in Rewasi, no small farmers in any of the villages surveyed owned a computer or laptop.

Data from these 16 villages substantiate the point made at the beginning, namely, the lack of inclusiveness of the extension system in India. If we want to address the constraints faced by small farmers, the reach of agricultural extension will have to be extended.

Appendix Table 2.3 *Number and proportion of small farmer households having television, mobile phone, transistor/radio and computer/laptop, study villages*

Village	State	Television		Mobile		Transistor/radio		Computer/laptop	
		No.	%	No.	%	No.	%	No.	%
Ananthavaram	Andhra Pradesh	143	63	4	2	12	5	0	0
Bukkacherla	Andhra Pradesh	65	61	0	0	6	6	0	0
Kothapalle	Telangana	55	56	2	2	12	12	0	0
Warwat Khanderao	Maharashtra	52	49	14	13	23	22	0	0
Nimshirgaon	Maharashtra	157	66	31	13	63	26	0	0
Harevli	Uttar Pradesh	18	40	2	4	13	29	0	0
Mahatwar	Uttar Pradesh	15	22	7	10	17	25	0	0
25F Gulabewala	Rajasthan	2	100	1	50	1	50	0	0
Rewasi	Rajasthan	31	29	78	73	15	14	1	1
Gharsondi	Madhya Pradesh	51	72	24	34	10	14	0	0
Alabujanahalli	Karnataka	74	65	69	61	24	21	0	0
Siresandra	Karnataka	38	68	29	52	10	18	0	0
Zhapur	Karnataka	10	38	12	46	3	12	0	0
Amarsinghi	West Bengal	20	42	28	58	0	0	0	0
Kalmandasguri	West Bengal	3	10	23	79	3	10	0	0
Panahar	West Bengal	58	48	61	50	0	0	2	2
Tehang	Punjab	51	52	43	43	0	0	5	5

Source: PARI survey data.

7

Fertilizer Use and Small-Scale Farms

Kamal Kumar Murari and T. Jayaraman, with Sanjukta Chakraborty

INTRODUCTION

Fertilizer is one of the key inputs for increasing agricultural productivity. It provides macronutrients – that is, nutrients consumed in large quantities – such as nitrogen, phosphorous, and potassium to plants. Lack of any one of these macronutrients affects the growth and productivity of plants. Of these nitrogen is a key macronutrient, with a significant proportion of it being absorbed from the soil during crop production. Field experiments suggest that about 28 kilograms of nitrogen are removed from the soil for 1 tonne of wheat production, while 10–31 kilograms of nitrogen are removed per tonne of rice (Prasad 1999). Crop cultivation without application of nitrogen may therefore lead to a decline in soil fertility in the long term, in turn affecting the sustainability of agricultural productivity.

India saw an increase in the rate of growth of food production during the period of the "green revolution." This was mainly accomplished by making available improved varieties of seed; expanding irrigation coverage (particularly through the exploitation of groundwater resources); and improving the application of fertilizers, particularly nitrogenous fertilizers. The expansion of fertilizer use has not been uniform; it is skewed in favour of greater use of nitrogen as compared to other macronutrients. Given that most Indian soils are nitrogen-deficient, use of nitrogen in areas of intensive agriculture has been very high, particularly in north-western India and the irrigated regions of southern India (Shindo *et al.* 2006). Fertilizer use intensity, measured in terms of the quantity of fertilizer applied per unit of gross cropped area, has crossed 300 kilograms per hectare in some districts of India (Government of India 2014). This is higher than many highly agriculturally productive regions of the world (Shindo *et al.* 2006). Excessive use of nitrogen on agricultural land is a serious concern from the viewpoint of environmental sustainability. Moreover, it is not clear whether excessive application of fertilizers (above the recommended dose) results in increased crop yields. Using the FAO's country-level data, Lassaletta *et al.* (2014) have argued that excess use of fertilizer has not resulted in increased productivity, particularly in India. Optimum application of fertilizer is necessary to maintain productivity, but the optimum rate of fertilizer application and the response in terms of crop yield vary across different agro-ecological zones.

Yield response to fertilizer application can be improved by employing better management practices and through better application of technology. However, farmers may be reluctant to employ these methods because of lack of adequate information (Mujeri *et al.* 2012). Most studies on fertilizer use focus on its effect on crop yield and the environment, and not so much on its costs and benefits for individual farmers. Improving the effectiveness of fertilizer application depends not only on physiological and technical factors (such as plot area, soil conditions, etc.), but also on socio-economic factors (such as assets and income of farmers, and access to credit). At present, the relative importance of these factors at the micro as well as macro level in India is poorly understood. Understanding this aspect is important because economic considerations influence a farmer's options, whether it be the choice of crops or fertilizer type or application rate (Ncube *et al.* 2006). Therefore, the ratio of fertilizer price to the value of the crop (in price terms) is an important indicator that informs us of the value of output from crop production that is realised per unit application of fertilizer.

Improvements in the efficiency of fertilizer application, measured by the ratio of fertilizer to gross value output (GVO) of the crop, reflect the farmer's decision regarding choice of crop, crop variety, irrigation practices, and fertilizer management practices.[1] However, the relationship is not straightforward and there are some studies that throw light on this issue, particularly in actual field conditions. To fill this gap in the literature, it is important to understand the nature of the relationship between fertilizer-use efficiency and economic factors by comparing indicators related to fertilizer use across socio-economic strata of farmers. Specifically, comparison of fertilizer use by small and large farmers in India is important, and can provide insight into these issues, because large farmers have a number of technological and economic options at their disposal whereas small farmers are generally choice-constrained.

Within this context, this chapter seeks to understand the pattern of nitrogen use and fertilizer input costs for small and large farmers, using PARI village-level survey data.[2] First, we review the current state of knowledge related to fertilizer use and its implications for productivity and economics. We then

---

[1] This chapter uses the partial factor productivity measure of fertilizer efficiency. Details of different measures of fertilizer efficiency are described in Table 10.
[2] Details of the PARI village surveys are given in chapter 2 of this volume.

offer an overview of the different types of fertilizer and their nutrient content. We examine fertilizer use and its nutrient content in terms of the application of NPK (where N stands for nitrogen, P for phosphorous, and K for potassium) per unit of land area, against the recommended NPK ratio for different crops in specific agro-ecological contexts. We analyse the fertilizer use patterns of small farmers using macro data, followed by an analogous discussion using data from village studies. The chapter ends with a summary of the key findings on fertilizer use by small farmers.

CURRENT FERTILIZER USE IN INDIAN AGRICULTURE

Soils in India generally have low to medium levels of nitrogen, phosphorous, and potassium, and are deficient in zinc (ICAR 2015). Application of inorganic fertilizers and manure is therefore necessary to improve agricultural productivity.

Inorganic fertilizers are made up of chemical and synthetic organic materials that are concentrated in character, and contain one or more water-soluble essential plant nutrients. As they are concentrated, inorganic fertilizers are less bulky and easier to handle than organic fertilizers during transportation and storage. Some of the commonly used chemical fertilizers in India are urea, diammonium phosphate (DAP), single super phosphate (SSP), and muriate of potash (MOP). In addition, various forms of other complex fertilizers that contain two or more essential nutrients are also used on Indian soils.[3]

Organic manures are derived from organic sources like plants, animals, and human residues. These are most commonly available in the form of farm yard manure (FYM) and compost, or in relatively concentrated forms such as oil-cake or neem-cake. FYM and compost are the most common organic manures used for agricultural purposes in India, but these are bulky (and therefore difficult to transport), and contain a low percentage of nitrogen, phosphate, and potash (NPK). In Indian conditions, FYM typically contains 0.3–1 per cent of N, 0.15–1 per cent of P, and 0.3–1 per cent of K (ICAR 2015). Green manure and bio-fertilizers are other sources of nutrients that are used on farms, but not as widely.

In order to sustain the growing food demand of the country, plant nutrients from both fertilizers and organic manure are required for agriculture. However, the choice of chemical fertilizer over the organic form is mainly driven by

---

[3] A complex fertilizer is referred to by its grade, in terms of the percentages of N, P, and K present in it. For example, a fertilizer with NPK grade of 10–26–26 indicates that it has 10 per cent of nitrogen, 26 per cent of phosphorous, and 26 per cent of potassium by weight.

two important factors: productivity and ease of application; and economic considerations. Chemical fertilizer is cheap relative to the value of increase in crop production that it stimulates. Application is also easier compared to the organic form. Table 1 shows the percentage share of various forms of chemical fertilizer used in India, and the nutrient content (in percentage of total weight) of different organic and inorganic forms of fertilizer. It is clear from the table that urea, a nitrogenous fertilizer, is the most consumed fertilizer in India. In addition to urea, DAP, MOP, and complex fertilizers are also used, although their percentage share is low compared to urea. Complex forms of fertilizers – that is, fertilizers containing two or more nutrients – are also used, but their share in total fertilizer consumption is very low.

NPK Recommendations for Varying Crop Types and Agro-Ecological Conditions

At the national level, it has been generally recognised that an NPK consumption ratio of 4:2:1 should be attained. The NPK needs of a farm are determined by the crop variety, the soil's capacity to supply nutrients, and the nutrient efficiency of a location. However, as we have already noted, a general recommendation to maintain an NPK ratio of 4:2:1 is largely in tune with the requirement of the two most important food crops in India, i.e. wheat and paddy. In addition, this ratio is a key benchmark adopted in promoting the use of nutrients (NAAS 2009). In fact, this ratio corresponds to the standard recommended dose rate of NPK of 120:60:30 in kilograms per hectare for paddy and wheat. Ideally, the recommendation for the NPK ratio should change according to the agro-climatic region, and the choice of crops and cropping pattern. Hence, varying NPK recommendations have been made (*ibid.*) that take into account the agro-climatic location and cropping patterns.

Table 2 shows crop-wise fertilizer use in India. Sugarcane has the highest fertilizer use per hectare of land in comparison to other crops; it has an almost equivalent proportion of use of nitrogenous, phosphorous, and potassium fertilizers. Here, and later in this chapter, we focus on NPK dose rates rather than the ratio, as the latter is confusing and may lead to misinterpretation, especially when the use of potassium is extremely low or negligible. For instance, the NPK ratio of total pulse production (a nitrogen-fixing crop) is about 9.25:4.1:1, which suggests a disproportionate use of nitrogen (relative to P and K), but this does not clarify whether the actual nutrient doses are within the permissible range. Therefore we will relate the observed NPK dose rates to the recommended dose of 120:60:30, and not the ratio of 4:2:1. All-India fertilizer use suggests that nitrogen application for most crops is within the recommended range. However, the application of P and K is much lower

Table 1 *Nutrient content of various organic and inorganic fertilizers*

Name of fertilizer	Nitrogen (N) content	Phosphorous (P) content	Potassium (K) content	Consumption in '000 tonnes (per cent of total)	Remarks
Urea	46	0	0	26651.3 (53.3)	Both domestically produced and imported. This fertilizer is subsidized and the price is controlled by the central government.
Calcium Ammonium Nitrate (CAN)	25–26	0	0	139.8 (0.3)	Domestically produced.
Ammonium Sulphate	20.6	0	0	280.4 (0.6)	Domestically produced.
Single Super Phosphate (SSP)	0	14.5–17	0	2400 (4.8)	The cost of the fertilizer is fixed every month for the entire country. Domestically produced.
Triple Super Phosphate	0	43	0	260.3 (0.5)	Totally imported. No domestic production.
Muriate of Potash (MOP)	0	0	50	4094.3 (8.2)	Totally imported. No domestic production.
Diammonium phosphate (DAP)	18	46	0	9422.6 (18.8)	Domestically produced and imported. This fertilizer is subsidised and the price is controlled by the central government.
Ammonium phosphate Sulphate	16–18	20	0	2665.9 (5.3)	Domestically produced.
Urea ammonium Phosphate	24–28	24–28	0	206.1 (0.4)	Domestically produced.
Other complex NPK fertilizers	Oct-19	15–35	14–26	3917.8 (7.8)	Domestically produced.
Farm yard manure (FYM)	0.5–1.5	0.4–0.8	0.5–1.9		
Compost and green manures	0.3–2	0.1–1	0.6–1.5		
Oil-cakes	2.5–7.2	0.9–2.9	1.0–2.2		
Manure of animal origin	0.6–10	0.15–30	0.2–1.0		

Note: The table also reports the percentage share (in percentage of total fertilizers consumed in India) of different chemical fertilizers.
Source: Partly derived from *Handbook of Agriculture* (ICAR 2015).

Table 2 *Crop-wise fertilizer use in India*

Crop	Fertilizer application in (kg/ha)					Farm yard manure (quintal/ha)
	DAP	Urea	Total	Total	Total	
			Nitrogen (N)	Phosphorous (P)	Potassium (K)	
Paddy	68.24	161.17	90.69	40	18.66	941.03
Sorghum	31.96	86.38	53.24	26.41	7.91	581.46
Maize	54.63	120.13	68.69	31.35	12.19	844.27
Wheat	96.6	218.08	119.21	49.68	9.06	486.65
Total Pulses	42.28	94.76	52.86	23.27	5.72	415.14
Sugarcane	137.08	298.09	174.71	94.21	50.66	1703.04
All Vegetables	97.33	153.64	95.56	64.52	37.62	1339.25
Total Oilseeds	48.94	91.95	53.55	31.09	7.97	517.65
Cotton	59.77	181.57	100.09	38.09	15.25	485.11

Source: "Agricultural Input Survey," Department of Agriculture and Cooperation, Government of India, 2011–12, available at http://inputsurvey.dacnet.nic.in

than the recommended dose for most crops, and particularly the application of P, which is very low for almost all crops except sugarcane and vegetables.

Environmental Implications of Fertilizer Use in Agriculture

There is large variation in fertilizer use across States in India. Among the major States, Punjab stands first in fertilizer consumption with 213 kilograms per hectare of gross cropped area, followed by Andhra Pradesh with 208 kilograms per hectare, Tamil Nadu with 193 kilograms per hectare, and Haryana with 187 kilograms per hectare (Fertilizer Statistics of India 2013). The levels of fertilizer use in Kerala, Madhya Pradesh, Chhattisgarh, Maharashtra, Rajasthan, Goa, Uttarakhand, Himachal Pradesh, Jammu and Kashmir, Jharkhand, Odisha, and the North Eastern States are below the national average of 119 kilograms per hectare (Government of India 2014). Despite awareness of the importance of correct application of fertilizer for boosting foodgrain production, its disproportionate use (in terms of excess application of fertilizer in fields, and application of nitrogen higher than the recommended dosage in comparison to potassium and phosphorous) is very common in regions of high agricultural productivity.

A number of studies have pointed out the serious environmental implications of excessive application of nitrogen (Vel Murugan and Dadhwal 2008; Pathak *et al.* 2010; Ladha *et al.* 2005). When a fertilizer is applied to a crop through the soil, it is subjected to various processes that result in changes in the nitrogen balance of the soil system. Some of the nitrogen from the applied fertilizer is absorbed by the crop, some of it may be fixed in or retained by the soil, some of it may be lost due to volatilisation, while the remainder is leached into the soil or washed away due to run-off. Excess nitrogen that is washed away by run-off has a tendency to pollute local water bodies, causing nitrate pollution in surface water and groundwater (Chandana *et al.* 2011; Vikas *et al.* 2015), and eutrophication of lakes and ponds (Divya and Belagali 2012). India has a serious problem of excess nitrogen in groundwater, with approximately 400 districts in the country having nitrate levels greater than the permissible limit of 45 milligrams per litre and so high levels of reactive nitrogen (CGWB 2009).

FERTILIZER USAGE OF SMALL FARMERS FROM MACRO DATA ANALYSIS

Small-scale farming generally has low productivity due to small farm size, lack of improved technologies and financial options, and also difficulties in coping with labour unavailability (Morton 2007). Three factors need to be taken into consideration when it comes to fertilizer application. These are: (1) economic factors (mainly the price of fertilizer); (2) physical and technological factors (related to the ability of the fertilizer to be released into the soil); and (3) institutional factors (Lal 2015). These factors are mainly related to the supply chain of fertilizer and the availability of credit facilities to purchase fertilizer. High costs of fertilizers, lack of credit, delays in the delivery of fertilizers, and poor transport and marketing infrastructure serve as disincentives for fertilizer use by smallholder farmers (Buresh and Giller 1998).

Small-scale farmers in developing countries mainly rely on experience, intuition, and comparisons with neighbours to arrive at farming decisions (Otoo *et al.* 2015). In the African context, the practices of small-scale farmers are generally characterised by low inputs of fertilizer, much less than world averages (Yengoh 2012). This may be due to the fact that small farmers tend to have cash after harvest but not at the time of planting. Unavailability of cash before planting forces small farmers to either apply less fertilizer than required or explore availability of credit facilities. Purchasing fertilizer on credit might increase the input price and impact overall profitability. In Asian countries, fertilizer application dosages are high even for small-scale farmers (Lal and Stewart 2014). For instance, in 2013, the consumption of chemical fertilizer per hectare of arable land was about 150 kilograms per hectare for Asian countries, which was higher than for all other geographical regions of the world (Takeshima *et al.* 2016).

In India, the latest agricultural input survey ("Agricultural Input Survey," 2016) suggests that fertilizer use per hectare of land is almost similar for

Table 3 *Fertilizer and nutrient application rates for small and large farmers in India*

Year	Group	Fertilizer application (kg/ha)		Nutrient application (kg/ha)			NPK ratio	Farm yard manure (quintal/ha)
		DAP	Urea	Total Nitrogen (N)	Total Phosphorous (P)	Total Potassium (K)		
1996–97	Small scale	55.59	138.73	83.03	38.04	11.97	6.94:3.17:1	7445.01
	Medium and large scale	63.98	161.06	92.31	39.77	8.41	10.97:4.73:1	5145.03
2001–02	Small-scale	54.60	125.94	73.65	37.20	18.05	4.08:2.06:1	4613.95
	Medium and large scale	65.73	157.89	88.22	37.39	11.80	7.47:3.16:1	5576.73
2006–07	Small scale	67.50	163.06	92.31	43.00	17.77	5.2:2.42:1	4140.74
	Medium and large scale	74.52	190.00	105.01	41.54	12.81	6.48:2.58:1	3964.99
2011–12	Small scale	80.76	181.71	105.82	49.64	19.70	5.32:2.4:1	3391.80
	Medium and large scale	86.54	180.96	103.91	47.49	14.37	7.23:3.3:1	3608.63

Source: "Agricultural Input Survey," Department of Agriculture and Cooperation, Government of India, various years, available at http://inputsurvey.dacnet.nic.in

small-scale farmers (farmers with landholdings less than 2 hectares), and medium and large farmers (Table 3).[4] This is true for the application rate of DAP and urea as well as for other NPK nutrients. The input surveys of 1996–97 and 2001–02 show higher use of fertilizer and NPK by medium and large farmers in comparison to small farmers, but data from the most recent survey (2011–12) show equal or higher application of fertilizers by small farmers. This suggests that despite limited access to key resources such as irrigation and credit facilities, small farmers strive to apply fertilizers in their fields at the same rate as farmers from other categories. One reason for this could be the government policy of providing subsidised fertilizer for both large and small farmers without any differential pricing based on farm size. Another explanation could be high price elasticity for a small farmer because of the lack of money at the time of sowing (Singh *et al.* 2002). The general belief that higher use of fertilizers leads to increased agricultural productivity and thereby income is also prevalent among small farmers, as they do not get appropriate extension advice in relation to the use of fertilizers (Howell 1984; Jat *et al.* 2015).

Other studies around the world have pointed out that recommendations regarding nutrient management, and in particular fertilizer use technologies, have rarely been implemented by small farmers (Dimes *et al.* 2002). This is true even for India according to Jat *et al.* (2015), who report that scientific recommendations are not aligned with the local situation of farmers, and therefore seldom meet the requirements of small and marginal farmers. As a result, small landholders may reduce the effectiveness of fertilizers through poor practices. Due to lack of knowledge or in an attempt to stretch their limited incomes, smallholders might apply an inappropriate type of fertilizer, insufficient quantity of fertilizers, or apply the fertilizer at the wrong time (IFC 2013). Data from the input surveys indicate that at an all-India scale, the rate of N, P, and K is less than the general recommended limit of 120, 60 and 30 kilograms per hectare, respectively. This is also true for farmers of other categories. The most recent input survey suggests that while the N application rate is closer to the recommended limit, the application rates of P and K are much lower than the recommended rates. This is true across farm-size

---

[4] The latest input survey, in the year 2011–12, was conducted in three phases. The first two phases provide vital information on the number and area of operational holdings, and their basic characteristics such as land use, cropping pattern, and irrigation status by size of holding and social group, and on the pattern of use of various agricultural inputs across the country by size of household, age, and education level of the holders ("Agricultural Input Survey," 2016). So far, there have been four input surveys, conducted in the years 1996–97, 2001–02, 2006–07, and 2011–12. Each survey provides information related to input usage of different agrarian classes.

classes. Input survey data also suggest that the application rate of K for other category farmers is much lower than that for small farmers. However, it will be interesting to study whether this trend is reflected at the household level, especially as it has been observed that the conclusions from the interpretation of input surveys vary. In the sections below, we use village-level data to explore the usage pattern of fertilizer application of small-scale farmers, together with comparisons with other categories of farmers.

FERTILIZER USAGE OF SMALL FARMERS IN VILLAGE STUDIES

This section discusses the findings on villages studied under PARI regarding household-level patterns of fertilizer application, specifically among households belonging to small farmers. It examines whether the reported costs of fertilizer inputs in the PARI villages is on a par with the costs of fertilizer inputs reported as part of the cost of cultivation studies by the Commission of Agriculture Costs and Pricing (CACP). This section also examines the relationship between partial factor productivity of fertilizer and the ratio of fertilizer cost to gross value of output (GVO) for different crops.

Description of Number of Small and Large Farmer
Households in Village Studies

The focus of analysis in this chapter is the villages surveyed under PARI. The details of these villages are given in chapter 2 of this volume. Here, we have used data from 16 study villages. Table 4 provides an overview of some characteristics of these villages. According to the PARI data, the number of households surveyed that belonged to the landlord/capitalist farmer class was small in all the villages, though relatively higher in the villages of Ananthavaram (Guntur district, Andhra Pradesh), Gharsondi (Gwalior district, Madhya Pradesh), and 25F Gulabewala (Sri Ganganagar district, Rajasthan). We therefore focus here on the categories of small and large farmers. 25F Gulabewala is an exception in that it had only two households in the small farmer category. On the other hand, Kothapalle (Karimnagar district, Telangana), Siresandra (Kolar district, Karnataka), Mahatwar (Ballia district, Uttar Pradesh), Amarsinghi (Malda district, West Bengal), Kalmandasguri (Koch Bihar district, West Bengal), and Panahar (Bankura district, West Bengal) had a very small number of households in the large farmer category. We have nevertheless included these villages in our analysis, while exercising caution in comparing fertilizer usage by small and large farmers in these villages. In the majority of villages (10 out of 16), a major proportion of small farmer households had irrigation facilities (chapter 2). Although there might be differences in the fertilizer application

Table 4 *Number of households by category, study villages*

Village	State	Total households surveyed	Small farmers	Large farmers	Landlords	Others
Ananthavaram	Andhra Pradesh	150	48	16	11	11
Bukkacherla	Andhra Pradesh	99	36	16	4	0
Kothapalle	Telangana	101	28	1	3	1
Harevli	Uttar Pradesh	109	45	21	3	11
Mahatwar	Uttar Pradesh	156	69	1	4	13
Warwat Khanderao	Maharashtra	250	106	13	3	0
Nimshirgaon	Maharashtra	137	33	12	3	1
25F Gulabewala	Rajasthan	204	2	36	20	0
Rewasi	Rajasthan	219	107	38	8	11
Gharsondi	Madhya Pradesh	263	71	59	12	2
Alabujanahalli	Karnataka	243	113	25	0	30
Siresandra	Karnataka	79	56	4	0	1
Zhapur	Karnataka	109	26	10	4	1
Panahar	West Bengal	248	144	1	7	16
Amarsinghi	West Bengal	127	54	0	0	10
Kalmandasguri	West Bengal	147	68	0	0	6

Source: PARI survey data.

patterns of small farmer households with irrigation and without irrigation, we do not provide this comparison in this chapter as it might reduce the sample of small farmers in our analysis. Table 5 gives a summary of cropping patterns in these villages. Details of cropping patterns and irrigation practices are given in earlier chapters.

In the following section, we show fertilizer application rates (in kilograms per hectare), the cost of fertilizer in comparison to total input costs and gross value of output (in percentages), fertilizer use efficiency (FUE), and economic efficiency for small and large farmer households. We present the median values of the indicators for comparison. The median is used because it is the best statistical measure when the size of the sample is small and it is a good measure of the central tendency in the presence of outliers, which is the case with this field data. A statistical test – two-sampled Kruskal–Wallis test – is also used to examine whether the distribution of samples for the categories of small and large farmers are the same or not.[5]

---

[5] The Kruskal–Wallis test is used to test whether the medians of two samples are significantly the same, with a null hypothesis that both samples have the same distribution. The median of two samples is considered to be significantly different if the null hypothesis is rejected at 0.05 and 0.1 significance levels.

Table 5 *Summary of cropping patterns of study villages*

Village	State	Irrigation index	Kharif crop(s)	Rabi crop(s)	Other crop(s)
Ananthavaram	Andhra Pradesh	86.5	Paddy	Maize, pulses	
Bukkacherla	Andhra Pradesh	9.2	Intercropped groundnut	Paddy	
Kothapalle	Telangana	42.2	Paddy, maize, cotton	Paddy, maize	
Harevli	Uttar Pradesh	72.1	Paddy	Wheat	Sugarcane
Mahatwar	Uttar Pradesh	98.6	Paddy, wheat	Wheat	
Warwat Khanderao	Maharashtra		Cotton intercropped with red gram and green gram		
Nimshirgaon	Maharashtra	30.2	Soybean, groundnut		Sugarcane, grape, vegetables
25 F Gulabewala	Rajasthan	88.4	Cotton, cluster bean, fodder crops	Wheat, rapeseed	
Rewasi	Rajasthan	39.7	Pearl millet intercropped with cluster bean, green gram	Wheat, mustard, onion, fenugreek	
Gharsondi	Madhya Pradesh	87.9	Soybean intercropped with sesame, paddy	Wheat, rapeseed, chickpea, lucerne grass	
Alabujanahalli	Karnataka		Paddy, finger millet	Paddy, finger millet	Sugarcane
Siresandra	Karnataka	25	Finger millet intercropped with sorghum, red gram, sesame		Vegetable, fodder crops
Zhapur	Karnataka	1.2	Red gram intercropped with maize, sesame, pearl millet, green gram		
Panahar	West Bengal		Paddy	Potato, mustard, rapeseed, wheat, paddy or sesame	
Amarsinghi	West Bengal		Paddy, jute	Paddy	
Kalmandasguri	West Bengal		Paddy, jute	Vegetables, sugarcane, potato	
Tehang	Punjab		Paddy	Wheat	

Source: PARI survey data.

Inputs Usage of Small and Large Farmers in Village Studies

As discussed in the previous section, small farmers may either apply fertilizer inefficiently or may incur higher costs because they buy fertilizer on credit at the time of sowing. This might result in them paying a higher share for fertilizer as a proportion of total input costs. We explore this using village-level data. Table 6 gives the cost of fertilizer as a percentage of total input costs as well as gross value of output (GVO) for small farmers. Information regarding the costs of cultivation in the PARI surveys refers to the costs of all material inputs (purchased and home-produced), the cost of hired labour, rental payments, the imputed value of interest on working capital, and the depreciation of owned fixed capital other than land. No cost is imputed for family labour and no rent is imputed for owned land. Broadly, the definition of cost of input refers closely to Cost A2 of the CACP. Gross value of output (GVO) of a crop is the sum of values of the produce and its by-products.

The village survey data suggest that for small farmers, the median value of fertilizer price ranged from 1 per cent to 15 per cent with respect to both total input cost and GVO, while it was 2 to 20 per cent for large farmers. For

Table 6 *Cost of fertilizer in comparison to total input cost or gross value of product*

Village	State	Share of fertilizer cost in total input costs (%)		Ratio of fertilizer cost to GVO (%)	
		Small farmers	Large farmers	Small farmers	Large farmers
Ananthavaram	Andhra Pradesh	13.7	13.4	15	10.3
Bukkacherala	Andhra Pradesh	5.3	5.8	6	5.3
Kothapalle	Telangana	16.5	18.5	12.1	20.9
Harevli	Uttar Pradesh	5.4	6	4.5	4.8
Mahatwar	Uttar Pradesh	10.9	12.5	11.5	11
Warwat Khanderao	Maharashtra	10	9.6	4.8	6.5
Nimshirgaon	Maharashtra	10.9	11	7.5	6.7
25F Gulabewala	Rajasthan	14.1	11.1	10.7	6.3
Rewasi	Rajasthan	1.1	3.2	1	2.2
Gharsondi	Madhya Pradesh	10	8.6	6.5	6.4
Alabujanahalli	Karnataka	15	13.2	11.3	11.2
Siresandra	Karnataka	9.9	8.8	8.4	5.7
Zhapur	Karnataka	7	4	5.9	3.5
Panahar	West Bengal	9.6	8.8	9.5	12.5
Amarsinghi	West Bengal	6.1	NA	4.3	NA
Kalmandasguri	West Bengal	6.1	NA	3.4	NA

Source: PARI survey data.

most villages, the cost of fertilizer per unit of input cost was higher for small farmers than large farmers. However, the median values of these indicators are not statistically different using the Kruskal–Wallis test for all villages. The same pattern was reflected in the ratio of cost of fertilizer to GVO, except in the case of Ananthavaram. For Ananthavaram, the percentage of fertilizer cost to total input cost did not show a significant difference in median values, but there was a significant difference in median values for percentage of fertilizer cost with respect to GVO. This might be because the GVO of a small farmer would be lower than that of a large farmer.

Tables 7 and 8 show the median value of fertilizer input cost per unit area for small and large farmers, and compare it with the per hectare cost of fertilizer as estimated by the cost of cultivation/production of principal crops (CCPC) of the CACP (Commission of Agricultural Costs and Prices). We have used CACP prices for the year before the survey period, which refers to the recall period of the survey. For most crops, the reported cost of fertilizer is higher than the CACP cost in almost all the villages (Tables 7 and 8). This suggests that input cost for fertilizer per hectare of land is higher than the

Table 7 *Cost of fertilizer per hectare of cultivated area of paddy and wheat*

Village	State	Crop	Cost of fertilizer (Rs/ha)		
			Small farmers	Large farmers	CACP price (Cost A2)
Ananthavaram[#]	Andhra Pradesh	Paddy	4200	3375	2579
Bukkacherla	Andhra Pradesh	Paddy	2500	2225 (*)	2579
Kothapalle	Telangana	Paddy	3750	NA	2579
Harevli	Uttar Pradesh	Paddy	1207	1139	1655
Harevli	Uttar Pradesh	Wheat	2373	2162	2085
Mahatwar	Uttar Pradesh	Paddy	1091	1666 (*)	1655
Mahatwar	Uttar Pradesh	Wheat	2063	2171 (*)	2085
25F Gulabewala	Rajasthan	Wheat	1953*	2107	1547
Gharsondi	Madhya Pradesh	Paddy	2750	2703	685
Siresandra	Karnataka	Paddy	4060	5000 (*)	3224
Alabujanahalli[##]	Karnataka	Paddy	4011	3645	3224
Panahar	West Bengal	Paddy	2089	NA	2331

Notes: * The value reflects the larger value of two households.
(*) Median of very few households in large farmer category.
[#] Median is statistically significant at 0.05 level of significance.
[##] Median is statistically significant at 0.1 level of significance.
Source: PARI survey data.

Table 8 *Cost of fertilizer per hectare of cultivated area of other crops*

Village	State	Crop	Cost of fertilizer (Rs /ha)		
			Small Farmers	Large farmers	CACP Cost A2
Ananthavaram	Andhra Pradesh	Maize	5278	4475	2353
Kothapalle	Telangana	Maize	3750	NA	2353
Kothapalle	Telangana	Groundnut	500	NA	962
Kothapalle	Telangana	Cotton	4025	5400	2082
Harevli	Uttar Pradesh	Sugarcane	2931	2677	2391
Mahatwar	Uttar Pradesh	Maize	937	NA	545
Warwat Khanderao	Maharashtra	Cotton	1297	1731	1525
Nimshirgaon	Maharashtra	Sugarcane	9500	14071	7498
Nimshirgaon	Maharashtra	Soybean	625	428	1107
25F Gulabewala	Rajasthan	Cotton	NA	1583	1064
25F Gulabewala	Rajasthan	Rapeseed/mustard	423*	1365	1077
Gharsondi	Madhya Pradesh	Rapeseed/mustard	1000	1250	1149
Alabujanahalli	Karnataka	Finger Millet	4320	4015	1361
Siresandra	Karnataka	Finger Millet	2000	1083(*)	1361
Panahar	West Bengal	Rapeseed/Mustard	3231	5225(*)	1846

Note: * The value reflects largest value of two households.
(*) Median of very few households under large farmer category.
Source: PARI survey data.

government-prescribed fertilizer input cost, which is a significant part of setting the minimum support price (MSP).

It is clear from an analysis of the village data that households from both small and large farmer categories paid higher prices for fertilizer per unit of cropped area as compared to the CACP input cost for fertilizer. This was true for most crops. The median values of fertilizer cost per unit of land were not statistically significant for small and large farmers, except in Alabujanahalli (Mandya district, Karnataka) and Ananthavaram (Guntur district, Andhra Pradesh) (Table 7). The data suggest that in both these villages, small farmers paid higher absolute values than large farmers. Some studies have concluded that an increase in fertilizer prices prompts farmers to offset the effect of the price hike by using a greater proportion of their land to cultivate less fertilizer-intensive crops such as pulses and mustard (Mujeri *et al.* 2012). This argument is supported by Dev (2012), who suggests that an increase in the price of fertilizer does not have a significant effect on fertilizer consumption, but farmers shift their cropping pattern from cultivating fertilizer-intensive crops like wheat to less fertilizer-intensive crops. However, the reverse also seems true as the data at an all-India level reveal that the area under cultivation of pulses has not increased significantly (Mohanty and Satyasai 2015). In fact, it was observed that there has been a significant shift in the area under pulses to high-yielding varieties of cereals, particularly in places like Punjab, Haryana, and western Uttar Pradesh (Dixit *et al.* 2014). This behaviour can be explained by the farmers' anticipation of higher returns. The PARI data support the second argument, as we do not observe any significant change in cropping pattern between small and large farmers in both Alabujanahalli and Ananthavaram. However, the data for other crops, such as for maize in Ananthavaram and finger millet in Alabujanahalli, do not suggest that the cost of fertilizer for small farmers is different from that for large farmers.

Nitrogen Input and Productivity of Small Farmers

Low input of fertilizer may result in changes in agricultural productivity, whereas excessive use of fertilizer may result in environmental pollution. Although the response of crop yield to use of fertilizer has evolved differently in various countries, the data from some countries show a linear trend in yield response to fertilizer use while other countries no longer show any such linear trend (Lassaletta *et al.* 2014). In a specific example, Lassaletta *et al.* (2014) observed that China and India show an increasing pattern of fertilizer application and gradual reduction in crop yield response. Additional use of fertilizers in these countries might lead to environmental loading, particularly of nitrogen.

Table 9 shows the nutrient application rate aggregated at the village level. It also shows the median nitrogen application rate for small and large farmers. Equivalent NPK values are estimated from application of inorganic fertilizers and manure, using standard conversion factors.

At the village level, on an aggregated scale, it is clear that nitrogen was the dominant nutrient in fertilizer application. Kothapalle (Karimnagar district, Telangana), Zhapur (Kalaburagi district, Karnataka), Gharsondi (Gwalior district, Madhya Pradesh), Warwat Khanderao (Buldhana district, Maharashtra), 25F Gulabewala (Ganganagar district, Rajasthan), Harevli (Bijnor district, Uttar Pradesh), and Mahatwar (Ballia district, Uttar Pradesh) had higher than recommended NPK ratio of application of nitrogenous fertilizer. Most of these villages had a paddy–wheat or cotton–wheat cropping pattern. Among them, Zhapur, Gharsondi, and Harevli reported the application of a negligible amount of potassium. Analysis of the village data suggests that villages in south India (except Kothapalle and Zhapur) had a balanced ratio of NPK application, but the amounts themselves were much higher than the recommended limit. The three villages in West Bengal

Table 9 *Village-wise nutrient application rate* in kg/ha

Village	State	Nutrient application rate, aggregate at village level			Nitrogen rate for small farmers	Nitrogen rate for large farmers	Recommended N–P–K ratio for dominating crop of agro-climatic region (in kg/ha)#
		Nitrogen (N) rate	Phosphorous (P) rate	Potassium (K) rate			
Ananthavaram[1,#]	Andhra Pradesh	215	39	50	195	145	60–40–20
Bukkacherala[2]	Andhra Pradesh	55	27	15	29	39	(120–60–60)
Kothapalle[2]	Telangana	197	39	15	196	159	(120–60–60)
Harevli[6]	Uttar Pradesh	219	33	0	216	213	150–60–60 (150–60–60)
Mahatwar[7]	Uttar Pradesh	223	37	12	210	198	(120–60–40)
Warwat Khanderao[5]	Maharashtra	71	11	6	66	60	100–50–50
Nimshirgaon[4]	Maharashtra	103	33	42	90	89	(250–115–115)
25F Gulabewala[5]	Rajasthan	126	21	6	90	96	NA
Gharsondi[1]	Madhya Pradesh	47	17	1	62	51	(120–60–40)
Alabujanahalli[3]	Karnataka	177	57	48	187	166	120–60–40 (225–60–120)
Siresandra	Karnataka	102	49	34	74	102	NA
Zhapur	Karnataka	30	11	1	10	83	NA
Panahar[1]	West Bengal	107	37	49	91	100	80–40–40
Amarsinghi[1]	West Bengal	63	21	21	51	NA	80–40–40
Kalmandasguri[1]	West Bengal	29	11	11	26	NA	80–40–40

Note: Dominant crops: [1] paddy during *kharif* season; [2] paddy during *rabi* season; [3] paddy during both *kharif* and *rabi* seasons; [4] sugarcane cultivation; [5] cotton cultivation; [6] paddy during *kharif* and wheat during *rabi*; [7] wheat during *rabi* season.
The benchmark refers to the ratio suggested in NAAS (2009) for a given crop and agro-climatic zone.
Source: PARI survey data.

showed very low application of all nutrients. However, the data suggest that the complex form of fertilizer was popular in these villages. This was true at the State level, as on an aggregated value, West Bengal showed a higher proportion mix of potassium nutrient than other States (except Kerala, Himachal Pradesh, and the North East). Interestingly, all the study villages showed a level of phosphorous application that was much less than the recommended dose.

In most of the villages (eight out of 13), the median value for the application rate of nitrogen among small farmers was higher than large farmers. Zhapur was the only village where small farmers' application rate of nitrogen was very low. This might be due to the fact that most small farmers in this village did not have irrigation facilities. Villages where the number of small farmers with irrigation facilities was large showed a higher median value of nitrogen application as compared to large farmers. This suggests that small farmers with irrigation facilities used an equal or higher application rate of nitrogen than large farmers, although this varied with the type of crop used and management practice at the farm level. However, the median values for the small farmer category was not statistically different from median values for large farmers, except in Ananthavaram. As seen earlier, Ananthavaram showed statistically higher median values for fertilizer price in addition to a higher nitrogen rate for small farmers.

Fertilizer Use Efficiency among Small and Large Farmers

As discussed, sustainable fertilizer management must be both efficient and effective to deliver the anticipated economic, social, and environmental benefits. Fertilizer use efficiency (FUE) is a critically important concept in the evaluation of crop production systems. FUE is greatly affected by fertilizer management as well as by soil and plant water management. It generally reflects crop productivity that is obtained per unit application of fertilizer on the field. "Higher FUE" is a key term to address the environmental implications of using fertilizer in agriculture. At the farmer's end, the choice of crop to be grown and type of fertilizer to be used are likely to be mainly influenced by economic factors. Therefore, the ratio of fertilizer to crop price, which is a reflection of economic efficiency of fertilizer use, is an important indicator that tells us how the value of output from crop production is realised per unit value of crop input.

It may appear that a plot or a household that has a higher FUE obtains higher economic benefits from fertilizer use. This might not be the reasoning of the farmer, however, especially the small farmer, because of the need for credit to finance the purchase of fertilizer. Inadequate availability of credit at an affordable cost is frequently mentioned as a major constraint on fertilizer use (Dev 2012; IFC 2013; Singh *et al.* 2002). In addition, farmers might have to pay a higher price for fertilizers bought on credit. This might influence the overall profitability or income of the farmer. Therefore, the farmer whose production has higher FUE might not derive a greater economic benefit, as this depends on his economic condition.

Using PARI village survey data, we have tried to explore the relationship between FUE and GVO per unit of fertilizer price for small and large farmers. Note that the term efficiency has different connotations for economic and agronomic aspects. Ideally, efficiency is a measure of the change in crop output (both in physical quantity and in monetary value) for unit change in input. For instance, we can estimate such changes if we have data on crop production with and without application of fertilizer for regions with similar geographical and economic characteristics. Data on such details are often difficult to obtain. Therefore, efficiency in this chapter is defined as a measure of the ratio of crop output to inputs to the crop, which is a partial factor productivity measure.

FUE appears to be a simple term; however, the literature suggests that its definition is complex due to the number of potential nutrient sources (soil, fertilizer, manure, atmospheric fixation, etc.), and the multitude of factors that influence crop nutrient demand (crop management, crop variety, and weather conditions) (Hou *et al.* 2012). Additionally, the intended use of the term FUE and the data that most appropriately allow determination of these indicators further complicate the scenario. An excellent review of measurements and calculation of nutrient use efficiency is available in Fixen *et al.* (2014). Table 10 gives a summary of the definitions of various measures of nutrient use efficiency, their usefulness and limitations, following Fixen *et al.* (*ibid.*). From this table, it is clear that partial factor productivity is a simple efficiency indicator that can be estimated from village-level data. We use partial factor productivity to understand the FUE of small and large farmers in a village. As stated in Table 10, partial factor productivity is not a useful measure for inter-regional comparisons. We do not, therefore, intend to use this indicator for inter-village comparisons, but only to study FUE for small farmers (and large farmers) within a village.

Tables 11 and 12 show a comparison of FUE and the ratio of GVO to fertilizer value for small and large farmers. Table 11 deals exclusively with wheat and paddy cultivation, while Table 12 gives the values for other crops across the study villages. FUE for most villages under paddy and wheat cultivation was very low, the exception being Harevli (Bijnor district, Uttar Pradesh). For cereal crops, the world average of FUE on partial factor productivity basis is

Table 10 *Indicators for measuring fertilizer/nutrient use efficiency and their applications*

Indicator	Definition	Advantage	Disadvantage
Partial factor productivity	Ratio of the quantity harvested crop per unit area to fertilizer applied per unit area.	It is a simple production efficiency expression, which can easily be calculated for any farm that keeps a record of inputs and crop yields.	It is not a good indicator for a comparison of different geographical regions and cropping systems, since crop response to fertilizer may be different.
Agronomic efficiency	The increase in yield per unit of fertilizer rate. Increase in yield is defined as a difference between yield with fertilizer applied and yield without any application of fertilizer.	A useful measurement to assess the extent to which fertilizer is responsible for enhancing productivity. Useful indicator at experimental plot scale.	It requires knowledge of yield without fertilizer input, which can be obtained in plot-level experiments. But information on yield without fertilizer application is difficult through household-level or macro-scale data.
Partial nutrient balance	The ratio of nutrient content of the harvested portion of the crop to the nutrient applied.	Useful measurement for nutrient efficiency for different regions and cropping patterns.	A complicated indicator that requires knowledge of crop nutrient availability. Also, different cropping systems have varying abilities to react with other forms of nutrient input, such as biological fixation, and nutrient recovery from manure and water.
Apparent recovery efficiency	The difference in nutrient uptake by the overground parts of the plant in fertilized and unfertilized conditions.	Useful indicator to address how much crop nutrient uptake has increased by fertilizer application.	A very complicated indicator to estimate. Usually measured in plot-level experiments.
Internal utilisation efficiency	The ratio of crop yield to nutrient uptake of crop per unit area.	Useful indicator to address the ability of the plant to transform nutrients acquired from all sources into economic yield.	It requires knowledge of the nutrients available in crop.
Physiological efficiency	The yield increase in relation to the increase in crop uptake of the nutrient in overground parts of the plant.	It measures the ability of the plant to transform nutrients acquired from all sources into crop productivity.	It requires information of crop yield for plots without application of nutrients. It also requires measurement of nutrient concentration in the crop. This indicator is mainly measured and used in research.

Source: Modified version of table adapted from Fixen *et al.* (2014), pp. 3–6.

Table 11 *Fertilizer use efficiency and economic efficiency of fertilizer application for paddy and wheat,* selected villages*

Village	State	Crop	FUE (kg/kg)		Ratio of GVO to fertilizer value (Rs/Rs)	
			Small farmers	Large farmers	Small farmers	Large farmers
Ananthavaram[##,$]	Andhra Pradesh	Paddy	6.2	5.0	7.5	8.5
Bukkacherla[#]	Andhra Pradesh	Paddy	5.9	8.1	13.3	14.0
Kothapalle	Telangana	Paddy	5.6	NA	7.93	NA
Harevli	Uttar Pradesh	Paddy	20	16.8	26.9	30.3
Harevli	Uttar Pradesh	Wheat	5.8	5.6	10.8	10.1
Mahatwar[#]	Uttar Pradesh	Wheat	6.7	1.3	6.6	6.0
Mahatwar[#]	Uttar Pradesh	Paddy	6.2	1.6	7.0	7.1
25F Gulabewala	Rajasthan	Wheat	10.0	10.7	13.6	13.5
Gharsondi	Madhya Pradesh	Paddy	7	4.7	13.6	13.7
Alabujanahalli[#]	Karnataka	Paddy	4	5.4	8.5	10.4
Siresandra	Karnataka	Paddy	10.0	6.8	10.8	10.4
Panahar	West Bengal	Paddy	7.6	5.4	16.7	17.8

Notes: * Values reflect median values of households under each category.
[#] Median for FUE is statistically significant at 0.05 level of significance
[##] Median is statistically significant at 0.1 level of significance.
[$] Median for economic efficiency is statistically significant at 0.05 level of significance.

44, while the South Asia average is 41 for nitrogen (Fixen *et al.* 2015).[6] We have seen that urea is the dominant form of nitrogenous fertilizer in India, as shown by both macro data and data from village studies. The corresponding world average and South Asia average figures (for urea use) are estimated to be close to 20 and 18.5 kilograms of output per kilogram of fertilizer.

The PARI data revealed a much lower FUE than this benchmark, except for the small farmers of Harevli. In most of the study villages, FUE for small farmers was higher than for large farmers, although in most cases the median of FUE for small farmers was not significantly different from the median of FUE for large farmers.

In terms of these considerations too, Ananthavaram (Guntur district, Andhra Pradesh) showed that the median of FUE (in the case of paddy) for small farmers was statistically greater than for the large farmer category. The data for Ananthavaram showed that the median of GVO to fertilizer price for small farmers was statistically lower than for the large farmer category. This

---

[6] The South Asia average appears particularly high, since the numbers from our data show far lower figures across India and India, clearly, must dominate the average for South Asia.

Table 12 *Fertilizer use efficiency and economic efficiency of fertilizer application for other crops based on median values*

Village	State	Crop	FUE (kg/kg)		Economic efficiency of fertilizer (Rs/Rs)	
			Small farmers	Large farmers	Small farmers	Large farmers
Ananthavaram[$]	Andhra Pradesh	Maize	6.0	5.0	5.7	8.3
Kothapalle	Telangana	Cotton	0.7	2.2	4.4	4.4
Kothapalle	Telangana	Groundnut	2	NA	11	NA
Harevli	Uttar Pradesh	Sugarcane	63	46	19	26
Warwat Khanderao[$]	Maharashtra	Cotton	8	5	21	12
Nimshirgaon	Maharashtra	Sugarcane	47	53	9	8
Nimshirgaon	Maharashtra	Soybean	1	1.0	20	14
25F Gulabewala	Rajasthan	Cotton	6.7	5	20	19
25F Gulabewala	Rajasthan	Rapeseed/mustard	5.3	4.9	22	19
Gharsondi	Madhya Pradesh	Rapeseed/mustard	10	5.5	24	23
Alabujanahalli	Karnataka	Finger millet	5	4.1	7.4	7.9
Siresandra	Karnataka	Finger millet	8	15	11.4	17.1
Panahar	West Bengal	Rapeseed/mustard	2.4	NA	5.0	NA

Note: [$] Median for economic efficiency is statistically significant at 0.05 level of significance.
Source: PARI survey data.

suggests that although the small farmer appears to have had higher fertilizer efficiency in the field, the gain in overall value due to fertilizer application was less than for the large farmer. This might be because the effective cost of fertilizer for small famers was higher than for large farmers, which is evident when comparing fertilizer costs for small and large farmers in the case of paddy cultivation in Ananthavaram (Table 7).

Our observations on Ananthavaram cannot be generalised because the statistics of other villages and crops did not show any statistical difference in the values of indicators for small farmers versus large farmers. For paddy cultivation in Alabujanahalli (Mandya district, Karnataka), small farmers had lower FUE than large farmers, and this is statistically significant. We also observed for paddy in Alabujanahalli, a lower median value of GVO to fertilizer price ratio for small farmers when compared to large farmers; however, this is not significant. Mahatwar (Ballia district, Uttar Pradesh) is another village that showed a pattern similar to that of Ananthavaram. We need to be cautious, however, in making generalisations, because in Mahatwar, the large farmer population size was very small in comparison to the number of small farmers.

Other crops did not show any significant difference of FUE and GVO to fertilizer price ratio between small and large farmers (Table 12). Here too, apart from sugarcane, FUE was much lower as compared to the world and South Asian averages.

CONCLUSIONS

There is no doubt that fertilizer has played an important role in raising crop productivity in India. However, the extent of use of fertilizer has not been uniform. Most of the intensive and productive agricultural regions of the country are witnessing the application of higher-than-recommended quantities of fertilizer, while there are regions that still do not utilise sufficient fertilizer inputs for raising crop productivity. Some districts in India have very high fertilizer intensity, comparable to the intensity in some of the most agriculturally productive regions of the world. FUE of Indian agriculture is generally considered to be very low. Moreover, inefficient use of fertilizer has adverse effects on the environment since the lost nutrients from excess use pollutes the hydrosphere, the biosphere, or both. For a farmer, the choice of fertilizer and related application practices are governed by economic considerations. Information related to the economic benefits of fertilizer application is useful in that it sheds light on farmers' decisions related to choice of crop, crop variety, irrigation, and fertilizer management practices. These decisions directly influence efficient application of fertilizer at the field level. Presently, there is limited evidence on these aspects because of the lack of required information at the macro level as well as micro-level data.

In this chapter, we have used PARI village survey data to understand the patterns of nitrogen use under different cropping patterns in the study villages. We have focused in particular on understanding fertilizer use among small farmers. We have also compared this to the patterns of fertilizer use by households in the category of large farmers in the study villages. FUE is an important concept to understand the role of fertilizer application in the improvement of crop production and crop productivity. We have used partial factor productivity as an indicator to measure FUE, and have also examined the relationship between FUE and the ratio of GVO to fertilizer price for small farmers. The following is a summary of some of the key conclusions derived from the analysis of macro data and PARI data.

Region-wise and crop-wise recommendations for optimal NPK ratios are available for India, and a ratio of 120–60–30 is broadly recommended across crops and regions. This is derived from the fact that the ratio of 120–60–30 is an appropriate standard for paddy and wheat cultivation. Input survey data suggest that, on a national scale, nutrient application of fertilizer is well within the recommended limits. For many crops, current use of nitrogen (N) is close to the recommended dose, while the application rates of phosphorous (P) and potassium (K) are much lower than the recommended doses. In particular, the application rate of K is much lower for almost all crops. This serious imbalance between application of nitrogen, and phosphorus and potassium is of course unsurprising.

The latest input survey for India shows that fertilizer use per hectare of gross cropped area is almost similar for small farmers (defined by the survey as farmers with operational landholdings of less than 2 hectares), and medium and large farmers. This is true in relation to the use of main fertilizers such as urea and DAP, and in terms of N, P, and K nutrients. This suggests that on a macro scale, the current fertilizer distribution policies have assisted the penetration of fertilizer use, as small farmers are not using less fertilizer than large farmers. The literature suggests that at the micro level, small farmers have rarely implemented recommendations related to nutrient management and utilisation of efficient fertilizer technologies (Dimes *et al.* 2002; Mujeri *et al.* 2012). However, this was not evident when we were analysing the input survey data. On the contrary, the input survey suggests that fertilizer application by medium and large farmers is skewed towards application of nitrogenous fertilizers.

The PARI village data suggest that, for small farmers, the median value of fertilizer price ranges from 1 per cent to 15 per cent with respect to both total input cost and gross value of output, while it is 2 to 20 per cent for large farmers. In most of the villages, there was no evidence to statistically confirm that the cost of fertilizer for small farmers is higher than for others, except in the case of Ananthavaram in Andhra Pradesh. However, when the absolute cost of fertilizer was compared with the costs of fertilizer as reported by the CACP, we found that the CACP costs were much lower than that reported in the village survey. The CACP costs were lower than the median of the cost for the small farmer household category.

Analysis of nitrogen application rates at the village level suggests that most of the study villages, except the ones that had paddy and wheat as a dominant part of their cropping pattern, had fertilizer nutrient application rates that were lower than the recommended limit. Villages with paddy and wheat as part of the cropping pattern, however, showed the application rate of nitrogen to be higher than the recommended dose of 120 kilograms per hectare. The PARI village data also suggest a higher median value of nitrogen application for small farmers as compared to large farmers for villages that had a higher population of small farmers with irrigation facilities. However, in most of the villages the median value of nitrogen application for small farmers was not statistically different from that for the large farmer category. Therefore, the village data (for most villages) do not offer significant confirmatory evidence that small farmers' usage of nitrogen is higher than that of large farmers.

Similar to macro data, village-level data also suggest lower application of potassium by small farmers in their fields. Data for the villages of Zhapur (Kalaburagi district, Karnataka), Harevli (Bijnor district, Uttar Pradesh), and Gharsondi (Gwalior district, Madhya Pradesh) show almost zero application of potassium. Zhapur is a dryland village in southern India with pulses as the major crops. The literature points to the usefulness of potassium for the cultivation of pulses. Harevli and Gharsondi are irrigated villages in northern India with paddy as a major crop. Here zero application of potassium points to sustainability issues for cultivation.

While analysing FUE of the study villages, we found that in most villages under paddy and wheat cultivation, FUE is much lower than the world and South Asia average. Comparing FUE with the ratio of GVO to price, we found that Ananthavaram (Guntur district, Andhra Pradesh) showed confirmatory evidence for small farmers having high FUE but low GVO to price. This finding, related to paddy cultivation in Ananthavaram, is contrary to the literature which suggests that fertilizer use of small farmers is not efficient due to poor application (in terms of type and quantity) and insufficient access to extension (Dimes *et al.* 2002; Mujeri *et al.* 2012; Morton 2007; Howell 1984; Jat *et al.* 2015; IFC 2013). This feature of paddy cultivation in Ananthavaram might be because almost all farmers in the village come under the small farmer category and have irrigation facilities. However, the second observation about the ratio of GVO to fertilizer price is very much in line with the common belief in the literature that insufficient access to cash at the time of sowing might result in higher fertilizer prices for small farmers (Buresh and Giller 1998; Singh *et al.* 2002; Dev 2012; Mujeri *et al.* 2012; IFC 2013). This observation about paddy in Ananthavaram cannot be generalised as a similar pattern is not observed for other villages and other crops.

In sum, we find that small farmers across the country, apart from one or two counter-examples that we have discussed in detail, use nitrogenous fertilizers heavily in wheat and paddy cultivation, especially in conjunction with the availability of irrigation. However, in both paddy and wheat, as well as in other crops, the agronomic efficiency of use is low compared to world averages. Absolute values of economic or agronomic efficiency of use do not give us as much insight as the relative significance of these values, whether in inter-country or intra-country and inter-regional comparisons. The particular features of small farmers' use of fertilizers also do not carry much significance in absolute quantitative terms unless compared with the efficiency use of other sections of farmers. The most significant overall conclusion from our analysis of village-level data is the broad convergence in fertilizer use and efficiency of fertilizer use among small farmers and large farmers, particularly when we test for statistically significant differences. Small farming is not characterised by a clear and evident distinction in fertilizer use and FUE, whether in agronomic or economic terms. This, of course, is not a conclusion that can be carried over to other inputs such as water or pesticides without independent analysis. Nor does this conclusion take away from the significance of features such as vast inequality in terms of accumulated wealth or access to amenities between small farmers and other farming households in India.

The authors would like to acknowledge Sreeja Jaiswal from the Tata Institute of Social Sciences, Mumbai, for her help in editing the chapter.

REFERENCES

"Agricultural Input Survey," Department of Agriculture and Cooperation, Government of India, various years, available at http://inputsurvey.dacnet.nic.in, viewed on 4 September 2017.

Buresh, R. J., and Giller, K. E. (1998), "Strategies to Replenish Soil Fertility in African Smallholder Agriculture," in S. R. Waddington, H. K. Murwira, J. D. T. Kumwenda, D. Hikwa, and F. Tagwira (eds.), *Soil Fertility Research for Maize-Based Farming Systems in Malawi and Zimbabwe*, Proceedings of the Soil Fert Net Results and Planning

Workshop held from 7–11 July 1997 at Africa University, Mutare, Soil Fert Net and CIMMYT–Zimbabwe, Harare, Zimbabwe, pp. 13–19.

Central Ground Water Board (CGWB) (2009), "Annual Report of Central Ground Water Board 2008–09," Ministry of Water Resources, available at http://cgwb.gov.in/NEW/Annual-Reports/Annual-Report-2008-09.pdf, viewed on 10 September 2016.

Chandna, P., Khurana, M. L., Ladha, J. K., Punia, M., Mehla, R. S., and Gupta, R. (2011), "Spatial and Seasonal Distribution of Nitrate-N in Groundwater Beneath the Rice–Wheat Cropping System of India: A Geospatial Analysis," *Environmental Monitoring and Assessment*, vol. 178, no. 1, pp. 545–62.

Dev, S. M. (2012), "Small Farmers in India: Challenges and Opportunities," Working Paper No. WP-2012-014, available at http://www.igidr.ac.in/pdf/publication/WP-2012-014.pdf, viewed on 10 July 2016.

Dimes, J., Twomlow, S., and Carberry, P. (2002), "Application of APSIM in Smallholder Farming Systems in the Semiarid Tropics," in T. S. Bontkes and M. C. S. Wopereis (eds.), *A Practical Guide to Decision-Support Tools for Smallholder Agriculture in Sub-Saharan Africa*, International Center for Soil Fertility and Agricultural Development (IFDC), Africa Division BP 4483, Lome Togo, pp. 85–99.

Dixit, G. P., Singh, J., and Singh, N. P. (eds.) (2014), "Pulses: Challenges and Opportunities Under Changing Climate Scenario," Proceedings of the National Conference on Pulses: Challenges and Opportunities Under Changing Climate Scenario, Indian Society of Pulses Research and Development.

Divya, J., and Belagali, S. L. (2012), "Effect of Extensive Use of Chemical Fertilizers on Nitrate and Phosphate Enrichment in Water Resources of Selected Agricultural Areas of T. Narasipura Taluk, Mysore District, Karnataka, India," Proceedings of International Conference SWRDM 2012, pp. 43–47, available at http://www.unishivaji.ac.in/uploads/journal/Journal_42/10.pdf, viewed on 10 September 2016.

Fixen, P., Bentrup, F., Bruulsema, T., Garcia, F., Norton, R., and Zingore, S. (2014), "Nutrient/Fertilizer Use Efficiency: Measurement, Current Situation and Trends," in H. F. Reetz (ed.), *Fertilizers and Their Efficient Use*, International Fertilizers Industry Association, Paris.

Government of India (GoI) (2014), "Indian Fertilizer Scenario 2013," available at fert. nic.in/sites/default/.../Indian per cent20Fertilizerper cent20SCENARIO-2014.pdf, viewed on 31 July 2016.

Hou, Y., Gao, Z., Heimann, L., Roekke, M., Ma, W., and Neider, R. (2012), "Nitrogen Balances of Smallholder Farms in Major Cropping Systems in a Peri-Urban Area of Beijing, China," *Nutrient Cycling in Agroecosystems*, vol. 92, pp. 347–61, available at DOI 10.1007/s10705-012-9494-0, viewed on 16 January 2017.

Howell, J. (1984), "Small Farmer Services in India: A Study of Two Blocks in Orissa State," Working Paper No. 13, Agricultural Administration Unit, Overseas Development Institute (ODI).

Indian Council of Agricultural Research (ICAR) (2015), *Handbook of Agriculture*, sixth edition, Indian Council of Agricultural Research, New Delhi.

International Finance Corporation (IFC) (2013), "Working with Smallholders: A Handbook of Firms Building Sustainable Supply Chains," Report of IFC Sustainable Business Advisory, Washington D. C.

Jat, M. L., Majumdar, K., McDonald, A., Sikka, A. K., and Paroda, R. S. (2015), *Book of Extended Summaries*, National Dialogue on Efficient Nutrient Management for Improving Soil Health, 28–29 September 2015, New Delhi.

Ladha, J. K., Pathak, H., Krupnik, T. J., and van Kessel, J. (2005), "Efficiency of Fertilizer Nitrogen in Cereal Production: Retrospects and Prospects," *Advances in Agronomy*, vol. 87, pp. 85–156.

Lal, R., and Stewart, B. A. (2015), "Sustainable Intensification of Smallholder Agriculture," in R. Lal and B. A. Stewart (eds.), *Sustainable Intensification of Smallholder Agriculture*, CRC Press, Boca Raton, pp. 385–92.

Lassaletta, L., Billen, G., Grizzetti, B., Anglade, J., and Garnier, J. (2014), "50 Year Trends in Nitrogen Use Efficiency of World Cropping Systems: The Relationship Between Yield and Nitrogen Input to Cropland," *Environmental Research Letters*, No. 9.

Mohanty, S., and Satyasai, K. S. (2015), "Feeling the Pulse: Indian Pulse Sector," Department of Economic Analysis and Research (DEAR), National Bank for Agriculture and Rural Development (NABARD).

Morton, J. F. (2007), "The Impact of Climate Change on Smallholder and Subsistence Agriculture," *Proceedings of the National Academy of Science*, vol. 104, no. 50, pp. 19680–85.

Mujeri, M., Shahana, S., Choudhary, T. T., and Haider, K. T. (2012), "Supporting Policy Research to Inform Agricultural Policy in Sub-Saharan Africa and South Asia," Report of Global Development Network, available at http://www.gdn.int/html/page2.php?MID=3&SCID=9&SID=24&SSID=5, viewed on 31 June 2016.

National Academy of Agricultural Sciences (NAAS) (2009), "Crop Response and Nutrient Ratio," Policy Paper No. 42, National Academy of Agricultural Sciences, New Delhi, available at http://naasindia.org/Policyper cent20Papers/policyper cent2042.pdf, viewed on 3 March 2015.

Ncube, B., Dimes, J. P., Twomlow, S. J., Mupangwa, W., and Giller, K. E. (2006), "Raising Productivity of Smallholder Farms Under Semi-Arid Conditions by Use of Small Doses of Manure and Nitrogen: A Case of Participatory Research," *Nutrient Cycling in Agroecosystems*, available at DOI 10.1007/s10705-006-9045-7, viewed on 17 January 2017.

Norton, R., Bruulsema, T., Roberts, T., and Snyder, C. (2015), "Crop Nutrient Performance Indicators," *Agricultural Science*, vol. 27, no. 2, November, pp. 33–38, available at http://search.informit.com.au.ezproxy.lib.monash.edu.au/documentSum mary;dn=685063951694712;res=IELAPA, viewed on 17 January 2017.

Otoo, J., Ofori, J. K., and Amoah, F. (2015), "Optimal Selection of Crops: A Case Study of Small Scale Farms in Fanteakwa District, Ghana," *International Journal of Scientific and Technology Research*, vol. 4, no. 5, pp. 142–46.

Pathak, H., Mohanty, S., Jain, N., and Bhatia, A. (2010), "Nitrogen, Phosphorous, and Potassium Budgets in Indian Agriculture," *Nutrient Cycling in Agroecosystems*, vol. 86, pp. 287–99, available at DOI 10.1007/s10705-009-9292-5, viewed on 17 January 2017.

Prasad, R. (1999), "Sustainable Agriculture and Fertilizer Use," *Current Science*, July, available at http://www.iisc.ernet.in/currsci/jul10/articles12.htm, viewed on 5 August 2015.

Shindo, J., Okamoto, K., and Kawashima, H. (2006), "Prediction of the Environmental Effects of Excess Nitrogen Caused by Increasing Food Demand with Rapid Economic Growth in Eastern Asian Countries, 1961–2020," *Ecological Modelling*, vol. 193, pp. 703–20.

Singh, R. B., Kumar, P., and Woodhead, T. (2002), "Smallholder Farmers in India: Food Security and Agricultural Policy," Food and Agriculture Organisation (FAO) Report No. 2002/2003, Rome.

Swaminathan, Madhura, and Rawal, Vikas (2011), "Is India Really a Country of Low Income-Inequality? Observations from Eight Villages," *Review of Agrarian Studies*, vol. 1, no. 1, available at http://ras.org.in/is_india_really_a_country_of_low_income_inequality_observations_from_eight_villages, viewed on 17 January 2017.

Vel Murugan, A., and Dadhwal, V. K. (2007), "Indian Agriculture and Nitrogen Cycle," in Y. P. Abrol, N. Raghuram, and M. S. Sachdev (eds.), *Agricultural Nitrogen Use and its Environmental Implications*, I. K. International, New Delhi, pp. 9–28.

Vikas, C., Kushwaha, R., Ahmad, W., Prasannakumar, V., Dhanya, P. V., and Reghunath, R. (2015), "Hydro-Chemical Appraisal and Geochemical Evolution of Groundwater with Special Reference to Nitrate Contamination in Aquifers of a Semi-Arid Terrain of NW India," *Water Quality, Exposure and Health*, vol. 7, no. 3, pp. 347–61.

Yengoh, G. T. (2012), "Determinants of Yield Differences in Small-Scale Food Crop Farming Systems in Cameroon," *Agriculture and Food Security*, pp. 1–12, available at http://www.agricultureandfoodsecurity.com/content/1/1/19, viewed on 17 January 2017.

8

Formal Credit and Small Farmers in India

Pallavi Chavan, with T. Sivamurugan

India has a long and fairly illustrious history of the development of a formal system of rural credit. It can be traced back to the formation of credit cooperatives in the early twentieth century under state sponsorship, with watershed events like bank nationalisation in 1969 and 1980 resulting in a massive expansion of public banking in the rural areas. These policy measures were centered around agriculture and within agriculture their focus was on small farmers, a predominant but relatively underprivileged segment of India's agrarian system.

The literature on credit-related issues concerning small farmers from developing economies, including India, contains concerns of efficiency and viability, vulnerability and equity. Although the definition of a small farmer differs across studies, the literature views access to formal credit to be an essential prerequisite for improving the efficiency and viability of small farms (FAO 2013; Berdegue and Feuntealba 2011). In the absence of a social security net, formal credit is also regarded as a means to tide over income fluctuations for small farmers, and thus address the concerns of vulnerability (Hulme and Mosley 1996; Clark and Dercon 2009).[1] Notwithstanding these arguments in favour of extending formal credit to small farmers, the literature also highlights that their access to formal credit has generally been low (IFPRI 2010). The low access is posed in the literature more as a supply-side issue, and hence is regarded as involuntary in nature.[2]

---

[1] There is also a vast literature dealing with micro-credit that regards access to such credit as a means to stabilise income and increase consumption among the poor; see Montgomery *et al.* (1996) and Morduch (1998).

[2] The issues relating to low access to credit for small farmers can be divided into two categories: supply-side issues concerning lenders (formal credit agencies) and demand-side issues concerning borrowers (small farmers themselves). Issues concerning lenders include: (a) high transaction costs in administering and monitoring small amounts of credit to a large group of borrowers (IFPRI 2010; Datta and Sriram 2010); (b) lack of credit histories for borrowers resulting in information asymmetries leading to risks of adverse selection (Narain 2010); (c) a focus – often associated with (b) in certain studies – on high-value, marketable collateral such as land by formal credit agencies (Swaminathan 1991). Issues concerning borrowers are relatively few and are associated more with the nature of operation of formal agencies. For instance, studies discuss the bureaucratisation of procedures resulting in high transaction costs for small farmers seeking formal credit (Dreze *et al.* 1998; Walker and Ryan 1990).

This chapter analyses the access to and adequacy of formal sources in meeting the credit needs, particularly agricultural credit needs, of small farmers in India with the help of banking data, and data on the borrowing profiles of these households collected through the village surveys of the Project on Agrarian Relations in India (PARI).

Three major institutions provide formal credit in the rural areas of India today: commercial banks, regional rural banks (RRBs), and credit cooperatives. Commercial banks and RRBs – primarily sponsored by public sector commercial banks – together control about three-fourths of total agricultural credit.[3] Despite being a late entrant into the field of rural credit, the contribution of commercial banks, after nationalisation, to shaping the history of development of rural and agricultural credit in India has been overwhelming. In course of time they have overtaken credit cooperatives, the oldest serving institution of rural credit, except in a few States where cooperatives continue to predominate even today.[4]

It may not be correct, however, to assume that the contribution of commercial banks to rural and agricultural credit has increased in a monotonic manner since their nationalisation. There have been changes in the role played by these institutions in the field of rural credit, following changes in banking policy. Whether these changes have also meant a change in the supply of bank credit to small farmers is an issue that is explored in this chapter along with certain other research questions.

This chapter attempts to answer the following questions using banking and village-level data sources:

1. What is the share of small farmers in formal agricultural credit, in particular bank credit, in India, and how has this changed over time?
2. How do small farmers meet their various credit needs, particularly agricultural credit needs?
3. What is the extent of access to formal credit (crop credit in particular) for small farmers? What are its determinants?
4. Do small farmers receive crop credit in line with the stipulated scales of finance or does the formal sector fail to adequately meet their credit needs?
5. What is the interest cost of credit, particularly agricultural credit, for small farmers?

[3] See Chavan (2013).
[4] See NSSO (2005; 2013). This point will also be illustrated later in the chapter while discussing village data.

The chapter is divided into six sections including this introductory section. The second section looks at the history of rural credit policy in India, particularly rural banking policy, through the prism of small farmers and their general credit needs, particularly agricultural needs. The third section examines certain conceptual and methodological issues relating to data analysis of credit for small farmers. The fourth section contains analysis of secondary data, although limited, on credit flow to small farmers. The fifth section provides insights into the debt profiles of small farmers, based on PARI village survey data. The sixth section provides concluding observations based on the earlier sections and discusses related policy implications.

RURAL CREDIT POLICY AND SMALL FARMERS

A review of the history of rural credit policy suggests that it can be divided into four broad phases: the phase of credit cooperatives prior to bank nationalisation; the phase of bank nationalisation; the phase of financial liberalisation since the early 1990s; and the phase of financial inclusion with a revival of some of the developmental measures with regard to rural credit since 2005.

In the first phase preceding bank nationalisation, credit cooperatives were envisaged as agencies of formal credit to farmers, particularly small farmers (GIPE 1975; Nicholson 1960). However, as shown by the All-India Rural Credit Survey (AIRCS) of 1951–52 – the first official nation-wide survey of rural credit – the degree of penetration of formal (read cooperative) credit in rural areas in general, and among small farmers in particular, remained low even after five decades of the development of cooperatives. Yet the AIRCS Committee argued for an Integrated System of Rural Credit with cooperatives as the central agency for financing agricultural production (RBI 1954). It also argued strongly in favour of the crop credit system, which later became the cornerstone of agricultural credit in India (*ibid.*). Credit for seasonal agricultural operations was provided to cultivators under crop loans, not against land owned by them but on the basis of their "repayment capacity" (GoI 1945, p. 48). By shifting the focus of cooperative credit from land to crops as security for credit, the crop credit system was expected to bring small farmers, particularly tenant farmers, into the fold of cooperative credit in a major way (GIPE 1975).

Under the Integrated Scheme, commercial banks were involved only at the periphery to provide finance for agricultural marketing (RBI 1954). This, however, changed radically with the nationalisation of 14 major banks in 1969, and six more banks later in 1980. From being urban-oriented credit institutions, commercial banks under public ownership, guided by branch licensing and priority sector lending policies, came to be regarded as the key agency of rural and agricultural credit (Narayana 2000).[5] The primary objective of banking policy during this phase – often referred to as the phase of "social and development banking" – was redistribution, at the expense, at times, of prudence and commercial considerations related to conventional banking (Wiggins and Rajendran 1987).[6]

Small farmers, being an integral part of India's agrarian system as shown in the earlier chapters of this book, figured prominently in the design of rural banking policy during this phase. First, priority sector guidelines were laid down in 1974 to direct a part of bank credit to certain sectors denoted as priority sectors. In 1980, two separate sub-targets were laid down for agriculture and "weaker sections" (constituting socio-economically backward sections) (RBI 2015). Small farmers explicitly figured in the list of weaker sections alongside landless labourers, Scheduled Castes (SCs), and Scheduled Tribes (STs) (*ibid.*). They also implicitly figured as those who were eligible for agricultural credit, particularly "direct" agricultural credit – that is, credit given directly to cultivators.[7]

Secondly, a policy of administered interest rates was followed to ensure adequate flow of bank credit to priority sectors at affordable rates (*ibid.*). The scheme of differential rate of interest (DRI) was introduced in 1972 to provide credit from public banks at a regulated rate of interest to individuals below the poverty line (RBI 2008). Thirdly, a new public banking institution was created in the form of the regional rural bank (RRB), to provide credit exclusively to the poor and underprivileged sections including small farmers (Regional Rural Banks' Act, 1976, cited in Maheswari 1995).

Fourthly, following the recommendations of the All-India Credit Review Committee (1969), a special agency in the form of the Small Farmers' Development Agency (SFDA) was created in select districts of the country in 1971 (it was merged with the Integrated Rural Development Programme (IRDP) in 1979–80 – a credit-based poverty alleviation programme). The objective of the SFDA was to enable small farmers to cope with the

---

[5] For the essentially urban-oriented nature of commercial banking in India prior to bank nationalisation, see Goyal (1967). The literature on rural credit discusses the striking increase in rural branches of commercial banks, and supply of bank credit to rural areas and agriculture during this phase (Shetty 2005; Chavan 2005).

[6] Wiggins and Rajendran (1987) have highlighted three aspects of conventional banking: "high recovery rates," "caution" in lending decisions, and "commercial stability through deposit mobilisation."

[7] The other part of agricultural credit was "indirect" credit, which was given to institutions/ organisations that supported agricultural production, such as input dealers and warehouse operators.

"non-neutral-to-scale" nature of markets, extension services, and credit institutions while adopting the new agricultural strategy (Desai 1979, p. 23).[8]

During the phase of financial liberalisation that began with economic reforms in 1991, however, the profitability and commercial viability of banks assumed more importance than redistribution. The Committee on the Financial Systems (CFS) of 1991, one of the first committees to argue strongly in favour of liberalisation of the banking sector, suggested that the objective of redistribution should be pursued only as part of the fiscal policy and not monetary policy (RBI 1991). As a result, this phase saw a reversal of many of the developmental measures that were introduced in the earlier phase, including complete liberalisation of the branch licensing policy.[9]

First, although the CFS argued for an initial reduction and eventual removal of priority sector targets, these targets were neither removed nor reduced. However, the definition of priority sectors, particularly in agriculture (under both the sub-categories, direct and indirect agricultural credit), was considerably widened during this phase (Ramakumar and Chavan 2007). These changes primarily took the form of: (a) inclusion of several new heads and sharp increases in the credit limits set for existing heads under indirect agricultural credit (*ibid.*). However, while making these changes, the cap of 25 per cent on indirect agricultural credit was retained, ensuring that direct credit/ credit to farmers received due priority in the overall supply of agricultural credit.[10] (b) Inclusion of several new heads even under direct agricultural credit. To cite an important change, a part of the loans to corporates involved in agricultural production was included under direct agricultural credit, thus essentially treating corporates on a par with farmers.[11]

It was argued that these changes were on account of a conscious shift in public policy relating to agriculture, to promote large-scale, commercial, capital-intensive forms of agricultural production and post-production

---

[8] These agencies were to act as coordinators between small farmers, credit institutions, and government departments. They were registered as societies, and were constituted by drawing members from credit institutions and governmental departments, and small farmers themselves (Singh 1979).

[9] See Chavan (forthcoming 2017) for illustrations of the various measures taken to liberalise the Indian banking sector during this phase, including liberalisation of the branch licensing policy, partial privatisation of public sector banks, and granting licenses to more domestic and foreign private sector banks.

[10] Also see RBI (2004) for arguments in favour of retaining the cap on indirect credit.

[11] All agricultural credit given to corporates up to a limit of Rs 10 million and one-third of the credit with a limit of over Rs 10 million were categorised as part of direct credit, while the rest was included under indirect credit. However, as per the modification in the guidelines in 2013, all agricultural credit to corporates up to a limit of Rs 20 million was considered as part of direct credit (www.rbi.org.in).

activities during the period of economic liberalisation (Ramakumar and Chavan 2007). However, this shift prompted questions about credit flow to farmers in general and small farmers in particular.

Secondly, this phase also witnessed several changes relating to RRBs, which had a direct impact on their contribution to rural credit, particularly to small farmers (Bose 2005). For instance, these institutions were allowed to lend to non-priority sectors from 1992.[12] Also, the rates of interest charged by them on loans were deregulated (*ibid.*).

Thirdly, the process of liberalisation affected credit cooperatives much later than commercial banks and RRBs, but it had far-reaching implications for the very survival of these institutions in many States. Prudential norms (excluding capital adequacy) were made applicable to State and district central cooperative banks in the second half of the 1990s.[13] However, despite the fact that cooperatives in most States had a weak capital base, they did not receive any recapitalisation support to enable them to meet the new norms, unlike commercial banks and RRBs.[14] Although the situation did change in 2008 when conditional capital support was granted to short-term credit cooperatives, the support was (a) delayed[15] and (b) not extended to long-term credit cooperatives. The latter continued to struggle with a low capital base and high levels of non-performing loans in most States.[16] Further, interest rates on advances charged by State- and district-level cooperative banks were also deregulated.[17]

Although the period after 2005 saw a continuation of the policy of financial liberalisation, there was a renewed commitment to bringing underserved sections into the ambit of banking as part of a policy of "financial inclusion."[18]

[12] Priority sector targets for these institutions, however, were set at levels higher than those applicable to commercial banks. See "Master Circular on Priority Sector Lending Targets and Classification for RRBs," available at www.rbi.org.in.

[13] See www.nabard.org for details. These norms were not applied to primary cooperative societies.

[14] This delay in recapitalisation was primarily because cooperation is a State subject (Sen 2005).

[15] The conditional capital support was a result of the government accepting the recommendations of the Task Force on Revival of Rural Cooperative Credit Institutions (headed by A. Vaidyanathan); for details, see www.nabard.org. The damage to these institutions, however, had already been done by then in terms of a fall in their credit growth; see Satish (2007), and Datta and Sriram (2010).

[16] See the *Report on Trend and Progress of Banking in India*, issues after 2007–08, for a discussion on the decline in the health and growth of long-term credit cooperatives in most States.

[17] The interest rates were almost entirely deregulated except for a floor interest rate of 12 per cent. See Subrahmanyam (1999).

[18] Officially, financial inclusion was defined as the "process of ensuring access to appropriate financial products and services needed by all sections of the society in general, and vulnerable groups, such as weaker sections and low income groups in particular, at an affordable cost in a fair and transparent manner by regulated, mainstream institutional players" (Chakrabarty 2013).

During this period, political changes at the centre with the formation of a government supported by Left parties opened up some space for inclusive economic policies (Ramachandran and Rawal 2010). The commitment to inclusive finance can be regarded as a part of these policies.

In terms of its objective, the policy of financial inclusion seemed similar to the policy followed after bank nationalisation. However, it was to be pursued in the broader context of financial liberalisation taking into account "business considerations" to ensure the "long-term sustainability of the process" (RBI 2008, p. 304). Hence, this phase witnessed the come-back of a few, but not all, policy mandates for the development of rural credit, but these mandates had to be pursued in an environment of greater operational autonomy of banks in the form of continuing interest rate deregulation and a broadening of the definition of priority sectors.

First, apart from a renewed mandate to commercial banks to open at least one-fourth of their total number of branches in a year in unbanked rural areas, they were asked to double the flow of agricultural credit over the three-year period from 2004–05 to 2006–07 as part of the "comprehensive credit policy" (Ramakumar and Chavan 2007). Secondly, although interest rates on bank advances were subjected to complete deregulation, an interest subvention scheme was introduced in 2006–07 for direct agricultural loans (to farmers) of up to Rs 0.3 million, which was expected to benefit small farmers given their small credit needs (*ibid.*). Thirdly, Kisan Credit Cards (KCCs), which were introduced as an innovative crop credit delivery mechanism during the earlier phase, received considerable impetus during the phase of financial inclusion.[19] Fourthly, the category of weaker sections was broadened to include joint liability groups (JLGs), for loans given by banks. This was expected to primarily benefit small farmers and tenant cultivators in gaining access to formal credit (NABARD 2014). Finally, in 2014, a separate sub-target of 8 per cent was introduced under agriculture for small farmers to ensure focused attention on this class of farmers.[20] However, while introducing this sub-target, the distinction between direct and indirect credit was removed. Hence, this change could be described as a mixed bag for small farmers. It, of course,

---

[19] These cards were originally designed to meet the short-term crop credit needs of a farmer for the entire year in one go, enabling the farmer to withdraw money as and when necessary. During the phase of financial inclusion, not only were these cards made smart cards, further easing the access for farmers, but the finance under these cards was broadened to include term credit and consumption credit components. See the RBI Circular on "Revised Kisan Credit Card Scheme," www.rbi.org.in.
[20] This target was applicable to domestic banks by 2017 and foreign banks with 20 branches or more from 2018; see "Master Circular on Priority Sector Lending Targets and Classification," July 2016, www.rbi.org.in.

recognised their due more formally than before; however, attainment of this target appeared doubtful given the greater presence of corporate and other entities under the combined category of agricultural credit. Moreover, even if the target was attained, given the poor representation of small farmers in agricultural credit in earlier years, it is difficult to regard this as a floor.

FORMAL CREDIT TO SMALL FARMERS:
ISSUES RELATED TO DATA, METHODOLOGY, AND LITERATURE

Data from secondary sources on formal credit available to small farmers are scant and scattered. Formal credit to small farmers can be analysed in two ways: by looking at (i) the share of formal sources in the debt profiles of small farmer households; (ii) the supply of credit to small farmer households by various formal credit agencies.

A key secondary source that provides household-level data over a long time-horizon is the decennial All-India Debt and Investment Survey (AIDIS). However, while this source provides data on the debt profiles of "cultivator households" (that operated more than 0.05 acre of land during the survey year), it offers no separate information on small cultivators/farmers.

It is possible to extract data from the Situation Assessment Survey (SAS) of Farmer Households of 2002–03 on the debt profiles of small farmer households, assuming a certain cut-off in terms of the size of landholding. However, this was a one-off survey. Although the National Sample Survey Organisation (NSSO) conducted another SAS in 2012–13, this survey was titled the SAS of Agricultural Households as it canvassed information on "agricultural households" – a category that was not comparable with "farmer households" canvassed in the earlier round.[21] Given this change in definition, any comparison between the two rounds is not appropriate. Hence, this chapter makes use of the 2002–03 round to offer limited insights into the debt profiles of small farmer households.

There are several data limitations as regards (ii) mentioned above too. First, there is no current source of secondary data on the credit flow to small

---

[21] Farmer households were defined as households that had at least one family member as a farmer. A farmer was defined as a person who possessed some land and was engaged in agricultural activities on any part of that land during the 365 days preceding the date of the survey (NSSO 2003). Agricultural activities were defined as cultivation of field and horticultural crops, growing of trees or plants, animal husbandry, fishery, bee-keeping, vermiculture, and sericulture, among others. Agricultural household, on the other hand, was a much narrower category and included households that annually produced agricultural output of more than Rs 3000 (NSSO 2013).

farmers from credit cooperatives.[22] Secondly, though the general availability of data on commercial banks has been better than on credit cooperatives, there are limitations again with regard to the data on small farmers. The key source of data on commercial banks (including RRBs) is the *Basic Statistical Returns of Scheduled Commercial Banks in India* (BSR), published by the Reserve Bank of India (RBI) annually since 1972, which does not offer data on small farmers. However, it offers data on small borrowal accounts (SBAs) under agriculture – accounts with a credit limit of Rs 0.2 million.[23] This category offers some insights into the credit going to cultivators with small credit requirements, who are expected to correspond closely but not exactly with small farmers.

In the literature, there are some studies based on secondary data from the BSR on the flow of credit to small farmers. These studies point to a marginalisation of small farmers in the distribution of formal credit to agriculture since the initiation of financial liberalisation (Ramakumar and Chavan 2007; Ramakumar and Chavan 2014). They point to a striking revival in the growth of bank credit to agriculture in the 2000s after a steep fall in the 1990s, and attribute it primarily to the comprehensive credit policy (Ramakumar and Chavan 2007; Ramakumar and Chavan 2014). However, given the various changes in the definition of agriculture, these studies raise questions about how far the revival has helped small farmers (*ibid.*; GoI 2010).

Annual priority sector returns are another source of data on commercial bank credit to small farmers. These returns also follow the conventional definition of small farmers (as having an operational holding of up to 5 acres). We have used these returns to bring out recent trends in the credit flow to small farmers.[24]

---

[22] Data on credit cooperatives were published through *Statistical Statements Relating to Cooperative Movement at a Glance*, an annual publication of the National Bank for Agriculture and Rural Development (NABARD), which included information on small farmers (defined as having an operational holding of up to 5 acres). However, over time, the lag in frequency of this publication has increased; and in order to minimise the lag, NABARD has often resorted to repetition of past data, which raises questions about the reliability of this data source. For a discussion of the data limitations related to cooperatives, also see GoI (2004). The last publication in this series, containing repetitions for certain States, was for the year 2003–04 (NABARD 2010).

[23] The limit was Rs 10,000 till 1983, Rs 25,000 between 1984 and 1998, and Rs 0.2 million since 1999.

[24] The RBI also publishes data on agricultural credit to small farmers in its *Handbook of Statistics on Indian Economy* (HBS). However, there are striking differences between the HBS data and data published in the various official committee reports on agricultural credit which have been sourced directly from the banks themselves (RBI 2015). As a result, in this chapter we have chosen not to use the HBS data but instead draw data from the annual priority sector returns submitted by banks and disseminated through the Database on Indian Economy by the RBI.

In sum, there are limitations to any analysis of secondary data on formal credit to small farmers, and primary surveys are a preferred option. This chapter does use secondary data sources but in a limited way, to address research questions (i) and (ii) raised earlier. Our focus here is the villages surveyed as part of the Project on Agrarian Relations in India (PARI). Village studies have been an important tool for understanding the rural credit system in India. As compared to the 1970s and 1980s, however, village studies on rural credit in the 1990s and 2000s are relatively few in number (Chavan 2012). Hence there appears to be a gap in the literature on the ground-level situation concerning formal credit with regard to small farmers in the phases of liberalisation of the banking sector and financial inclusion, which this chapter sets out to address.[25]

Out of 15 PARI study villages (after excluding 25F Gulabewala, which is a large farmer-dominated village with only two small farmer households), this chapter makes use of data on 13 villages.[26] The key criterion in selecting these villages was that the survey year was a normal agricultural year for a given village. This is because for analysing formal credit to small farmers, particularly for agricultural purposes, it is essential that the given village had a normal agricultural activity cycle during the survey year. Based on this criterion, we shortlisted 13 villages; Gharsondi (Gwalior district, Madhya Pradesh) and Rewasi (Sikar district, Rajasthan), which reported a drought-like situation during the survey year, were excluded. We have used the data from these 13 villages to address questions (2), (3), (4), and (5).

However, for answering specific questions concerning crop credit in relation to the crop credit estimate based on the scale of finance, we have used case studies of small farmer households, and have reduced our sample further to 11 villages – after excluding Siresandra (Kolar district, Karnataka) and Zhapur (Kalaburagi district, Karnataka). This was because these two villages showed a relatively small incidence of crop credit (from formal sources) among small farmer households during the survey year, with only four and three households, respectively, reporting a fresh loan from formal sources. These

---

[25] The few studies available for these decades, however, suggest a diminishing role for formal credit and the strengthening of informal lenders, giving rise to more oppressive debt-related practices (Ramachandran and Surjit 2005; Rawal and Mukherjee 2005; Gill 2005; Ramachandran et al. 2010). Interestingly, however, village studies from Maharashtra and Kerala, where credit cooperatives have been a more dominant source of formal credit, have continued to show a remarkably high share of formal sources in the total debt of small farmers (Ramakumar 2005; Rao and Charyulu 2007; Chavan 2012).

[26] All other details about these villages have been discussed in Chapter 2 of this volume.

accounted for 22 per cent and 7 per cent, respectively, of the total size of small farmer households in these villages.

While we have used data on debt outstanding (including interest and principal outstanding) on the survey date to illustrate the stock of debt, the key variable of analysis is fresh borrowings during the survey year (including amount borrowed and repaid during the year itself). This reflects the flow of credit and it can then be compared with other flow variables associated with agricultural activity. To compare the flow of credit with crop credit estimate, we separated out the investment credit component and worked out the crop credit taken for seasonal agricultural production from formal sources by these households.

While we present village-level data to illustrate various trends in formal credit, we have used pooled data (across all 13 villages) for the analysis of access to formal credit. We resorted to pooling of data to overcome the constraint of small-sized samples of small farmers across villages. Given that the surveys were not conducted in one year, while pooling: (a) all variables in value terms were normalised by a demographic or economic variable, and were expressed as proportions; (b) State dummies were used to control for State-specific factors affecting the access to formal credit. The access to formal credit was modelled using a logistic model, details of which are given below under the section titled "Analysis of Village Data."

Apart from data on borrowing profiles extracted from the village surveys, we also sourced data on crop-wise scales of finance for the respective survey years. These data were collected from the lead banks for the various districts to which the selected villages belonged.

TRENDS IN FORMAL CREDIT TO SMALL FARMERS: SOME INSIGHTS FROM SECONDARY DATA

Secondary data on the debt profiles of small farmers can be drawn from SAS 2002–03, taking a land size cut-off of 5 acres. This is the standard definition followed in most official sources in India, but is different from the carefully designed definition followed in this volume of 5 standard (irrigated) acres. It shows that in 2003: (a) about half of the total debt of small farmer households was raised from formal sources leaving the rest to be met by informal sources; (b) the share of formal sources was lower for these households than for large farmer households; (c) moneylenders accounted for about one-third of the total debt of small farmer households; (d) commercial banks were the most important formal source of credit for them (Table 1).

Data drawn from the annual priority sector returns of banks offer more current insights into the flow of bank credit for agriculture to small

Table 1 *Share of debt by source, small farmer and other farmer households, 2003*

Source	Small farmer households	Other farmer households	All farmer households
All formal sources	51.6	65.9	57.7
Government	3.1	1.7	2.5
Commercial banks	31.1	41.7	35.6
Credit cooperatives	17.5	22.5	19.6
All informal sources	48.4	34.1	42.3
Moneylenders	29.5	20.6	25.7
Traders	4.5	6.1	5.2
Relatives and friends	10.7	5.5	8.5
Others	3.6	1.9	2.9
Total	100.0	100.0	100.0

Source: Calculated from NSSO (2003).

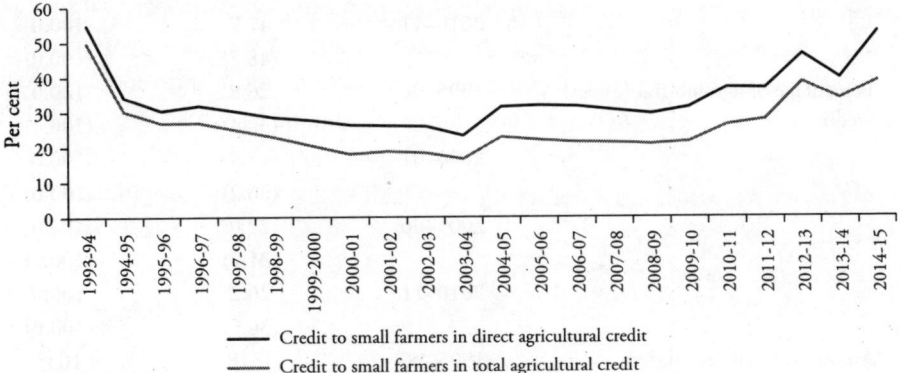

——— Credit to small farmers in direct agricultural credit
——— Credit to small farmers in total agricultural credit

Figure 1 *Share of small farmers in total agricultural credit from banks*
Source: Database on Indian Economy, Reserve Bank of India (RBI).

farmers. In 2014–15, small farmers accounted for about 39 per cent of the total agricultural credit from banks. Their share in direct agricultural credit was 53 per cent. There was a falling trend in the share over the 1990s and 2000s except for a spurt in 2004–05, the year the comprehensive credit policy was introduced. Their share was on a recovery path only after 2009–10.

Notwithstanding these changes, it is evident that the share of small farmers in total agricultural credit from banks has rarely been above 50 per cent during this entire period. Further, their share in credit has lagged behind their share in land operated. The availability of credit per hectare for small farmers too has been much lower than for all farmers (Table 2).

Table 2 *Share of small farmers in land operated and agricultural credit, 1995–96 to 2010–11*

Size-class	Year	Small farmers	All farmers
Percentage of landholding	1995–96	80.3	100.0
	2000–01	81.8	100.0
	2005–06	83.3	100.0
	2010–11	85.0	100.0
Percentage of area operated	1995–96	36.0	100.0
	2000–01	38.9	100.0
	2005–06	41.1	100.0
	2010–11	44.3	100.0
Percentage of agricultural (direct) accounts	1995–96	60.6	100.0
		(61.7)	(100.0)
	2000–01	40.5	100.0
		(41.1)	(100.0)
	2005–06	43.6	100.0
		(47.0)	(100.0)
	2010–11	47.7	100.0
		(48.7)	(100.0)
Percentage of agricultural (direct) credit	1995–96	26.8	100.0
		(30.4)	(100.0)
	2000–01	17.9	100.0
		(26.2)	(100.0)
	2005–06	22.0	100.0
		(31.7)	(100.0)
	2010–11	26.2	100.0
		(36.9)	(100.0)
Amount of deflated direct agricultural credit per hectare (Rs), (Compound annual rate of growth (%))	1995–96	1378	1635
		–	–
	2000–01	2017	2987
		(3.9)	(6.2)
	2005–06	6248	8253
		(12.0)	(10.7)
	2010–11	12808	15396
		(7.4)	(6.4)

Source: Agricultural Census of India, Database on Indian Economy, Reserve Bank of India (RBI).

ANALYSIS OF VILLAGE DATA

General Features of Village Credit Systems

Wide variation in development of formal sources

There was a wide variation across villages in terms of the development of formal sources of credit.[27] We used two indicators of development of formal sources: (a) number of formal sources available to village households; (b) percentage share of formal sources in fresh borrowings (columns 3 and 5 in Table 3). The first indicator captured the availability of formal sources, while the second showed the effective contribution of available formal sources to the flow of credit in a village. Taking the two indicators together, Nimshirgaon (Kolhapur district, Maharashtra), Warwat Khanderao (Buldhana district, Maharashtra), Alabujanahalli (Mandya district, Karnataka), Bukkacherla (Anantapur district, Andhra Pradesh), and Mahatwar (Ballia district, Uttar Pradesh) had a well-developed formal credit system.

The remaining villages could be divided into the following two categories. (1) Villages with fewer formal sources but contributing more than half to total credit. These included Kothapalle (Karimnagar district, Telangana), and Harevli (Bijnor district, Uttar Pradesh). (2) Villages having a large presence of formal sources but these contributing less than half to the flow of credit. These included Zhapur (Kalaburagi district, Karnataka), Siresandra (Kolar district, Karnataka), Ananthavaram (Guntur district, Andhra Pradesh), Panahar (Bankura district, West Bengal), Kalmandasguri (Koch Bihar district, West Bengal), and Amarsinghi (Malda district, West Bengal). Given the superiority of indicator (b) over (a), it could be said that villages in category (1) had a relatively well-developed formal system of credit as compared to villages in category (2). The backward nature of credit systems in Zhapur

---

[27] Following the SAS and AIDIS, our definition of formal sources includes commercial banks, credit cooperatives, regional rural banks, insurance companies, provident funds, government agencies, and transport finance companies. Informal sources include all other sources, including micro-finance institutions (MFIs) and self-help groups (SHGs). As illustrated in greater detail in Box 2, MFIs include a few non-banking financial companies (NBFCs) which are registered as NBFC–MFIs with the Reserve Bank of India, as well as a large number of non-governmental organisations, trusts, and Section 25 companies, which are not under the purview of any uniform regulation. And hence, all institutions operating in this sector have been treated as informal/semi-formal.

Table 3 *Basic indicators of borrowing profiles of all households, 13 PARI villages*

Village	State	Number of formal sources available to village households	% of formal sources in total debt outstanding	% of formal sources in fresh borrowings	% of formal sources in fresh borrowings for agriculture	Largest formal source in total fresh borrowing from all formal sources (% share)	Largest formal source in total fresh borrowing from all formal sources for agriculture (% share)
(1)	(2)	(3)	(4)	(5)	(6)	(7)	(8)
Nimshirgaon	Mah	31	92.6	75.9	97.3	Credit cooperative (73.1)	Credit cooperative (70.5)
Warwat Khanderao	Mah	20	84.2	66.8	69.1	Credit cooperative (73.8)	Credit cooperative (84.7)
Alabujanahalli	Kar	18	50.1	54.2	88.4	Commercial bank (68.4)	Commercial bank (77.8)
Mahatwar	UP	10	56.5	44.4	97.8	Commercial bank (96.2)	Commercial bank (98.6)
Zhapur	Kar	9	28.9	21.5	50.8	Commercial bank (97.6)	Commercial bank (100.0)
Bukkacherla	AP	8	45.0	61.1	87.6	Commercial bank (80.3)	Commercial bank (77.0)
Ananthavaram	AP	8	33.1	36.6	35.0	Commercial bank (83.6)	Commercial bank (97.8)
Panahar	WB	7	45.6	21.5	48.7	Commercial bank (85.8)	Commercial bank (78.9)
Kalmandasguri	WB	7	37.8	33.9	76.2	Commercial bank (81.9)	Commercial bank (72.5)
Harevli	UP	6	82.9	72.1	93.2	Commercial bank (86.7)	Commercial bank (84.7)
Kothapalle	TE	6	57.9	74.7	94.3	Commercial bank (99.3)	Commercial bank (100.0)
Siresandra	Kar	6	33.5	11.5	–	Commercial bank (90.2)	–
Amarsinghi	WB	3	57.0	43.5	73.9	Commercial bank (89.9)	Commercial bank (75.8)

Notes: 1. Agricultural credit refers to crop and investment credit taken for cultivation by a household.
2. Villages are arranged in descending order of column 3.
3. – Nil.
Source: PARI survey data.

and Siresandra was also evident from the prevalence of interlinked credit and labour transactions (Bhavani 2016).[28]

In most villages, the shares of formal sources in total debt outstanding were observed to be higher than their shares in borrowings (columns 4 and 5 in Table 3). This implies that formal sector contributed more to the stock than to the flow of credit in these villages.

Importance of formal sources in meeting agricultural credit needs

In every village, without exception, formal sources were more active than informal sources in meeting the agricultural credit needs of households.[29] Of the total amount borrowed for agriculture, more than 70 per cent was from formal sources (column 6 in Table 3). This finds resonance with the literature that discusses the focus of the formal sector on meeting production credit needs, and thereby creating space for the informal sector to grow to cater to consumption credit needs.

Commercial banks as key agencies of formal credit

In all except two villages, commercial banks (including RRBs) were the key agency of formal credit. They accounted for a major part of the fresh flow of formal credit, including the flow of formal credit to agriculture (columns 7 and 8 in Table 3). The two exceptions were Nimshirgaon and Warwat Khanderao in Maharashtra, where credit cooperatives were a more dominant source of formal credit in general, and formal credit for agriculture in particular (Box 1).

---

[28] Such transactions were primarily seen among manual labour households, but also in the case of a few small farmer households. These loans were taken for consumption purposes, including house construction, marriage and medical expenses. The borrower had to work either for free or at a lower wage rate than the prevalent village wage towards repayment of these loans. The lender was a landlord/rich capitalist farmer either from the village or a neighbouring village. An implicit rate of interest was tied to these loans (Bhavani 2016).

[29] The "purpose" of credit in the PARI surveys, as in all credit-related surveys, is the "stated purpose." It is difficult to arrive at the exact amount of credit that actually gets used for agriculture. This is because credit suffers from fungibility and hence, the "stated purpose" may be different from the "actual purpose" of credit. In other words, some part of credit raised for agriculture can be used for other purposes. Likewise, a part of the credit raised for other purposes can get diverted to agriculture.

Box 1: Credit Cooperatives in Maharashtra

Maharashtra ranks high in terms of the number of cooperatives and village coverage of cooperatives – both credit and non-credit (production and distribution) societies – among all Indian States (Chavan 2012). Studies have attributed the development of cooperatives in the State to its pioneering role in taking two major steps: adoption of the crop loans scheme and promotion of cooperative sugar factories (Dalaya and Sabnis 1973; Majumdar 1993).

In consecutive rounds of the AIDIS, Maharashtra has consistently remained in the top brackets when compared with other States in the share of formal credit, spearheaded by cooperative credit (RBI 1965; RBI 1977; RBI 1989b; NSSO 1998b; NSSO 2006a). Moreover, following banking sector liberalisation, when the share of formal sources in rural debt showed a fall in most States between the two AIDIS rounds of 1991 and 2002, Maharashtra reported an exceptional rise on account of an increase in the share of cooperative credit.

Among the villages discussed in this chapter, Nimshirgaon and Warwat Khanderao in Maharashtra stood out on account of a remarkably high share of formal credit in the borrowing profiles of village households in general, and small farmer households in particular. This was attributed mainly to cooperative credit. This commonality was despite the fact that Nimshirgaon belongs to Kolhapur district in western Maharashtra, which is historically known for a well-developed cooperative sector, while Warwat Khanderao is a dry, cotton-growing village from Buldhana district in eastern Maharashtra where cooperative development has received a setback since the early 1990s (Chavan 2016). The share of cooperatives in total borrowing by small farmer households in Nimshirgaon and Warwat Khanderao was 72 and 83 per cent, respectively.

There were in all 20 cooperatives in Nimshirgaon (including four urban cooperative banks, one DCCB, and seven non-agricultural societies/*pat sansthas*) and in Warwat Khanderao the total was nine (including one urban cooperative bank, one DCCB, and one *pat sanstha*).

Credit cooperatives in both villages could be classified into two broad categories based on the nature of their operations (Swaminathan 2012). In the first category, there were primary agricultural credit societies and DCCBs mainly catering to agricultural (crop) credit needs of the landed sections through Kisan Credit Cards (KCCs). Given their reliance on land as collateral, the agricultural (crop) credit needs of the landless sections, including tenant cultivators, could not be met by these societies. This distinction could be seen even among landed and landless small farmer households, as shown later in the chapter while discussing the issue of adequacy of crop credit.

As regards non-agricultural needs, including non-agricultural production (setting up of telephone booths and shops, and purchasing trucks), house construction/repair, and ceremonial expenses, village households relied mainly on the second category of cooperatives, i.e. urban cooperative banks and non-agricultural credit societies or *pat sansthas*. These societies relied on a host of securities including land, property, and fixed deposits. Given the focus on land by both types of cooperatives, their primary beneficiaries in both villages were landed farmer households. The landless sections, including manual workers, also borrowed from the cooperative societies, but loans taken by these households were fewer in number and were from cooperatives belonging to the second category. While the accessibility to cooperatives in the second category seemed better for the landless sections, the rates charged by them were perceptibly higher than the societies in the first category.

Key Features of the Borrowing Profiles of Small Farmer Households

Low incidence of formal sector borrowing

In each of the villages, at the most one-third of the small farmer households reported a fresh formal sector loan during the survey year, indicating very low incidence of formal credit (column 4 in Table 4).[30] The proportion was even lower for agriculture. At the most one-fourth of the small farmer households reported a fresh agricultural loan from the formal sector in all villages (column 6 in Table 4).

The incidence of formal sector borrowing among small farmer households was lower than, or at best comparable with, the average for all farmer households in each of the villages (columns 3 and 4 in Table 4). A similar trend was discernible when the incidence of formal sector borrowing for agricultural purposes of small farmer households was compared with all farmer households (columns 5 and 6 in Table 4).

More prominent presence of informal sources in meeting credit needs other than agriculture

The percentage shares of formal sources in fresh borrowing by small farmer households was less than or equal to 50 per cent, leaving the rest to be met by informal sources in all villages excepting Nimshirgaon, Harevli, Bukkacherla, Amarsinghi, and Warwat Khanderao (column 4 in Table 5).

Informal sources, thus, were a more dominant source of credit for small farmer households in most villages. Details about the various types of informal sources operational in these villages are discussed in Box 2.

---

[30] A similar point is also noted in the literature; see villages studies discussed in Dreze *et al.* (1998); Swaminathan (1991); Sarap (1990); Ramachandran (1990).

Table 4 *Incidence of borrowing, all farmer and small farmer households, 13 PARI villages*

Village	State	% of households reporting at least one fresh borrowing from formal sources		% of households reporting at least one fresh borrowing from formal sources for agriculture	
		All farmer households	Small farmer households	All farmer households	Small farmer households
(1)	(2)	(3)	(4)	(5)	(6)
Bukkacherla	Andhra Pradesh	42.9	33.3	28.6	27.8
Ananthavaram	Andhra Pradesh	32.0	27.1	30.7	25.0
Alabujanahalli	Karnataka	27.5	25.7	16.7	15.9
Warwat Khanderao	Maharashtra	27.9	25.5	24.6	22.6
Nimshirgaon	Maharashtra	35.4	24.2	27.1	15.2
Harevli	Uttar Pradesh	33.3	22.2	24.6	13.3
Kalmandasguri	West Bengal	22.1	22.1	19.1	19.1
Kothapalle	Telangana	21.9	21.4	12.5	10.7
Amarsinghi	West Bengal	14.8	14.8	13.0	13.0
Panahar	West Bengal	13.8	11.8	12.5	11.1
Siresandra	Karnataka	8.3	8.9	–	–
Mahatwar	Uttar Pradesh	9.5	7.2	9.5	7.2
Zhapur	Karnataka	12.5	3.8	2.5	–

Notes: 1. Agricultural credit refers to crop and investment credit taken for cultivation by a household.
2. Villages are arranged in descending order of column 4.
3. – Nil.
Source: PARI survey data.

Table 5 *Basic indicators of borrowing profiles of small farmer households, 13 PARI villages*

Village	State	% of formal sources in fresh borrowings by all farmer households	% of formal sources in fresh borrowings by small farmer households	% of formal sources in fresh borrowings for agriculture by all farmer households	% of formal sources in fresh borrowings for agriculture by small farmer households	% of formal sources in fresh borrowings for agriculture and allied activities by small farmer households
(1)	(2)	(3)	(4)	(5)	(6)	(7)
Nimshirgaon	Maharashtra	95.9	77.2	97.0	76.0	76.0
Harevli	Uttar Pradesh	77.1	57.8	98.4	96.3	96.3
Bukkacherla	Andhra Pradesh	64.8	55.6	82.5	84.3	85.1
Amarsinghi	West Bengal	50.3	50.3	77.2	77.2	75.2
Warwat Khanderao	Maharashtra	73.1	50.0	76.9	67.2	67.2
Alabujanahalli	Karnataka	64.2	46.8	91.0	87.4	87.4
Kalmandasguri	West Bengal	39.9	39.9	74.7	74.7	74.7
Panahar	West Bengal	22.5	38.9	48.4	47.9	47.9
Kothapalle	Telangana	84.6	36.7	89.0	53.4	43.0
Mahatwar	Uttar Pradesh	33.9	23.3	97.4	95.3	95.3
Ananthavaram	Andhra Pradesh	38.9	17.9	40.3	19.6	18.5
Siresandra	Karnataka	12.3	13.2	–	–	–
Zhapur	Karnataka	24.6	12.4	41.7	–	–

Notes: 1. Agricultural credit refers to crop and investment credit taken for cultivation by a household.
2. Villages are arranged in descending order of column 4.
3. – Nil.

Source: PARI survey data.

Box 2: System of Informal/Semi-formal Credit for
Small Farmer Households

In the case of informal sources of credit, as the name suggests, the terms and conditions of credit transactions lack any formalisation or standardisation. While there may be certain State-specific laws regulating usury and bonded labour, most informal sources, in practice, are not bound by any legislative control (Swaminathan 1986). Similarly, they are outside the purview of monetary or prudential regulations, such as rules governing interest rates. This gives them complete freedom in negotiating the terms of credit. Hence, every informal credit transaction turns out to be unique in nature (*ibid.*). It is influenced by a variety of factors, such as "prior relation" between the lender and the borrower, and asset-holding, caste, and gender of the borrower (Nagaraj 1981; Harriss-White and Colatei 2004). This makes the credit relation invariably personalised and, in many cases, power-based (Bhaduri 1982). Thus the persistence of an informal sector is considered to be a feature of backwardness of a credit system. And hence, the objective of rural credit policy in India since Independence has been to countervail this sector (RBI 1956).

In the PARI study villages, informal sources operated in a number of guises, including landlords, rich/medium peasants, professional moneylenders, input dealers, and friends/relatives. While the presence of friends/relatives as an interest-free informal source of credit for small farmer households could be seen in every village, there was a distinct State-wise pattern that emerged with regard to the nature of other informal sources. First, the control of landlords and rich/middle peasants was prominent in villages of Andhra Pradesh, Telangana, Karnataka, and Uttar Pradesh: Ananthavaram and Bukkacherla in Andhra Pradesh; Kothapalle in Telengana; Alabujanahalli, Siresandra, and Zhapur in Karnataka; and Harevli in Uttar Pradesh. Secondly, contrary to general belief, the presence of professional moneylenders was less widespread and they were seen mostly in the States of Andhra Pradesh and Karnataka (in Siresandra village, which had a higher share of 20 per cent in total loans of small farmer households). Thirdly, input dealers/traders were an important informal source for small farmer households in all three villages from West Bengal (their shares in total loans ranging between 48 and 76 per cent). Fourthly, in Nimshirgaon, Maharashtra, with the significant spread of formal sources, informal sources operated primarily in the form of friends/relatives, and there was very little presence of other types of informal sources as seen elsewhere. In Warwat Khanderao (Maharashtra) too, small farmer households mainly borrowed from friends/relatives, but some loans were also reported from input dealers (both friends/relatives and input dealers having shares of 22 per cent each in total loans).

The terms and conditions of most informal sources except friends/relatives were onerous. The rate of interest charged by professional moneylenders went up to 120 per cent per year in some villages. There was one case of a small farmer household in Alabujanahalli (Karnataka) which had borrowed from landlords/rich peasants and committed to work for them at lower than the market rate.[31] This suggested the presence of interlinked credit transactions among small farmer households.[32] Similarly, interlinkages with the product market could also be seen in the West Bengal villages, where input dealers sold inputs at higher than market prices to those small farmer households who bought them on credit. These small farmer households also had to sell the produce to the same input dealer immediately after the harvest. Although the prices were not fixed in advance, small farmer households generally received prices lower than the market prices (Bhavani 2015).

In some of the villages, micro-finance institutions (MFIs) and self-help groups (SHGs) were also a part of the credit system. We define the micro-finance sector as an informal or, at best, a semi-formal sector. This is because, pending the passage of the Micro Finance Development and Regulation Bill, the vast micro-finance sector comprising non-banking finance company (NBFC)–MFIs, not-for-profit companies (Section 25 companies), not-for-profit non-governmental organisations (NGOs)/charitable or investment trusts, and SHGs, is not regulated by any uniform legislation. Moreover, except for NBFC–MFIs, comprising large-sized MFIs which are registered with and regulated by the RBI, the financials of other institutions are not regulated or supervised regularly by a public authority.

Following the introduction of micro-finance in India in the early 1990s, there emerged a concentration of MFIs in the southern region of the country, particularly in the States of Andhra Pradesh and Karnataka. Accordingly, we observed that MFIs and SHGs were an important source of credit for small farmer households in all three study villages from Karnataka and in Kothapalle from Telangana (earlier Andhra Pradesh), with the share ranging between 18 per cent and 40 per cent of total loans taken by small farmer households. The MFIs operational in these villages were SKS Micro Finance, Shri Chaitanya Finance, and Spandana Spoorthy Financial Limited. We also observed that an MFI (Bandhan Micro Finance) was a popular source of credit among small farmer households in West Bengal villages, with the share in total loans ranging from 31 per cent in Kalmandasguri to 13 per cent in Panahar (Bhavani 2015).

---

[31] The market wage rate was reported to be Rs 100 per day, while they received Rs 80 per day with the rest deducted towards interest on the loan.

[32] Such interlinked transactions were more commonly observed among manual worker households in the village.

A closer look at the operations of MFIs brought out their aggressive marketing strategies, reliance on a wide network of agents, and carefully designed loan and insurance products, which set them apart from other formal and informal sources of credit.[33] First, the agents of these companies played a vital role in making assessments of the payment capacity and histories of the borrowers. MFIs also employed collection agents who visited the villages for collection of repayments. In case a borrower failed to repay, other borrowers from the village were required to mobilise money for repayment. There were cases where borrowers had to sell assets, including land, to make their weekly payments. Even if a borrower fled the village, the MFI was able to trace him through its network of branches and collection agents, and ensure recovery. Secondly, interest rates ranged between 27 per cent and 56 per cent per annum. Moreover, the repayment schedule was designed in a manner such that the real interest rate worked out to be much higher than when the entire loan was repaid at the end of the term of the loan. Interestingly, many borrowers were not even aware of the actual rates and often perceived the rates to be very low, underlining the role played by agents in marketing the products. Loans were repaid through equalised weekly instalments. Thirdly, there was a debt trap that the MFI engineered in such a manner that once a loan was repaid, the borrower had to take another loan immediately; failing to do so could result in cancellation of the membership of the borrower. Finally, through compulsory savings schemes in which the borrower was forced to invest, MFIs cushioned themselves against defaults.

Importantly, a comparison of the shares of formal sources in total borrowings of small farmer households and all households in a village suggested that the extent of development of a village credit system was neither a necessary nor a sufficient condition for the formal sector to be more involved in meeting the credit needs of small farmer households (read Table 3 with Table 5).

When compared with all farmer households too, the shares of formal sources turned out to be lower for small farmer households (columns 3 and 4 in Table 5). There were three exceptions to this observation: the villages of Amarsinghi, Kalmandasguri, and Panahar – all small farmer-dominated villages in West Bengal (column 3 in Table 8).

Formal sources were more important for meeting the agricultural credit needs of small farmer households. In nearly all the study villages except Panahar (Bankura district, West Bengal) and Ananthavaram (Guntur district, Andhra Pradesh), formal sources met more than half of the agricultural credit needs

---

[33] These observations primarily relate to MFIs in West Bengal, as discussed in Ramachandran (2013). However, we have also tried to include certain features of MFIs working in Andhra Pradesh and Karnataka in our discussion.

of these households (column 6 in Table 5). In the case of Ananthavaram, the share of formal sources in total borrowings for agriculture was only about 20 per cent, mainly because of the large presence of tenant cultivators in the village (Ramachandran *et al.* 2010). Small farmer households in Siresandra (Kolar district, Karnataka) and Zhapur (Kalaburagi district, Karnataka) reported no fresh borrowing for agriculture during the survey years. In Siresandra, apart from agriculture, small farmer households were also involved in sericulture. However, even for sericulture, no small farmer household in the village reported any fresh formal sector loan. In other villages too, formal sector credit for allied activities, including sericulture, dairying, poultry, and fishery, was almost close to zero; this can be discerned from the fact that there is hardly any difference between columns 6 and 7 of Table 5 for most villages. Only in Amarsinghi (Malda district, West Bengal), Kothapalle (Karimnagar district, Telangana), and Ananthavaram small farmer households reported some borrowing for dairying (purchase of cows and buffaloes), but this borrowing was entirely from informal sources.

The shares of formal sources in total borrowing for agricultural purposes by small farmer households were either lower than or comparable with those for all farmer households (columns 5 and 6 in Table 5). Evidently, small farmer households relied more on informal sources for meeting their agricultural credit needs than other farmer households. However, the gap between columns 3 and 4 was generally wider than that between columns 5 and 6 across the villages. This implied that the reliance of small farmer households on informal sources was greater in the case of consumption credit needs.

Higher interest cost of borrowing
The weighted average interest rate for small farmer households was higher than for all farmer households except in the three West Bengal villages and Siresandra in Karnataka, where small farmer households made up nearly the whole of farmer households (columns 3 and 4 in Table 6). The weighted average was worked out by weighing the interest rate on every fresh loan contracted during the survey year by the share of its amount in total amount borrowed irrespective of the source.[34] Similarly, in the case of borrowing for agriculture too, the weighted average interest for small farmer households was generally higher than for all farmer households (columns 3 and 4 in Table 7). A higher interest cost for small farmer households reflected the dominance of informal sources in their borrowing profile as compared to other categories of farmer households.

---

[34] As the rates are nominal and pertain to different survey years, we refrain from a direct comparison of the interest rates across villages here.

Table 6 *Weighted average interest rates on borrowings by all farmer and small farmer households, 13 PARI villages*

| Village | State | Weighted average interest rate on | |
		Fresh borrowings by all farmer households	Fresh borrowings by small farmer households
(1)	*(2)*	*(3)*	*(4)*
Amarsinghi	West Bengal	5.6	5.6
Panahar	West Bengal	13.2	6.5
Kalmandasguri	West Bengal	10.0	10.0
Mahatwar	Uttar Pradesh	10.1	10.9
Nimshirgaon	Maharashtra	11.5	13.8
Warwat Khanderao	Maharashtra	12.9	14.2
Harevli	Uttar Pradesh	14.0	15.9
Bukkacherla	Andhra Pradesh	16.5	17.7
Ananthavaram	Andhra Pradesh	17.4	19.0
Kothapalle	Telangana	13.8	19.4
Alabujanahalli	Karnataka	19.2	21.4
Siresandra	Karnataka	25.4	24.6
Zhapur	Karnataka	23.9	26.8

Note: Villages are arranged in ascending order of column 4.
Source: PARI survey data.

The interest rates in the formal sector, even if deregulated, are governed by set standards, as against completely non-transparent and arbitrary ways of setting rates in the informal sector. As a result, borrowing from the formal sector is expected to be cheaper than from the informal sector. Accordingly, the weighted average rates on formal sector borrowings for agriculture were also found to be significantly lower than the rates on total borrowings for agriculture by small farmer households (columns 4 and 5 in Table 7).

Access to Formal Credit for Small Farmer Households

The access to crop credit from formal sources for small farmer households was modelled using a logistic model. The binary dependent variable used in the model is given below:

Crop Credit Access$_i$ = {1 if a small farmer household reported at least one crop loan from a formal source (commercial bank/credit cooperative/any other formal source) during the survey year; {0 otherwise.

The results from the baseline model (model summarised in Appendix 1 and results in Appendix 2) indicated a positive and statistically significant role of

Table 7 *Weighted average interest rates on borrowings by all farmer and small farmer households for agriculture, 13 PARI villages*

Village	State	Weighted average interest rate for		
		All farmer households on fresh borrowings for agriculture	Small farmer households on	
			Fresh borrowings for agriculture	Fresh borrowings for agriculture from formal sources
(1)	*(2)*	*(3)*	*(4)*	*(5)*
Panahar	West Bengal	38.7	7.9	7.0
Kalmandasguri	West Bengal	7.9	7.9	7.0
Amarsinghi	West Bengal	8.0	8.0	7.0
Alabujanahalli	Andhra Pradesh	8.8	8.7	6.2
Harevli	Uttar Pradesh	12.9	13.2	13.3
Bukkacherla	Andhra Pradesh	14.2	14.2	12.3
Mahatwar	Uttar Pradesh	12.0	14.2	12.5
Nimshirgaon	Maharashtra	11.4	14.5	11.4
Kothapalle	Telangana	13.2	17.1	12.0
Warwat Khanderao	Maharashtra	15.8	17.6	12.4
Ananthavaram	Andhra Pradesh	17.5	19.2	12.0
Zhapur	Karnataka	25.9	–	–
Siresandra	Karnataka	6.9	–	–

Notes: 1. Agricultural credit refers to crop and investment credit taken for cultivation by a household.
2. Villages are arranged in ascending order of column 4.
Source: PARI survey data.

land operated in determining the access to formal credit.[35] This underlined the point about collateral and a possible risk aversion on the part of formal sources while extending credit to small farmer households. Similarly, the social group status of a small farmer household also significantly determined the access to formal credit; if a small farmer household belonged to a backward social group (Scheduled Caste/Scheduled Tribe/Muslim), it was less likely to gain access to formal credit, *ceteris paribus*. Apart from a social bias, the issue of social group also relates to the lack of access to land as collateral for backward social groups. It thus brings to the fore the debilitating role of caste in economic development.

[35] There have been very few attempts to model the access to formal credit for small farmers in India. Among these, the study by Sarap (1990) using data on small farmers (defined as households operating less than or equal to 5 acres) from rural Odisha, similar to our study, found land operated to be the most significant determinant of access to formal credit. He also found that caste had a negative but not significant impact on access.

The village-specific number of small farmer households was significant in determining the access to crop credit; the higher the number of small farmer households in a village, the higher was the average access to crop credit for these households. This suggested the possibility of small farmer households organising themselves into cooperatives/joint liability groups in order to gain greater access to credit, if they were in large numbers in a given village.

The baseline model was augmented to test different hypotheses:

(1) We tested for the impact of tenancy status of small farmer households on access (column 2 of Appendix 2). It showed that if a small farmer household operated self-owned land, the probability of gaining access to crop credit was higher than when it was a partly or fully tenant cultivator household.

(2) We then tested the hypothesis that greater crop diversification by a small farmer household through better irrigation and commercial cropping augured well for the access to crop credit. We observed that the controls for crop diversification, although positive, were not significant in determining access (column 3 in Appendix 2).

(3) The hypothesis of diversification outside agriculture in the form of allied activities and non-agricultural employment facilitating access to crop credit for small farmer households was also tested. This was because employment outside agriculture, apart from offering newer avenues of income, could also open up newer formal credit facilities to these households. While non-agricultural employment had a positive impact on access, the presence of allied activities diminished the access to crop credit (column 4 in Appendix 2). While crop credit is short-term in nature, credit to allied activities figures under long-term agricultural credit. Hence, this negative association could purely be a result of the accounting of agricultural credit. However, the impact of both these variables on access was not statistically significant.

(4) Finally, we tested the hypothesis of alternative/informal sources of agricultural credit bringing down the access to crop credit from formal sources. We observed that the incidence of informal credit for agriculture brought down the access to formal credit, suggesting a substitution effect, but the impact was not found to be statistically significant (column 5 in Appendix 2). In running each of these models, our baseline results held true and each of the models passed the goodness of fit test.

Some Pointers about the Adequacy of Formal Credit for Small Farmer Households

The question of adequacy of formal credit for small farmer households was addressed using two basic indicators. First, following from the literature (Randhawa and Sundaram 1990), the share of small farmer households in total borrowings from the formal sector in a village was compared with their share in total land operated (columns 4 and 5 in Table 8). We also compared their share in total crop credit from the formal sector with their share in total land operated (columns 4 and 6 in Table 8). The shares in formal credit in general, and crop credit in particular, for small farmer households were lower than their share in land operated. The notable exceptions to this observation were Amarsinghi, Panahar, and Siresandra, villages that had a high concentration of small farmer households among farmers (column 3 in Table 8).

Secondly, using the scale of finance as a rough benchmark for the amount of crop credit a household was entitled to, the crop credit estimate for a small farmer household was worked out taking its crop-mix (excluding intercrops) in the survey year. This estimate was then compared with the actual amount of crop credit received by/sanctioned to the small farmer household, to work out a ratio of actual crop credit to the crop credit estimate (column 6 in Table 9). For this comparison, small farmer households in a given village were arranged in an ascending order of the size of land operated, and the smallest two fully owner-cultivator households were selected as case studies.

The comparison of actual crop credit with crop credit estimate brings out a wide variation across villages irrespective of the crop-mix. However, the broad observation was that small farmer households generally received an amount that was lower than what they were entitled to, based on crop-mix and area operated. The divergence in these two variables could have been on account of many factors, including the lack of willingness of the formal agency to sanction the amount that was due owing to the past credit history of a small farmer household and the lack of adequate collateral in the form of land, among others.

We also worked out the ratio of actual crop credit to crop credit estimate for the smallest partly tenant and fully tenant small farmer household (column 6 in Table 10). It showed that the ratio was of course lower than 100 per cent, but it was also significantly lower than that of fully owner cultivators as shown in Table 9. Moreover, for fully tenant cultivator households, the ratio was zero in all villages. This implied that notwithstanding the focus of the crop credit scheme on crops cultivated and not land owned, landless tenant cultivators did not receive any crop credit from formal agencies.

---

[36] The literature uses the share of small farmers in total land operated as an indicator of the adequacy of agricultural credit, particularly crop credit, to this segment (Randhawa and Sundaram 1990). This is because given the low capacity of self-financing among small farmers, their credit requirements for seasonal agricultural operations are expected to be comparable with the land operated by them (*ibid.*).

Our crop credit estimate was based only on the prevailing scale of finance and area under cultivation of a given crop. However, as per the crop credit scheme, formal agencies are expected to also add a component for post-harvest/consumption requirements, maintenance expense on assets, and small investment credit requirements while fixing the crop credit limit over and above the scale of finance and area under cultivation.[37] Hence, in all likelihood, the ratio that we obtained might have been an overestimate, and the gap between the actual crop credit and crop credit that was due could have been even wider.

Whether or not the crop credit received is adequate to meet the credit needs of small farmer households is a much broader issue. It may require more granular data to ascertain whether or not the scales of finance fixed by formal agencies are adequate to cover the going costs of cultivation, an exercise that cannot be attempted here.[38] Further, as we did not attempt a comparison for large farmer households, it was also not possible to judge whether small farmer households were being treated more unfairly than large farmer households in the provision of crop credit. However, given that small farmer households are an integral part of India's agriculture, the inadequacy of the crop credit scheme in meeting their credit requirements suggests the need for a relook at the scheme.

Conclusions and Policy Implications

This chapter attempts to analyse the formal credit system in rural India from the point of view of small farmers and their credit needs. Small farmers have been at the centre of the design of credit policy even before Independence, when credit cooperatives, institutions aimed at catering to the production credit needs of small farmers, were formed. After bank nationalisation, when commercial banks were roped into the provision of rural credit, small farmers were identified as a sector of priority for these institutions as well. A number of supporting policy measures were undertaken during this phase, including the creation of RRBs and SFDA, to ensure enhanced flow of formal credit to small farmers for agriculture at regulated rates of interest.

The initiation of banking sector liberalisation in the early 1990s altered the social orientation of the banking system, and, as a result, a number of definitional changes were brought about in the priority sectors identified

[37] See "Revised Scheme for Issue of Kisan Credit Card (KCC) – May 2012," www.rbi.org.in.
[38] The existing field-level evidence on this issue suggests that scales of finance are themselves underestimates and need to be raised to adequately cover the credit needs of farmer households; see Samantara (2010).

Table 8 *Basic indicators of small farmer households, 13 PARI villages*

Village	State	Small farmer households as % of total number of farmer households	% of land operated by small farmer households	% of formal sector borrowings by small farmer households	% of formal sector borrowings for crop cultivation by small farmer households
(1)	(2)	(3)	(4)	(5)	(6)
Amarsinghi	West Bengal	100.0	87.1	95.7	100.0
Kalmandasguri	West Bengal	100.0	79.0	78.3	70.3
Panahar	West Bengal	94.7	59.9	45.7	65.5
Siresandra	Karnataka	93.3	69.5	91.7	–
Mahatwar	Uttar Pradesh	93.2	48.3	17.0	45.8
Kothapalle	Telangana	87.5	51.3	7.7	9.8
Warwat Khanderao	Maharashtra	86.9	47.2	32.2	57.3*
Alabujanahalli	Karnataka	81.9	48.8	25.7	46.1
Nimshirgaon	Maharashtra	68.8	27.5	10.7	8.9
Harevli	Uttar Pradesh	65.2	22.4	21.3	15.2
Zhapur	Karnataka	65.0	25.3	9.0	–
Bukkacherla	Andhra Pradesh	64.3	40.2	31.2	30.1
Ananthavaram	Andhra Pradesh	64.0	20.9	12.2	14.8

Notes: 1. – no crop credit by small farmer households during the survey year.
2. Villages are arranged in descending order of column 3.
3. * This share is high on account of one small farmer household which had reported three formal sector loans for agriculture. After excluding this outlier, the share was 43.3 per cent.
Source: PARI survey data.

Table 9 *Comparison of crop credit estimate with total crop credit received by small farmer households, 11 PARI villages, two case studies of smallest owner cultivators (in terms of size of land operated)*

No.	Village	State	Crop credit estimate based on scale of finance (Rs)	Total crop credit received during the survey year (Rs)	Ratio (%)	Crop-mix in the survey year
(1)	(2)	(3)	(4)	(5)	(6=5/4)	(7)
1	Alabujanahalli	Karnataka	32,450	15,000	46.2	Sugarcane, mulberry, paddy
2	Alabujanahalli	Karnataka	48,000	20,000	41.7	Paddy, sugarcane, mulberry
3	Ananthavaram	Andhra Pradesh	9,500	9,000	94.7	Paddy, black gram
4	Ananthavaram	Andhra Pradesh	12,700	8,000	63.0	Paddy, maize
5	Bukkacherla	Andhra Pradesh	26,250	30,000	114*	Groundnut, red gram
6	Bukkacherla	Andhra Pradesh	15,000	12,000	80.0	Groundnut, red gram, green gram, cow pea
7	Kothapalle	Telangana	20,575	6,000	29.2	Paddy, maize
8	Kothapalle	Telangana	39,200	20,000	51.0	Cucumber, groundnut, paddy, maize
9	Nimshirgaon	Maharashtra	32,625	20,000	61.3	Sugarcane, coriander, cluster beans
10	Nimshirgaon	Maharashtra	12,970	10,000	77.1	Soybean, jowar
11	Warwat Khanderao	Maharashtra	23,300	10,000	42.9	Cotton, sorghum
12	Warwat Khanderao	Maharashtra	4,550	3,500	76.9	Cotton
13	Panahar	West Bengal	104,045	40,000	38.4	Paddy, potato, vegetables
14	Panahar	West Bengal	9,765	6,000	61.4	Paddy, potato, sesame

15	Amarsinghi	West Bengal	26,532	10,000	37.7	Paddy, mustard/rapeseed, paddy
16	Amarsinghi	West Bengal	46,347	10,000	21.6	Paddy, mustard, potato, jute
17	Kalmandasguri	West Bengal	41,827	15,000	35.9	Paddy, potato
18	Kalmandasguri	West Bengal	10,000	10,000	100.0	Paddy, potato
19	Harevli	Uttar Pradesh	56,430	20,000	35.4	Sugarcane, wheat
20	Harevli	Uttar Pradesh	72,270	11,000	15.2	Sugarcane, wheat, sorghum
21	Maharwar	Uttar Pradesh	26,000	25,000	96.2	Paddy, pulses, wheat
22	Maharwar	Uttar Pradesh	37,198	4,500	12.1	Paddy, sugarcane, maize, pulses

Note: * The household reported two crop loans during the survey year.
Source: PARI survey data.

Table 10 *Comparison of crop credit estimate with total crop credit received by small farmer households, 11 PARI villages, case study of the smallest partly and fully tenant cultivator (in terms of size of land operated)*

No.	Village	State	Crop credit estimate based on scale of finance (Rs)	Total crop credit during the survey year (Rs)	Ratio (%)	Crop-mix
(1)	(2)	(3)	(4)	(5)	(6 = 5/4)	(7)
			Partly tenant cultivator households			
1	Alabujanahalli	Karnataka	51,500	10,000	19.4	Paddy, sugarcane
2	Amarsinghi	West Bengal	31,870	18,000	56.5	Paddy, potato, jute
3	Ananthavaram	Andhra Pradesh	59,530	7,000	11.8	Paddy, maize
4	Bukkacherla	Andhra Pradesh	82,357	10,000	12.1	Groundnut, paddy
5	Harevli	Uttar Pradesh	84,843	6,500	7.7	Sugarcane, wheat, paddy, lentil
6	Kalmandasguri	West Bengal	87,381	20,000	22.9	Paddy, potato, jute
7	Kothapalle	Telangana	29,000	0	0	Paddy
8	Mahatwar	Uttar Pradesh	30,800	0	0	Sugarcane, paddy
9	Nimshirgaon	Maharashtra	9,000	0	0	Sugarcane, soybean
10	Panahar	West Bengal	17,950	3,000	16.7	Paddy, potato
11	Warwat Khanderao	Maharashtra	42,400	16,000	37.7	Cotton, green gram, sorghum

Fully tenant cultivator households

1	Alabujanahalli	Andhra Pradesh	32,000	0	Paddy, sugarcane, paddy
2	Amarsinghi	West Bengal	8,000	0	Paddy
3	Ananthavaram	Andhra Pradesh	26,000	0	Paddy, maize
4	Bukkacherla	Andhra Pradesh	38,000	0	Paddy, groundnut
5	Harevli	Uttar Pradesh	6,700	0	Paddy
6	Kalmandasguri	West Bengal	24,400	0	Paddy, potato, mustard
7	Kothapalle	Telangana	8,500	0	Paddy
8	Mahatwar	Uttar Pradesh	23,200	0	Paddy, wheat
9	Nimshirgaon	Maharashtra	7,000	0	Tomato
10	Panahar	West Bengal	12,300	0	Potato, paddy

Note: There was no fully tenant cultivator household in Warwat Khanderao.
Source: PARI survey data.

for banks. Small farmers continued to be a part of agriculture under priority sectors. However, changes in the definition of agriculture, with a thrust on commercial and capital-intensive agricultural activities, seems to have weakened the focus on small farmers. Also, a decline in the rural branch network and fall in the growth of agricultural credit during this phase raised questions about the actual flow of bank credit to small farmers.

The policy of financial inclusion pursued since 2005, along with the comprehensive credit policy for agriculture, arrested the decline in the rural branch network and revived the flow of formal credit to agriculture. Moreover, credit to small farmers as part of priority sector lending was formalised by setting a separate sub-target for this group of cultivators in 2015.

Whether these changes in rural credit policy also meant a change in the flow of formal credit to small farmers on the ground requires to be established through credible time-series data, which are not available from secondary sources. The available banking data indicate that the growth of credit to small farmers and their share in total agricultural credit were on a decline since the early 1990s. The decline was arrested after 2005 and a rise in these variables could be seen after 2010.

Notwithstanding these favourable changes over the second half of the 2000s, the PARI village data collected between 2005 and 2010 suggest a general inadequacy of, and limited access to, formal credit for small farmer households. In none of the villages surveyed did more than one-third of small farmer households report fresh borrowing from formal sources during the survey year. The weak access to formal credit for small farmer households was also reflected in the relatively high interest costs for these households.

We drew data from a diverse mix of 13 villages belonging to five States – Andhra Pradesh, Karnataka, Maharashtra, Uttar Pradesh, and West Bengal – which showed a wide variation in the overall development of the formal credit system. However, except for Nimshirgaon and Warwat Khanderao in Maharashtra, in no other village did we find a direct correlation between the overall development of formal sources in a village and the extent of involvement of these sources in providing credit to small farmer households. In Maharashtra, the fairly extensive reach of credit cooperatives – both agricultural and non-agricultural societies – ensured better access to formal credit for small farmer households. Hence, the overall development seemed neither a necessary nor a sufficient condition to ensure the provision of formal credit to small farmer households in a village. The most striking case in point was in the villages of West Bengal, which had a relatively underdeveloped formal credit system. However, the share of formal sources for small farmer households was higher than that for all farmer households, suggesting better availability of formal credit for small farmer households.

The key to access to formal credit for small farmer households in all the surveyed villages was the availability of owned land and the social group status of the household. The larger the extent of land, the higher was the probability of gaining access to formal credit for a small farmer household, *ceteris paribus*. Juxtaposing this finding with the observation made in the foregoing paragraph, it could be said that land reforms and the availability of land as collateral were possibly factors that made a difference to the access of formal credit for small farmer households in West Bengal.

Across villages, we observed that formal sources were mainly involved in lending for agriculture, subject to, of course, an element of fungibility. Their involvement in lending for the consumption-related needs of small farmer households was limited. Notwithstanding the involvement of small farmer households in allied activities, including dairying and sericulture, there was hardly any formal sector credit reported even for these activities. This implied that apart from the availability of collateral, the purpose of seeking credit was also responsible for limited access to formal credit by small farmer households, and a consequent perpetuation of informal sources in village credit systems. Not only was access to formal credit limited for small farmer households, but the extent of crop credit was also inadequate when compared with the prevailing scales of finance.

This chapter, thus, suggests the need to expand the reach of formal credit to small farmer households. Redistribution of land through land reforms is a means of ensuring collateral for formal credit to small farmer households, apart from the other economic and social benefits entailed in these reforms. However, there is also a need to reform the formal credit system in general, and crop credit system in particular, to broaden their reach and adequacy. Apart from reducing the exclusive thrust on land as collateral and widening the reach of formal credit to meet the consumption-related needs of small farmer households, there is also a need to ensure that crop credit reaching small farmer households is in line with the existing scales of finance.

The authors thank Madhura Swaminathan, Judith Heyer, Aparajita Bakshi, Shamsher Singh, and Arindam Das for useful comments on an earlier draft. The views expressed in this chapter are the personal views of the authors and not of the organisations to which they are affiliated.

REFERENCES

Berdegue, J. A., and Fuentealba, R. (2011), "Latin America: The State of Smallholders in Agriculture. IFAD Conference on New Directions for Smallholder Agriculture," Rome, available at http://www. fad.org/events/agriculture/doc/papers/Berdegue.pdf, viewed on 14 August 2017.

Bhavani, R.V. (2015), "Indebtedness in the West Bengal Study Villages," presentation at the Symposium on Results from Village Surveys, Durgapur.

Bhavani, R. V. (2016), "Indebtedness of Households," in Madhura Swaminathan and Arindam Das (eds.), *Socio-economic Surveys of Three Villages in Karnataka: A Study of Agrarian Relations,* Tulika Books, New Delhi.

Bose, Sukanya (2005), "Regional Rural Banks: The Past and the Present Debate," available at www.macroscan.com, viewed on 14 August 2017.

Chavan, P. (2005), "Banking Sector Reforms and Growth and Distribution of Rural Banking in India," in V. K. Ramachandran and Madhura Swaminathan (eds.), *Financial Liberalisation and Rural Credit,* Tulika Books, New Delhi.

Chavan, P. (2013), "Credit and Capital Formation in Agriculture: A Growing Disconnect," *Social Scientist,* September–November.

Chavan, P. (forthcoming 2017), "Public Banks and Financial Intermediation in India: The Phases of Nationalisation, Liberalisation, and Inclusion," in Christoph Scherrer (ed.), *Public Banks in the Age of Financialisation: A Comparative Perspective,* Edward Elgar Publishing Limited, Gloucestershire.

Clarke, Daniel, and Dercon, Stefan (2009), "Insurance, Credit and Safety Nets for the Poor in a World of Risk," DESA Working Paper No. 81 ST/ESA/2009/DWP/81, October.

Dalaya, Chandra, and Sabnis, Ravindra (1973), *Cooperation in Maharashtra, Review and Perspectives,* Centre for the Study of Social Change, Bombay.

Datta, S., and Sriram, M. S. (2010), *Towards a Perspective on Flow of Credit to Small and Marginal Farmers in India,* Publication No. 240, Centre for Management in Agriculture, Ahmedabad.

Desai, B. M. (1979), "Interventions for Rural Development: Experiences of the SFDA," Indian Institute of Management (IIM), Ahmedabad.

Dreze, Jean, Lanjouw, Peter, and Sharma, Naresh (1998), "Credit," in Peter Lanjouw and Nicholas Stern (eds.), *Economic Development in Palanpur over Five Decades,* Oxford University Press, Oxford.

Food and Agriculture Organisation (FAO) of the United Nations (2013), *Investing in Smallholder Agriculture for Food Security: A Report by the High Level Panel of Experts on Food Security and Nutrition of the Committee on World Food Security,* Rome, available at http://www.deza.admin.ch/ressources/resource_en_225682.pdf, viewed on 14 August 2017.

Gokhale Institute of Politics and Economics (GIPE) (1975), *Writings and Speeches of Prof. D. R. Gadgil on Cooperation,* Orient Longman, New Delhi.

Goyal, S. K. (1967), "Banking Institutions and Indian Economy," ISID Working Paper 1, available at http://isidev.nic.in/pdf/banking.PDF, viewed on 14 August 2017.

Hoda, Anwarul, and Terway, Prerna (2015), "Credit Policy for Agriculture in India: An Evaluation," Working Paper No. 302, Indian Council for Research in International Economic Relations (ICRIER) .

Hulme, D., and Mosely, P. (1996), *Finance against Poverty*, Vols. 1 and 2, Routledge, London.

Kloeppinger-Todd, R., and Sharma, M. (2010), "Innovations in Rural and Agricultural Finance," *Focus 18*, International Food Policy Research Institute (IFPRI), July.

Maheshwari, S. (1995), *Rural Development in India: A Public Policy Approach*, Sage Publications, New Delhi.

Majumdar, N. A. (1993), "Cooperative Banking in Maharashtra," in Malati Anagol and B.D. Ghonasgi (eds.), *Banking in Maharashtra: A Regional Profile*, Himalaya Publishing House, Mumbai.

Montgomery, R., Bhattacharya, D., and Hulme, D. (1996), "Credit for the Poor in Bangladesh: The BRAC Rural Development Programme and the Government Thana Resource Development and Employment Programme," in D. Hulme and P. Mosely (eds.), *Finance against Poverty*, Vols. 1 and 2, Routledge, London.

Morduch, J. (1998), "Does Microfinance Really Help the Poor? New Evidence from Flagship Programs in Bangladesh," Working Paper, MacArthur Foundation Project on Inequality, Princeton University.

Narain, S. (2010), "Gender and Access to Finance," Analytical Paper, World Bank, Washington D. C.

Narayana, D. (2000), "Banking Sector Reforms and the Emerging Inequalities in Commercial Credit Deployment in India," Working Paper No. 300, Centre for Development Studies, Thiruvananthapuram.

National Bank for Agriculture and Rural Development (NABARD) (2014), *JLGs: Sharing Liabilities and Creating Assets*, NABARD, Mumbai.

National Sample Survey Organisation (NSSO) (2005), *Household Indebtedness in India as on 30–06–2002: Report No. 501*, NSSO, New Delhi.

Nicholson, Frederick (1960), *Report Regarding the Possibility of Introducing Land and Agricultural Banks into Madras Presidency, 1895: Vol. 1*, reprint, Reserve Bank of India, Bombay.

Raj, Krishna (1979), "Small Farmer Development," *Economic and Political Weekly*, vol. 14, no. 21.

Ramachandran, V. K. (2013), "Agrarian Relation and Village Studies," Radha Kamal Mukherjee Memorial Lecture, Indian Society of Labour Economics, New Delhi.

Ramachandran, V. K., and Rawal, Vikas (2010), "The Impact of Globalisation and Liberalisation on India's Agrarian Economy," *Global Labour Journal*, vol. 1, no. 1.

Ramachandran, V. K., Rawal, Vikas, and Swaminathan, Madhura (2010), *Socio-economic Surveys of Three Villages in Andhra Pradesh*, Tulika Books, New Delhi.

Ramakumar, R., and Chavan, P. (2014), "Agricultural Credit in the 2000s: Dissecting the Revival," *Review of Agrarian Studies*, vol. 4, no. 1, February–June, available at http://ras.org.in/bank_credit_to_agriculture_in_india_in_the_2000s, viewed on 8 September 2017.

Ramakumar, R., and Chavan, Pallavi (2007), "Revival in Agricultural Credit in the 2000s: An Explanation," *Economic and Political Weekly*, vol. 42, no. 52, 29 December–4 January.

Reserve Bank of India (RBI) (1954), *All India Rural Credit Survey: The General Report – Vol. 2*, RBI, Bombay.

Reserve Bank of India (RBI) (1991), *Report of the Committee on the Financial System*, RBI, Mumbai.

Reserve Bank of India (RBI) (2004), *Report of the Advisory Committee on Flow of Credit to Agriculture and Related Activities from the Banking System* (Chairman: V. S. Vyas), RBI, Mumbai.

Reserve Bank of India (RBI) (2008), *Report on Currency and Finance: 2006–08*, RBI, Mumbai.

Reserve Bank of India (RBI) (2015), *Report of the Internal Working Group to Revisit the Existing Priority Sector Lending Guidelines*, RBI, Mumbai.

Satish, P. (2007), "Agricultural Credit in the Post-Reform Era," *Economic and Political Weekly*, vol. 42, no. 26, 30 June.

Shetty, S. L. (2005), "Regional, Sectoral and Functional Distribution of Bank Credit," in V. K. Ramachandran and Madhura Swaminathan (eds.), *Financial Liberalisation and Rural Credit*, Tulika Books, New Delhi.

Swaminathan, Madhura (1991), "Segmentation, Collateral Undervaluation, and the Rate of Interest in Agrarian Credit," *Cambridge Journal of Economics*, vol. 15, no. 2.

Swaminathan, Madhura (2012), "Who Has Access to Formal *Credit* in Rural India?" *Review of Agrarian Studies*, vol. 2, no. 1, available at http://ras.org.in/who_has_access_to_formal_credit_in_rural_india_evidence_from_four_villages, viewed on 8 September 2017.

Walker, T. S., and Ryan, James G. (1990), *Village and Household Economies in India's Semi-arid Tropics*, John Hopkins University Press, Baltimore.

Wiggins, Steve, and Rajendran, S. (1987), "Rural Banking in Southern Tamil Nadu: Performance and Management," Final Research Report No. 3, The University of Reading, U. K.

APPENDIX 1
BASELINE MODEL FOR ACCESS TO CROP CREDIT

Crop Credit Access
$= C + \beta_1$ Land operated$_i + \beta_2$ social group$_i + \beta_3$ Average schooling years$_i + \beta_4$ Number of small farmer households$_i + \eta_i$

where,

Number of small farmer households$_i$ captures the number of small farmer households in a given village. It thus controls for the dominance of small farmer households in a given village, which can have implications for the bargaining power of these households in the credit system.

η_i = household-specific effect for idiosyncratic characteristics of household 'i'.

The illustration of each of the variables used in the model and its descriptive statistics are given in Appendix 3 and 4 respectively. To minimise the presence of multi-collinearity in the model, the correlation coefficients for each pair of variables are worked out and are found to range between 0 and (+/–)0.5, suggesting weak-to-moderate degree of correlation. See Jain *et al.* (2011).

Appendix 2 *Modelling access to crop credit from the formal sector for small farmer households, 13 PARI villages*

Explanatory variable	Dependent variable: Crop Credit Access$_i$ (Access to crop credit from formal sector during the survey year – 0 if no access; 1 if access)				
	(1)	(2)	(3)	(4)	(5)
	Baseline model	Model II: Control for tenancy	Model III: Control for crop diversification	Model IV: Control for diversification outside crop cultivation	Model V: Control for alternative sources of agricultural credit
Land operated$_i$	0.231**	0.233**	0.209**	0.178*	0.185**
	(0.094)	(0.094)	(0.096)	(0.099)	(0.100)
Social group$_i$	1.073***	0.899**	0.933**	0.916**	0.908**
	(0.364)	(0.372)	(0.372)	(0.375)	(0.377)
Average schooling years$_i$	0.017	0.002	0.0008	0.003	0.003
	(0.047)	(0.047)	(0.048)	(0.048)	(0.048)
Number of small famer households$_i$	0.790*	1.006*	1.744*	1.681*	1.650*
	(0.450)	(0.528)	(0.938)	(0.960)	(0.959)
Tenancy status$_i$		−0.667***	−0.633***	−0.637***	−0.628***
		(0.228)	(0.231)	(0.231)	(0.232)
Irrigation index$_i$			−0.039	−0.011	−0.035
			(0.372)	(0.381)	(0.390)
Commercial crop cultivation$_i$			0.622**	0.636	0.665**
			(0.311)	(0.425)	(0.343)
Share of allied activities GVO$_i$				−1.084	−1.036
				(0.789)	(0.778)
Share of non-agricultural employment$_i$				0.047	0.018
				(0.742)	(0.735)

Incidence of informal sector borrowing‡					
Constant					−0.260 (0.347)
State dummies	Yes	Yes	Yes	Yes	Yes
	Yes	Yes	Yes	Yes	Yes
No. of observations	744	744	744	744	744
P-value (Hosmer-Lemeshow χ)^	0.40	0.61	0.76	0.57	0.52

Notes: Figures in parentheses are robust standard errors.

*** $p<0.01$, ** $p<0.05$, * $p<0.1$.

^ gives the p-value for the null hypothesis that the specified model fits the data.

Source: Estimated from PARI survey data.

Appendix 3 *Description of variables used in the logistic model*

Nomenclature	Description of the variable
Land operated	Total area operated by a household during the survey year
Social group	SC/ST/Muslim households = 0; Others = 1
Average schooling years	Average years of schooling of all members in a household
Number of small farmer households	Log (number of small farmer households in a given village)
Tenancy status	Owner cultivator = 0; partly tenant cultivator = 1; fully tenant cultivator = 2
Commercial crop cultivation	If not cultivated commercial crop during the survey year (cotton, sugarcane, jute, coffee) = 0; if cultivated = 1
Irrigation index	Total area irrigated by a household/gross cropped area of the household in the survey year
Share of allied activities GVO	Amount of gross value of output (GVO) from allied activities including animal husbandry and fishery/total GVO
Share of non-agricultural employment	Number of adults employed in non-agricultural employment (manual, salaried, self-employment/business)/total adults in the household
Incidence of informal sector borrowing	If loan not taken for agriculture from informal sources during the survey year = 0; if taken = 1

Appendix 4 *Descriptive statistics of variables in the logistic model*

Variable	Mean	Standard deviation	Minimum	Maximum
Land operated	1.633	1.307	0	5
Average schooling years	4.937	2.786	0	14.5
Number of small farmer households	1.893	0.235	1.4	2.2
Share of allied activities GVO	0.203	0.198	0	0.9
Share of non-agricultural employment	0.080	0.163	0	1
Irrigation index	0.666	0.420	0	1

9

Climate Change, Agriculture, and the Small Farmer

T. Jayaraman

There is now widespread scientific consensus, based on detailed analysis of a vast body of data from multiple sources and multiple theoretical directions of enquiry, that global warming is a scientific fact and that it is ongoing. However, despite broad recognition of the phenomenon in the policy literature, there are real differences, and indeed a great degree of confusion, regarding the measures and initiatives that help human society's production systems and the organisation of human activity to cope with the consequences of global warming, and the effects and impact of such warming on the earth's geosphere and biosphere.

For many developing countries, the presence of a large rural population that is significantly dependent on agriculture for income and livelihoods makes the impact of climate change on agriculture an issue of particular concern.[1] As is well known, agriculture is dependent on weather and climate. However, while much of agricultural science has developed with a clear understanding of climatic conditions and variability, and its relevance to crop production, livestock, and other agriculture-related activities, equal attention has not been paid to the consequences of a changing climate. The new challenge presented by the phenomenon of global warming is to understand the relationship between the intrinsic variability of climate, and the impact of changes in the levels and variability of climate indicators as a result of global warming.

---

[1] In this chapter, we do not deal with climate mitigation in the agricultural sector, where mitigation refers to the effort to limit the rise in global temperatures by limiting, and eventually possibly eliminating, the emissions of greenhouse gases – which are the ultimate cause of global warming – into the atmosphere. The focus of this chapter is climate adaptation in agriculture.

But it is the significance of their impact on human and social well-being that converts objective, natural processes into matters of environmental issues.[2] Thus, global interest in the matter of climate change and agriculture stems from the importance of agriculture for food production and supply, and additionally, in developing countries, because agriculture is the overwhelmingly significant source of livelihoods and incomes for a large section of the population (even if it contributes only a relatively low share to national income). It is in this specific context of the impact of global warming on agriculture in developing countries that various perspectives on the strategy required to deal with this question fundamentally differ.

Within the rapidly growing policy-oriented literature on climate change and agriculture, two broad streams of thinking may be distinguished. Some policy researchers and activists of climate science suggest that global warming has already led to widespread negative consequences for agricultural production. The hallmark of such arguments is the juxtaposition of scientific conclusions regarding *future* climate change with examples drawn from the impact of *current* climate variability, without clarifying the relationship between the two. The problem here is enhanced by the fact that such examples are typically drawn from case studies of small and marginal farmers, for whom the vulnerability of agricultural production to climate variation, even in the absence of climate change, leads to a serious loss of well-being.

Another line of argument, though more circumspect about asserting that the negative consequences of climate change are already apparent, nevertheless regards climate change as *the* issue of overriding concern for agriculture. In this view, all problems related to agricultural production have to be examined primarily in the context of global warming. This line of argument often pays little attention to the complexity of the interplay between the socio-economic factors affecting agricultural production, and the environmental and climatic conditions under which such production occurs.

The first line of argument cited above has been popular in particular with non-governmental and social work organisations, both international and

---

[2] We are indeed suggesting that the notion of environment is clearly anthropocentric in nature. We would go further to suggest that the very notion of an environmental science draws its significance from its anthropocentric character. Otherwise it could be easily be subsumed as part of the study of the biosphere and the geosphere. Unfortunately, we will not be able to pursue this line of thought further in this chapter. Another aspect of anthropocentricity with particular reference to global warming is that the current warming is anthropogenic in origin. However this does not take away from the fact that the rise in global temperatures is an objective process due to man-made emissions.

national.[3] The second line of argument has been put forth by multilateral institutions of various kinds, agencies from the United Nations system, and aid agencies of individual developed countries.[4] A crucial problem with both sets of arguments is that they do not adequately study data on agricultural production over time, and across regions, crops, and socio-economic strata of producers, in order to understand the interplay of the environmental and socio-economic dimensions of the climate sensitivity of agriculture. Thus, studies that arise from these two viewpoints routinely conflate problems of current climate vulnerability with problems of adaptation to climate change in the future. Some studies also proceed to conflate a variety of issues of environmental degradation with climate change.

Both these arguments have in common an underlying vision that sees issues of environmental degradation as overriding socio-economic realities in the rural sphere, even when there is striking evidence that the latter is the base on which the former is realised. In refusing to recognise the role of socio-economic differentiation in the countryside, the first line of argument would seem to subscribe to the romantic notion of a peasant community. The second line of argument, on the other hand, is clearly apologist in character, brushing aside the evidence that the manner in which different sections of farmers react to the shocks posed by the weather or climate depends very much on their location within the larger system of relations of production in agriculture.

A more nuanced and precise account of the impact of climate change on agricultural production that takes note of differentiation within the category of farmers is not, however, easy to develop. Except in rare instances, detailed panel data required from observations over several years, that include socio-economic, agronomic, and climate variables, are simply not available. Secondary data from official sources are of course extremely useful in determining, to some extent, the impact of climate variability and climate change on overall agricultural production. But the detailed panel data that one really needs is not available from such sources. In the absence of such information, the kind of

---

[3] A good illustration is provided by a report of the Working Group on Climate Change and Development, a consortium of 23 non-governmental organisations (see Working Group on Climate and Development 2007).

[4] There are many reports that follow this second line of argument. Among some of the recent ones are World Bank (2012) and World Bank (2013). The latter reference has specific discussions of the impact of climate change on agriculture in Southeast Asia, South Asia, and Sub-Saharan Africa. However, it does not refer to yield gaps. There is little discussion in this report of how and why current agricultural production is sub-optimal in many regions, and how dealing with production deficits can help cope with climate change in the future. The report also tends to overemphasise the negative effects of climate change in a one-sided way, without careful discussion of the uncertainties involved.

detailed cross-sectional data that studies like those undertaken by the Project of Agrarian Relations in India (PARI) can provide are critical for drawing some inferences at least at a general level, which can also inform future efforts to obtain panel data.

This chapter, which is part of an ongoing attempt to bring together results from two distinct areas in the scientific and social science literature, discusses current climate variability and its consequences as well as the impact of climate change in the future. It argues that the distinction between current and future climate variability is fundamental to understanding the differential impact of climate change on different classes of producers, and that this differential impact stems directly from the distinctly poorer economic status of small farmers in contrast to farmers who are better-off. To neglect this difference between current and future climate is, in effect, to deny the true significance of the differential impact of climate change and the basis of such differentiation.

In the next section of the chapter, we briefly make a note of some key observations on current climate variability and climate change, and comment on some aspects of these in relation to India. These include some results regarding the current impact of mean temperature and precipitation changes on agriculture globally. Careful statistical analyses show that ongoing temperature and precipitation changes indeed have had a negative impact on agricultural production and yields, but that these have been more than overtaken by the overall trend of growth in agriculture due to other factors. We also briefly discuss some specific details of the temperature and precipitation dependence of Indian agriculture, and the impact of climate variability.

In the third section of the chapter, we deal with the issue of differential impact of climate change on the livelihoods and incomes of different classes of agricultural producers. We do not, of course, have explicit time-series data on crop production linked to climate variables across different strata of farmers, which would best illustrate the differential impact of climate variability. However, we argue, such differentiation is strongly suggested by the cross-sectional data available on agricultural production, yield, and incomes from the PARI surveys of villages on which this volume is based. This part of the chapter also provides a critique of the two dominant approaches in the literature that we have referred to earlier.

CLIMATE CHANGE, CLIMATE VARIABILITY, AND AGRICULTURE IN INDIA

Before entering into details, we take into account here some of the highlights of the scientific findings of the Fifth Assessment Report of the Intergovernmental Panel on Climate Change (IPCC) (2014). These findings, which relate to temperature increase and temperature variability, are the most robust on the global scale, while the findings relating to precipitation are subject to much greater uncertainty. Sea-level rise is another area in the field of global warming studies where significant and robust results are available.

Global temperatures are rising, as also the occurrence of extreme temperature events when temperatures rise well above the mean. Such events are "very likely" to be attributable to anthropogenic global warming. Global warming is also contributing to the increase in extreme rainfall events in terms of frequency of occurrence, intensity, and quantum of precipitation, though the connection to anthropogenic global warming is not as certain as in the case of temperature.

In general, there are more studies now available on the impact of higher average temperatures on crop production than at the time of the Fourth Assessment Report of the IPCC (2007). It now appears that in the absence of nitrogen stress, the impact of climate change on crop production is not as severe as estimated earlier. However, if nitrogen stress is included, the results are similar to those outlined in the Fourth Assessment Report.

With regard to the role of climate variability, we note that simulation models provide evidence that greater climate variation alone can lower yields to an extent comparable to (or greater than) the impact of increased mean temperatures. There is empirical evidence in this regard: more days of exposure to extreme temperatures of wheat in northern India resulted in correspondingly earlier onset of senescence in wheat (Lobell 2012). While Lobell's study was based on remote sensing data, Mahato *et al.* (2014) confirmed this through an analysis of the dependence of district-level yields on extreme temperatures for the same region over approximately the same time-period. This analysis shows clearly that greater exposure to extreme temperatures lowers yield.

Ongoing climate change due to rising temperatures and changes in precipitation levels has had a negative impact on crop production in different parts of the world, though this impact has been more than offset in practice by improved management techniques and other factors (Lobell 2011). Predictions on the basis of climate models of temperature trends in the future indicate that a greater proportion of global crop production will be exposed to heat stress, potentially leading to lowered yields and decreased production.

WHO IS AFFECTED BY THE SENSITIVITY OF AGRICULTURE TO CLIMATE?

Jayaraman (2011) has reviewed the different mathematical techniques and models that have been used to measure the impact of environmental stresses and shocks on aggregate agricultural production, and the consequent impact on commodity supplies and prices. A succinct criticism of such contemporary mainstream models and their inability to study the differential impact of climate change has been made by the report of the High Level Panel of Experts (HLPE) on Food Security and Climate Change, of the Food and Agriculture Organisation (FAO): "None of these global scenario efforts attempts to address distributional issues within countries and the possibility that climate change affects the vulnerable disproportionately" (HLPE 2012, p. 47).[5] This criticism also applies in accounting for the differential impact of climate change across developed and developing countries. The report also notes, even if somewhat elliptically, in relation to food security and the World Trade Organisation (WTO), that governments need to build a transparent, rules-based, and accountable multilateral trading system. More specifically, it states that these rules need to give a larger place to public policy concerns regarding food security, better account for the heterogeneity of WTO member-states, and take into account the special needs of poor and vulnerable countries or social groups (*ibid.*)

It is generally acknowledged that globally, the poor and sections of the population that are most vulnerable to climate change are categories with a substantial overlap. As a consequence, when agricultural production and food supply are affected by environmental shocks, these categories are the most likely to suffer the consequences. Overall social and economic development that guarantees an adequate supply of food, nutrition and health, education, and access to basic amenities to the broad masses of the population constitutes, therefore, the first line of defence against climate change.

In a pioneering econometric study, Guiteras (2009) attempted to quantify the impact of future climate change on yields (in terms of value of output per hectare) under various scenarios, by regressing observed yields against current temperature and precipitation trends (taking careful account of variability in both), and then utilising these results for future predictions. A limited number of economic variables were also included in the analysis to ensure that the relationship between yields, and temperature and precipitation was accurately estimated. This study predicts that climate change will affect Indian agriculture significantly, and that the reduction in yield in the long term (that is, by 2070–99) will be of an order of 20 to 30 per cent, and in the medium term (that is, by 2040), of an order of 10 per cent. It is possible, Guiteras suggests, to estimate the distributional outcomes of such reductions in the value of agricultural output using various methods, including those based on

[5] For a brief review of this report and other reports of the HLPE, see Sridhar (2012) and Sridhar (2013).

the social accounting matrix (SAM). This study and a few other similar studies of agriculture in the United States and Europe have not received adequate appreciation in the literature; nevertheless they offer what appears to be a fruitful methodology that is worth exploring further.

The HLPE of the FAO has pointed out both the crucial role played by small farmers in food production and food security, and their particular responsibility in climate adaptation. At the same time, the Panel recognises that we perhaps know too little yet about variations in agricultural production and livestock-rearing methods across different scales of production and economic activity.

> We know too little about how crops and livestock (are) grown and management practices change with scale to identify global patterns consistently, but it is commonly assumed that small-scale farms are more likely to engage in diversified crop and livestock agriculture, which might be more resilient to climate change. On the other hand, small-scale operations are less likely to have access to extension services, markets for new inputs and seeds, and loans to finance operations. Gaining a better understanding of the differences in farm activities, and vulnerabilities to climate change is critical, both to finding ways to improve food security and to deal with the challenges which climate change poses to agricultural productivity and stability. (HLPE 2012)

It is clear that when households are classified by any criterion of *economic size* – that is, by size-class of landholding or size-class of asset-holding or size-class of annual total (of crop) income earned – or by other methods of socio-economic classification, the overwhelming reality is of *very sharp economic inequality*. The point being made is that small farmers – that is, farmers at the lower end of the scale with regard to household holdings of land, other assets, and incomes – are more vulnerable to different kinds of environmental and economic shocks, and to fluctuations in livelihoods and incomes. Further, they are likely to be worse-off in respect of crop output (in value and physical terms), scientific equalisation of inputs, other aspects of technology, and land-tenure arrangements.

In what follows, we re-examine these observations in the light of PARI data, focusing especially on the differentiation between small farmers and large farmers. This re-examination draws heavily on the analysis of PARI data already undertaken by other colleagues and reported in other chapters in this volume. We restrict ourselves here to drawing out some stylised facts suggested by these analyses in relation to climate change and agriculture, and the vulnerability of small farmers to climate change and climate variability. In what follows we rely mostly on the analyses in chapter 4, on cropping pattern, productivity, and incomes.

One of the first important results derived from the data is the substantial variability in mean incomes from crop production across different villages in various parts of India. The results discussed in chapter 4 show that net income per hectare varied considerably, ranging from Rs 3,000 per hectare to Rs 50,000 per hectare. The data show that the low-income villages were either dependent on rainfed agriculture or the survey years were ones of moisture stresss. High-income villages, on the other hand, enjoyed irrigation, multiple cropping, and commercial crops to some degree. On the whole, production data across the villages suggest that small farmers were more given to foodgrain production than large farmers, though this does not imply, as the authors note, that small farmers were typically subsistence farmers. In the context of the impact of climate change on crop production, it is evident that the relatively worse-off farmers in drier regions would be disproportionately vulnerable to increased temperatures, both on account of the potential loss of income as well as the loss of ability to grow their own food.

The second set of results relevant to our argument here are the relative differences in the productivity of small farmers versus large farmers. The data show that these differences were not always substantial and varied across crops. In the case of wheat, small farmers reported lower yields than large farmers across almost all the PARI study villages. However, the relative absence of variations in yield should not obscure the fact that small farmer households have small holdings and hence absolute production levels will be low. This undoubtedly has consequences for their net overall income, and the consequent capacity for savings and accumulation of assets.

One of the most striking consequences of the overall instability of incomes from crop production that is brought out uniquely by PARI data is the regular occurrence of negative incomes from crop production. While negative incomes (see Table 1 below) occur among both small farmer and large farmer households, clearly, a larger number of small farmers are at risk of making losses. As global warming proceeds in a situation of instability of crop production, the regular prevalence of negative incomes suggests that income losses affect a significant number of small farmer households across India.

Table 2 shows the drastic divergence of the mean of incomes from crop production between the worst-off 20 households and the top 20 households in eight selected villages surveyed as part of PARI. As can be seen, the earnings of the top 20 households are a multiple of the mean, while the earnings of the bottom 20 households are a fraction of the mean. In seven out of eight cases, the figure is *negative* for the bottom 20 households; that is to say, on average, they ran at a loss from crop production during the survey year.

Table 1 *Number and proportion of households with negative incomes from crop production by farmer category, selected villages*

Village	State	Small farmers		Large farmers		Combined (%)
		Number	Percentage	Number	Percentage	
Ananthavaram	Andhra Pradesh	80	35	0	0	20
Bukkacherla	Andhra Pradesh	42	39	18	31	20
Kothapalle	Telangana	24	24	0	0	17
Harevli	Uttar Pradesh	6	13	1	4	7
Mahatwar	Uttar Pradesh	13	19	0	0	12
25F Gulabewala	Rajasthan	0	0	0	0	0
Rewasi	Rajasthan	56	52	2	4	24
Gharsondi	Madhya Pradesh	16	23	10	14	11
Nimshirgaon	Maharashtra	3	1	22	41	11
Warwat Khanderao	Maharashtra	3	3	1	6	4
Siresandra	Karnataka	11	21	0	0	11
Zhapur	Karnataka	13	50	4	29	19
Alabujanahalli	Karnataka	9	8	1	4	7
Amarsinghi	West Bengal	1	1	NA	NA	1
Kalmandasguri	West Bengal	6	8	NA	NA	8
Panahar	West Bengal	43	29	3	33	30
Tehang	Punjab	2	3	2	4	3

Source: Chapter 4, Table 9.

Table 2 *Average annual net incomes from crop production per acre of operational holding of bottom 20 households and top 20 households in eight PARI survey villages, surveyed between 2006 and 2010*

State	District	Village	Bottom 20 households	Mean	Top 20 households
Andhra Pradesh	Anantapur	Bukkacherla*	−5027	1049	6648
Telangana	Karimnagar	Kothapalle*	−1801	3091	8015
Uttar Pradesh	Bijnor	Harevli	−4965	6343	16350
	Ballia	Mahatwar	−3016	2665	9017
Maharashtra	Buldhana	Warwat Khanderao	−782	6301	15893
	Kolhapur	Nimshirgaon*	−72	10598	26253
Rajasthan	Sri Ganganagar	25F Gulabewala	3553	7737	12024
Madhya Pradesh	Gwalior	Gharsondi	−5172	5338	20081

Notes: Incomes are estimated at 2008–09 prices using state-level CPIAL (Consumer Price Index for Agricultural Labour).
* Bottom 20 and top 20 households of villages marked with an asterisk are averages of sample households.
Source: PARI survey data, as reported in Ramachandran (2011).

Table 3, which pools data from the same eight villages, further illustrates the inequalities of income.

Households in the lowest deciles or the bottom 20 households need not, of course, refer solely to smallholders. Households may belong to this category because of the instability of agricultural production under current conditions, particularly higher exposure to risk in some crops and the instability of dryland agriculture in some villages. Table 4 reports on median net incomes per acre among different socio-economic classes for the same villages reported in Tables 2 and 3. This analysis of PARI data (Ramachandran 2011) uses a finer classification of the socio-economic strata, in terms of classes, as compared to the one that is the focus of this volume. Broadly speaking, we may consider the categories Peasant 1 and part of Peasant 2 as analogous with large farmers, while parts of Peasant 2, Peasant 3, and hired manual workers may be considered to be analogous to small farmers. However, the actual identification of these categories and of the class nature of every household is a complex exercise, and the reader is referred to the original reference for these details.

The inability of small farmer households to keep production costs down, to use inputs efficiently, and to obtain a higher income per unit of production, is a reflection of the socio-economic inequalities of rural society – particularly inequalities in the ownership of land and other productive assets, payment of rents for land and machinery which small farmers have to make, the higher costs at which they gain access to inputs, and their lack of access to markets.

Table 3 *Average incomes from crop production of households operating land by decile of agricultural income, pooled data from PARI survey villages, at 2008–09 prices*

Decile of household ranked by income from crop production	All villages
1	–19161
2	–2397
3	859
4	3296
5	6419
6	11788
7	19427
8	33338
9	60661
10	323049
D10/D9	5.32

Note: This table is based on data from nine survey villages: two in Andhra Pradesh (Ananthavaram and Bukkacherla); one in Telangana (Kothapalle); two in Uttar Pradesh (Harevli and Mahatwar); two in Maharashtra (Warwat Khanderao and Nimshirgaon); one in Rajasthan (25F Gulabewala); and one in Madhya Pradesh (Gharsondi). For the purpose of comparison, incomes of all households were converted to 2008–09 prices using state-level CPIAL (Consumer Price Index for Agricultural Labour).
Source: PARI survey data, as reported in Ramachandran (2011).

The conditions of labour and livelihood for manual workers, and small and medium farmers also translate into serious human development deficits for a significant proportion of rural households.

Similar results, which speak to the issue of scale raised by HLPE (2012), are reported from a study commissioned by the Planning Commission, Government of India, on agriculture in eastern India (Haque *et al.* 2013). This study uses a sample drawn from several districts across the States of Uttar Pradesh, Bihar, Jharkhand, Odisha, and West Bengal. It shows that crop yields from ownership holdings of marginal and small farmers are significantly lower than corresponding yields from operational holdings of large farmers. It also reports lower input–output ratios as well as income–input ratios on marginal and small landholdings, than on large holdings.

The considerable differentials in output and incomes suggest that small and marginal farmers are more susceptible to climate variability and climate change than large farmers. Climate change is effectively an immediate threat to small and marginal farmers, though this is not to forget that their susceptibility to environmental stresses and shocks is more a consequence of socio-economic conditions than any form of "environmental poverty."

With reference to the impact of climate change on agriculture, the Central Research Institute for Dryland Agriculture (CRIDA), India has brought out

Table 4 *Median net income from crop production per acre of operational holding by class, pooled data from PARI survey villages, at 2005–06 prices*

State	District	Village	Landlord	Peasant 1 (Rich)	Peasant 2 (Middle)	Peasant 3 (Poor)	Hired manual worker
Andhra Pradesh	Guntur	Ananthavaram	7534	15022	3238	485	993
	Anantapur	Bukkacherla	–274	1134	894	207	2159
Telangana	Karimnagar	Kothapalle	4839	2210	3188	3523	2039
Uttar Pradesh	Bijnor	Harevli	6636	8627	6640	2134	1634
	Ballia	Mahatwar	3458	6957	2656	952	1745
Maharashtra	Buldhana	Warwat Khanderao	9576	7594	5515	5660	1358
	Kolhapur	Nimshirgaon	16231	13001	9449	5888	–58
Rajasthan	Sri Ganganagar	25F Gulabewala	7077	6004	5890	nil	nil
	Sikar	Rewasi	3304	3299	469	517	–572
Madhya Pradesh	Gwalior	Gharsondi	7031	5634	3924	3035	1258

Source: Rawal (2014).

a number of publications[6] dealing with drought, drought management, and various strategies and coping plans for drought and its impact on different sub-sectors. These publications are undoubtedly of scientific value from the point of view of agricultural science. But they make little or no reference to issues of scale, and to the potential and actual implementation of their recommendations by farmers belonging to different socio-economic categories.

We can now understand the deficiencies of the two perspectives, both emerging from an "environmentalist" standpoint that we had referred to at the beginning of this chapter. It is evident that to claim that climate change is already affecting the poor and medium peasantry is to ignore, on the one hand, the evidence that the dominant classes in the countryside are by no means affected by climate change today. These classes are no doubt susceptible to disruptions in agricultural production due to extreme climate events and climate variability in general, but their dominant economic status ensures their ability to recover from such losses, and also to utilise the years of good production to build reserves that would enable them to cope with such disruptions. Clearly, it is not climate change *per se* that in any way affects the poor and medium peasantry.

On the other hand, by ignoring the source of the poverty of the bulk of the rural population, this view also misses the point that climate adaptation must begin with the fundamental social and economic transformation of the countryside, without which the inability of the poor to deal with climate and weather shocks cannot be reduced significantly.[7] We may point out here that such environmentalist arguments began more than forty years ago, when environmental groups began to campaign on the slogan that environmental issues play a more significant role than class relations in understanding the poverty of the poor and medium peasantry, and of agricultural labour. The case of global warming has merely provided a fresh lease of life to such arguments, even though the seeming radicalism of early environmentalism has almost vanished thanks to the adoption of "environmentalism" by even mainstream/ establishment academic thinking.

The other line of argument that sees all problems related to enhancing agricultural production in a purely climate perspective, of course, belongs to a much older viewpoint, especially in India, which has constantly eschewed any talk of social and economic transformation of the countryside, and has always insisted that improvement of the conditions of production and technological

---

[6] Some of these publications can be accessed on the web at www.crida.in

[7] This is of course not to argue that various forms of relief to the poor cannot provide some short-term benefits. Nevertheless, the basic causes for the deprivation of the poor will not be eliminated by such relief.

change are the key issues. Such a view is little more than an apologist argument for the continuation of unequal and oppressive social and economic conditions in the countryside, which, by focusing solely on the technological aspect of modernisation of agricultural production, actually seeks to perpetuate these oppressive conditions in renewed forms. Environmental issues, especially of the kind posed by global warming, are, according to this perspective, to be dealt with by merely technological means. Climate adaptation in this perspective is exclusively a matter of new technologies for agriculture, with social and economic policies being purely secondary means to enable such technological transformation.

It is indeed true that despite the considerable penetration of modern production techniques in agriculture, there is still considerable room for the expansion of productive forces in the Indian countryside. This is evident, for instance, from the considerable yield gaps in several crops in several parts of the country. As we have argued earlier in this section, however, only radical transformation of agrarian relations would enable expansion in such a manner that substantially benefits the small farmer. Typically, both lines of the environmentalist argument we have discussed earlier insist that the question of productive forces in the current period is basically one of transformation to environmental-friendly technologies, rather than seeking to ensure environmental sustainability while simultaneously expanding the productive forces.

SOME CONCLUSIONS FROM A CRITICAL PERSPECTIVE

As has been clearly articulated elsewhere (see, for instance, Ramachandran 2011), a basic characteristic of Indian agriculture is the development of capitalism in agriculture without a corresponding radical transformation of agrarian relations. The Indian state has sought to promote capitalism in agriculture, transforming traditional landlords into capitalist landlords and creating a stratum of rich peasants. These sections have drawn the overwhelming bulk of the benefits of modernisation of agriculture, including access to new technologies and inputs, access to markets, and access to credit, while retaining in many respects their traditional hold on social and economic relations in the countryside. On the other hand, the bulk of the rural population has faced growing inequalities and persistent poverty. While peasant production is increasingly subjugated to markets, it is also marked by increasing loss of land and other productive assets such as livestock, access to modern inputs, and continued dependence on usurious credit. Agricultural production for the majority of the peasantry, the bulk of whom constitute the small farmers that this volume deals with, is therefore unremunerative, crisis-prone, and increasingly unable to provide more than bare subsistence for an ever-larger number while accelerating their dependence on wage incomes (from agricultural and other forms of labour too) across the board (Ramachandran 2011).

This, of course, does not imply that a resolution of the agrarian question in India would immediately lead to a resolution of all environmental questions. Resolution of environmental issues, which cover a wide range as a consequence of increasing scientific knowledge, cannot be achieved without further application of science and technology, and the general development of productive forces. But, as we have attempted to point out in this chapter, this cannot be achieved under the current social and economic conditions prevailing in the countryside. Without transforming these conditions, environmental crises would at best be periodically and partially resolved at the expense of the vast majority of the working population in the countryside.

We believe that the issues under consideration in this chapter have wider relevance to many developing societies, even if specific social and economic considerations vary. One may also anticipate that developing countries where the agrarian question has been resolved will be much better placed to deal with the issues raised by the impact of global warming on agriculture.[8] Developing societies, their peoples, and even their governments and elites are of course keenly aware of this problem. They are also aware that a solution to the global challenge of climate change involves overcoming the inequalities existing between nations on the global scale, especially in assuring an equitable and just solution to the problem of sharing the burden of climate mitigation.

While this understanding is undoubtedly correct, it is our argument in this chapter that radical socio-economic transformation that overcomes the gross inequalities within developing societies is as important when dealing with the question of climate adaptation – and climate change in agriculture and the fate of the small farmer in the era of global warming offer particularly telling examples.

---

[8] A detailed comparative study of environmental issues, including climate variability, in relation to agricultural production in the case of two countries, such as Bangladesh and Vietnam, would be interesting in this context.

REFERENCES

Chaturvedi, R. K., Joshi, J., Jayaraman, M., Bala, G., and Ravindranath, N. H. (2012), "Multi-model Climate Change Projections for India under Representative Concentration Pathways," *Current Science*, vol. 103, no. 7, pp. 1–12.

Cutter, Susan, Emrich, Christopher, Webb, Jennifer J., and Morath, Daniel (2009), *Social Vulnerability to Climate Variability Hazards: A Review of the Literature*, Final Report to Oxfam America from the Hazards and Vulnerability Research Institute (HVRI), Department of Geography, University of South Carolina, June, available at http://adapt. oxfamamerica.org/resources/Literature_Review.pdf, viewed on 18 December 2013.

Easterling, W. E., and Aggarwal, Pramod (2007), *Food, Fibre and Forest Products*, in M. L. Parry, O. F. Canziani, J. P. Palutikof, P. J. van der Linden, and C.E. Hanson (eds.), *Climate Change 2007: Impacts, Adaptation and Vulnerability: Contribution of Working Group II to the Fourth Assessment Report of the Intergovernmental Panel on Climate Change*, Cambridge University Press, Cambridge, pp. 273–313.

Gadgil, S. (2003), "The Indian Monsoon and Its Variability," *Annual Review of Earth Planet Science*, 31, pp. 429–67, doi:10.1146/annurev.earth.31.100901.141251.

Gadgil, S., and Kumar, K. R. (2006), "The Asian Monsoon, Agriculture and Economy," in B. Wang, (ed.), *The Asian Monsoon*, Springer, Praxis, U. K., pp. 651–83.

Gadgil, S., and Srinivasan, J. (2012), "Monsoon Prediction: Are Dynamical Models Getting Better than Statistical Models?" *Current Science*, 103, pp. 257–59.

Goswami, B. N., Venugopal, V., Sengupta, D., Madhusoodanan, M. S., and Xavier, Prince K. (2006), "Increasing Trend of Extreme Rain Events over India in a Warming Environment," *Science*, vol. 314, no. 5804, pp. 1442–45.

Gourdji, S. M., Sibley, A. M., and Lobell, D. B. (2013), "Global Crop Exposure to Critical High Temperatures in the Reproductive Period: Historical Trend and Future Projections," *Environmental Research Letters*, vol. 8, no. 2, available at doi:10.1088/1748-9326/8/2/02404, viewed on 6 September 2017.

Guiteras, R. (2009), "The Impact of Climate Change on Indian Agriculture," University of Maryland, available at http://www.econ.umd.edu/research/papers/34, viewed on 15 December 2013.

Gulati, A., Saini, S., and Jain, S. (2013), "Monsoon 2013: Estimating the Impact on Agriculture," Commission on Agricultural Costs and Prices (CACP), Department of Agriculture and Cooperation, Ministry of Agriculture, Government of India, October, available at http://cacp.dacnet.nic.in/DP8_Monsoon_2013.pdf, viewed on 15 December 2013.

Gustafson, David I., Collins, Michael, Fry, Jonna, Smith, Saori, Matlock, Marty, Zilberman, David, Shryock, Jereme, Doane, Michael, and Ramsey, Nathan (2013), "Climate Adaptation Imperatives: Global Sustainability Trends and Eco-efficiency Metrics in Four Major Crops – Canola, Cotton, Maize, and Soybean," *International Journal of Agricultural Sustainability*, available at doi:10.1080/14735903.2013.84601, viewed on 4 September 2017.

Hafner, S. (2003), "Trends in Maize, Rice and Wheat Yields for 188 Nations over the Past 40 Years: A Prevalence of Linear Growth," *Agriculture, Ecosystems and Environment*, 97, pp. 275–83.

Haque, T., Bhattacharya, Mondira, Sinha, Gitesh, Kalra, Purtika, and Thomas, Saji (2010), *Constraints and Potentials of Diversified Agricultural Development in Eastern India*, Council for Social Development, New Delhi, available at htttp://planningcommission. nic.in/reports/sereport/ser/ser_agridiv1102.pdf, viewed on 15 December 2013.

High Level Panel of Experts (HLPE) (2012), *Climate Change and Food Security: A Report by the High Level Panel of Experts on Food Security and Nutrition of the Committee on World Food Security*, Food and Agriculture Organisation (FAO), Rome.

Iizumi, Toshichika, Sakuma, Hirofumi, Yokozawa, Masayuki, Luo, Jing-Jia, Challinor, Andrew J., Brown, Molly E., Sakurai, Gen, and Yamagata, Toshio (2013), "Prediction of Seasonal Climate-induced Variations in Global Food Production," *Nature Climate Change*, available at doi: 10.1038/nclimate1945, viewed on 5 September 2017.

Intergovernmental Panel on Climate Change (IPCC) (2000), *Special Report on Emission Scenarios*, Cambridge University Press, Cambridge, available at https://www.ipcc.ch/ pdf/special-reports/spm/sres-en.pdf, viewed on 6 December 2013.

Intergovernmental Panel on Climate Change (IPCC) (2001), *Climate Change 2001: The Scientific Basis. Contribution of Working Group I to the Third Assessment Report of the Intergovernmental Panel on Climate Change*, J. T. Houghton, Y. Ding, D. J. Griggs, M. Noguer, P. J. van der Linden, X. Dai, K. Maskell, and C. A. Johnson (eds.), Cambridge University Press, Cambridge and New York.

Intergovernmental Panel on Climate Change (IPCC) (2012), *Managing the Risks of Extreme Events and Disasters to Advance Climate Change Adaptation. A Special Report of Working Groups I and II of the Intergovernmental Panel on Climate Change*, C. B. Field, V. Barros, T. F. Stocker, D. Qin, D. J. Dokken, K.L. Ebi, M. D. Mastrandrea, K. J. Mach, G. K. Plattner, S. K. Allen, M. Tignor, and P. M. Midgley (eds.), Cambridge University Press, Cambridge and New York.

Intergovernmental Panel on Climate Change (IPCC) (2013), "Summary for Policymakers," *Climate Change 2013: The Physical Science Basis. Contribution of Working Group I to the Fifth Assessment Report of the Intergovernmental Panel on Climate Change*, T. F. Stocker, D. Qin, G. K. Plattner, M. Tignor, S. K. Allen, J. Boschung, A. Nauels, Y. Xia, V. Bex, and P. M. Midgley (eds.), Cambridge University Press, Cambridge and New York.

Intergovernmental Panel on Climate Change (IPCC) (2013), *Climate Change 2013: The Physical Science Basis. Contribution of Working Group I to the Fifth Assessment Report of the Intergovernmental Panel on Climate Change*, T. F. Stocker, D. Qin, G. K. Plattner, M. Tignor, S. K. Allen, J. Boschung, A. Nauels, Y. Xia, V. Bex, and P. M. Midgley (eds.), Cambridge University Press, Cambridge and New York.

Jayaraman, T. (2011), "Climate Change and Agriculture: A Review Article with Special Reference to India," *Review of Agrarian Studies*, vol. 1, no. 2, available at http://www. ras.org.in/climate_change_and_agriculture/, viewed on 6 September 2017.

Lobell, D. B., Schlenker, W. S., and Costa-Roberts, J. (2011), "Climate Trends and Global Crop Production since 1980," *Science*, 333, pp. 616–20.

Lobell, D. B., Sibley, A., and Ortiz-Monasterio, J. I. (2012), "Extreme Heat Effects on Wheat Senescence in India," *Nature Climate Change*, 2, pp. 186–89.

Mahato, Sandeep, Murari, Kamal K., and Jayaraman, T. (2015), "Empirical Evidence of the Impact of Extreme Temperatures on Yield in a Major Wheat Growing Region in India," presentation at National Climate Science Conference, Bengaluru.

Menon, A., Levermann, A., Schewe, J., Lehmann, J., and Frieler, K. (2013), "Consistent Increase in Indian Monsoon Rainfall and its Variability across CMIP-5 models," *Earth System Dynamics Discussion*, vol. 4, no. 2, pp. 1–24, available at doi:10.5194/esdd-4-1-2013, viewed on 6 September 2017.

Murari, Kamal K., Ghosh, Subimal, Pal, Sujan, Sengupta, Agniv, and Karmakar, Subhankar (2013), "Evaluations of Superensemble and Multi Model Average of CMIP5 Simulations for Indian Rainfall," unpublished paper.

Natcom II (2012), *India: Second National Communication to the United Nations Framework Convention on Climate Change*, Ministry of Environment and Forests, Government of India, available at http://envfor.nic.in/, viewed on 10 December 2013.

Nelson, Gerald C. *et al.* (2009), *Climate Change: Impact on Agriculture and Costs of Adaptation*, IFPRI Food Policy Report, International Food Policy Research Institute, Washington D. C., available at http://www.ifpri.org/publications/.climate-change-impact-agriculture-and-costs-adaptation, updated October 2009, viewed on 15 December 2013.

Ramachandran, V. K. (2011), "Notes on the State of Agrarian Relations in India Today," unpublished.

Rao, C. A. Rama, Raju, B. M. K., Rao, A. V. M. Subba, Rao, K. V., Rao, V. U . M., Ramachandran, K., Venkateswarlu, B., and Sikka, A. K. (2013), *Atlas on Vulnerability of Indian Agriculture to Climate Change*, Central Research Institute for Dryland Agriculture, Hyderabad, available at http://www.nicra-icar.in/nicrarevised/index.php?option=com...edit&id, viewed on 15 December 2013.

Rawal, Vikas (2014), "Cost of Cultivation and Farm Business Incomes in India," paper presented at the Tenth Anniversary Conference of the Foundation for Agrarian Studies, Kochi, 9–12 January.

Rosenzweig, C. *et al.* (2013), "Assessing Agricultural Risks of Climate Change in the 21st Century in a Global Gridded Crop Model Intercomparison," in *Proceedings of the National Academy of Sciences*, pp. 1–6, available at doi:10.1073/pnas.1222463110, viewed on 6 September 2017.

Schellnhuber, Hans Joachim, Hare, Bill, Serdeczny, Olivia, Schaeffer, Michael, Adams, Sophie, Baarsch, Florent, Schwan, Susanne, Coumou, Dim, Robinson, Alexander, Vieweg, Marion, Piontek, Franziska, Donner, Reik, Runge, Jakob, Rehfeld, Kira, Rogelj, Joeri, Perette, Mahe, Menon, Arathy, Schleussner, Carl-Friedrich, Bondeau, Alberte, Svirejeva-Hopkins, Anastasia, Schewe, Jacob, Frieler, Katja, Warszawski, Lila, and Rocha, Marcia (2013), "Turn Down the Heat: Climate Extremes, Regional Impacts, and the Case for Resilience – Full Report," World Bank, Washington D. C.

Semenov, M. A., and Porter, J. R. (1995), "Climatic Variability and the Modeling of Crop Yields," *Journal of Agriculture of Forest Meteorology*, vol. 73, pp. 265–83.

Singh, D., and Singh, S. (2011), "Crop Responses to Climate Variation," in V. U. M. Rao (ed.), *Agricultural Drought: Climate Change and Rain-fed Agriculture*, Central Research Institute for Dryland Agriculture, pp. 145–55.

Sridhar, V. (2012), "Globalisation and the Determinants of Food Security," *Review of Agrarian Studies*, vol. 2, no. 2, available at http://ras.org.in/globalisation_and_the_determinants_of_food_security, viewed on 8 September 2017.

Sridhar, V. (2013–14), "Biofuels, Smallholder Agriculture, and Food Security," *Review of Agrarian Studies*, vol. 3, no. 2, available at http://ras.org.in/biofuels_smallholder_agriculture_and_food_security, viewed on 8 September 2017.

Teixeira, E. I., Fischer, G., Velthuizen, V. G., Walter, C., and Ewert, F. (2013), "Global Hot-spots on Heat Stress on Agricultural Crops Due to Climate Change," *Agriculture and Forest Meteorology*, 170, pp. 206–15.

Turner, A. G., and Annamalai, H. (2012), "Climate Change and the South Asian Summer Monsoon," *Nature Climate Change*, 2, pp. 587–95, doi: 10.1038/nclimate1495.

Working Group on Climate Change and Development (2007), *Up in Smoke: Asia and the Pacific: The Threat from Climate Change to Human Development and the Environment*, New Economics Foundation, November.

10

Access to Education

Madhura Swaminathan and Rakesh Kumar Mahato

In this chapter we briefly examine the access to education and educational attainments among small farmers in India, with a focus on adults and not children.

A set of reports prepared for the UNICEF by the Foundation for Agrarian Studies (FAS) examined in detail the data collected by the Project on Agrarian Research in India (PARI) to arrive at an understanding of educational deprivation among children in the villages surveyed by the Project. In no village did we find *all* children below the age of 18 attending educational institutions, and in only one village (Siresandra, Kolar district, Karnataka) did we find universal attendance even in the 6 to 14 age-group (FAS 2014). The proportion of girls out of school exceeded that of boys in most of the villages; the proportion of Scheduled Caste (SC) children out of school generally exceeded that of all children; and children with major disabilities were invariably out of school (*ibid.*). Further, as discussed in Chapter 3 of this volume, child labour persisted in most of the villages and particularly among small farmer households, where children worked on the family's operational holding.

One limitation of these data on children's deprivation is worth noting. As might be expected, since the data pertain to different survey years for different villages, educational attainments were found to be higher in villages surveyed more recently, such as Tehang (Jalandhar district, Punjab) in 2011, as compared to villages surveyed almost a decade ago, such as Ananthavaram (Guntur district, Andhra Pradesh) in 2006. Nevertheless, taken together, the data for a cross-section of villages brought out the extent of educational deprivation among children in small farmer households.

EDUCATIONAL ATTAINMENTS AMONG ADULTS

We begin with the rates of literacy among persons aged 7 and above in the study villages (Table 1). Strikingly, overall literacy rates were less than 60 per cent for females and less than 75 per cent (with a few exceptions) for males in

Table 1 *Average literacy rate for females, males, and persons aged 7 years and above, by size-class of operational holdings by farmer category in per cent*

Village	State	Small farmers			Large farmers		
		Females	Males	Persons	Females	Males	Persons
Ananthavaram	Andhra Pradesh	58	61	60	86	86	86
Bukkacherla	Andhra Pradesh	46	63	55	70	89	81
Kothapalle	Telangana	52	70	61	67	75	71
Harevli	Uttar Pradesh	48	61	55	71	89	81
Mahatwar	Uttar Pradesh	43	71	58	69	92	80
Nimshirgaon	Maharashtra	63	85	76	80	87	84
Warwat Khanderao	Maharashtra	66	86	76	85	95	90
Gharsondi	Madhya Pradesh	47	73	61	53	82	69
25F Gulabewala	Rajasthan	60	60	60	70	85	78
Alabujanahalli	Karnataka	58	72	65	79	84	82
Siresandra	Karnataka	49	71	60	63	82	73
Zhapur	Karnataka	45	50	48	38	61	51
Rewasi	Rajasthan	46	76	60	50	75	62
Amarsinghi	West Bengal	66	81	73	–	–	–
Kalmandasguri	West Bengal	54	72	63	–	–	–
Panahar	West Bengal	50	61	55	81	89	85
Tehang	Punjab	75	78	76	76	81	79

small farmer households. The exceptions were the two villages in Maharashtra (Warwat Khanderao, Buldhana district and Nimshirgaon, Kolhapur district) and Tehang, the most recently surveyed village in Punjab. Further, with one exception each, the rates of male and female literacy for persons in small farmer households were lower than for persons in large farmer households in the same village – the exceptions were Zhapur (Kalaburagi district, Karnataka) for females, and Rewasi (Sikar district, Rajasthan) for males. In Harevli village (Bijnor district, Uttar Pradesh), for example, the male literacy rate for persons in small farmer households was 61 per cent as compared to 89 per cent for males in large farmer households. The corresponding figures for females were 48 per cent and 71 per cent.

MEDIAN YEARS OF SCHOOLING

A useful measure of adult achievement with respect to formal education is the median years of schooling among adults, that is, the number of years of schooling completed by one-half of all persons. The distribution of median years of schooling for the population aged above 16 years for small farmers as well as large farmers, and landlords and capitalist farmers, across all the study villages is presented in Table 2.

Among females in small farmer households, the median number of years of schooling was zero in nine villages. In other words, in these villages, 50 per cent of women above the age of 16 had not completed even one year of schooling. In the remaining villages, with one exception, the median was less than five; these were Ananthavaram, Nimshirgaon, Warwat Khanderao, 25F Gulabewala (Sri Ganganagar district, Rajasthan), Alabujanahalli (Mandya district, Karnataka), Amarsinghi (Malda district, West Bengal) and Kalmandasguri (Koch Bihar, West Bengal). The exception was Tehang, where the median was 10 years.

The situation was better among males, but not significantly so. The median number of completed years of schooling for males was less than 10 in all the villages, and it was less than or equal to five in five villages. In Zhapur village, the median number of years of education among males in small farmer households was two.

The median years of schooling for both males and females in small farmer households are generally lower than in large farmer households, and the gap is even bigger when compared to landlord and capitalist farmer households. For example, among females belonging to small farmer households in Harevli, Mahatwar (Ballia district, Uttar Pradesh), and Zhapur the median years of schooling was zero, whereas in the case of landlord and capitalist farmer households in the same villages it was 10 or more years of schooling. Similarly,

Table 2 *Median years of schooling for females and males aged 16 years and above, by village and farmer category*

Village	State	Small farmers		Large farmers		Landlords and capitalist farmers	
		Females	Males	Females	Males	Females	Males
Ananthavaram	Andhra Pradesh	2	4	7	7	8	12
Bukkacherla	Andhra Pradesh	0	5	5	9	7	11
Kothapalle	Telangana	0	7	–	–	9	12
Harevli	Uttar Pradesh	0	5	5	10	12	15
Mahatwar	Uttar Pradesh	0	8	8	15	10	13
Nimshirgaon	Maharashtra	5	9	7	9	11	15
Warwat Khanderao	Maharashtra	4	8	7	10	10	11
Gharsondi	Madhya Pradesh	0	8	0	8	5	10
25F Gulabewala	Rajasthan	3	6	7	8	8	10
Alabujanahalli	Karnataka	4	9	10	10	10	11
Siresandra	Karnataka	0	8	4	10	–	–
Zhapur	Karnataka	0	2	0	3	10	12
Rewasi	Rajasthan	0	6	0	6	0	8
Amarsinghi	West Bengal	4	7	–	–	–	–
Kalmandasguri	West Bengal	2	6	–	–	–	–
Panahar	West Bengal	0	5	10	6	7	12
Tehang	Punjab	10	8	8	8	–	–

Notes: – represents no or few observations.
For the village Tehang, identification of the landlord and capitalist farmer category, though sizeable, is not complete.

in the case of males, the median years of schooling for small farmer households in Zhapur and Ananthavaram were two and four, respectively, whereas for landlord and capitalist farmer households it was 12 years in both villages.

It is clear that the median years of schooling for both males and females in small farmer households were much lower than in landlord and capitalist farmer households across all villages. Further disaggregation shows that the least attainment among females was to be found among those belonging to Scheduled Caste (SC), Scheduled Tribe (ST), Muslim, and Backward Class (BC) small farmer households.

PERSONS WITH 10 YEARS OF SCHOOLING

We now turn to an indicator at the other end of the spectrum, namely, the number and proportion of males and females who had completed 10 years of schooling among those above the age of 25 in the survey year (Table 3).

In the case of females in small farmer households, educational achievement was low in almost all the villages. The proportion of females aged 25 years and above with 10 completed years of schooling among small farmer households was zero in Zhapur and Bukkacherla (Anantapur district, Andhra Pradesh), whereas in Tehang, Alabujanahalli, and Nimshirgaon, it was 36 per cent, 16 per cent, and 14 per cent, respectively. Educational achievement was higher among women who resided in irrigated villages – such as Ananthavaram, Nimshirgaon, Alabujanhalli, and Tehang – than in other villages. This may be on account of more differentiation among small farmers in high-productivity, irrigated villages.

The picture was better in the case of males aged 25 years and above, but in no village did more than one-third of the adult males in small farmer households report 10 years of schooling. The highest proportion was 36 per cent in Tehang; it must be noted, of course, that Tehang was surveyed more recently (in 2011).

Not surprisingly, and quite consistently, men and women in small farmer households had lower educational attainment than those in large farmer households.

HOUSEHOLDS WITHOUT A SINGLE ADULT LITERATE

The presence or absence of literate adults in a household can influence many decisions, including that of seeking access to scientific information on farming practices. Table 4 shows the percentage of small farmer households without a single literate adult across all the study villages. All the villages had

Table 3 *Proportion of females and males aged 25 years and above with 10 completed years of schooling by farmer category, study villages in per cent*

Village	State	Small farmers		Large farmers		Landlords and capitalist farmers	
		Females	Males	Females	Males	Females	Males
Ananthavaram	Andhra Pradesh	10	14	27	35	40	71
Bukkacherla	Andhra Pradesh	0	16	13	34	24	53
Kothapalle	Telangana	5	27	0	0	13	70
Harevli	Uttar Pradesh	8	18	12	54	63	89
Mahatwar	Uttar Pradesh	1	32	0	67	56	89
Nimshirgaon	Maharashtra	14	30	30	45	82	100
Warwat Khanderao	Maharashtra	10	28	11	47	58	50
Gharsondi	Madhya Pradesh	2	17	6	28	29	59
25F Gulabewala	Rajasthan	0#	0#	15	32	34	43
Alabujanahalli	Karnataka	16	31	41	47	75	60
Siresandra	Karnataka	7	25	23	53	–	–
Zhapur	Karnataka	0	15	0	11	33	89
Rewasi	Rajasthan	1	13	3	24	8	39
Amarsinghi	West Bengal	2	20	–	–	–	–
Kalmandasguri	West Bengal	5	7	–	–	–	–
Panahar	West Bengal	4	20	0	0	13	58
Tehang	Punjab	36	36	32	31	–	–

Notes: – represents no or few observations.

For the village Tehang, identification of the landlord and capitalist farmer category, though sizeable, is not complete.

Table 4 *Proportion of small farmer households without any adult literates* in per cent

Village	State	Small farmers
Ananthavaram	Andhra Pradesh	73
Bukkacherla	Andhra Pradesh	81
Kothapalle	Telangana	82
Harevli	Uttar Pradesh	82
Mahatwar	Uttar Pradesh	93
Nimshirgaon	Maharashtra	63
Warwat Khanderao	Maharashtra	65
Gharsondi	Madhya Pradesh	79
25F Gulabewala	Rajasthan	100*
Alabujanahalli	Karnataka	81
Siresandra	Karnataka	96
Zhapur	Karnataka	92
Rewasi	Rajasthan	90
Amarsinghi	West Bengal	70
Kalmandasguri	West Bengal	85
Panahar	West Bengal	81
Tehang	Punjab	65

Note: * There were only two small farmer households in 25F Gulabewala.

a significant number of such households. It can be clearly seen from the table that the percentage of small farmer households without a literate adult ranged from 63 per cent to 96 per cent across all villages – with the exception of 25F Gulabewala which had only two small farmer households. In Nimshirgaon, Tehang, and Warwat Khanderao, the percentage of households without a literate adult was lower, around 63 to 65 per cent, than in the other villages. In four study villages, more than 90 per cent of small farmer households did not have even one literate adult.

CONCLUSION

Two methodological points concerning data on educational indicators for adults in small farmer households need to be highlighted here. First, the 17 study villages were surveyed between 2006 and 2011. Improvements in educational attainment happen over time, and therefore villages that were more recently surveyed presented a better picture of education as compared to those surveyed earlier. Secondly, a comparison between small farmer and large farmer households was not very relevant in villages with very few small farmers (such as 25F Gulabewala), or in villages where small farmers predominated (such as the three West Bengal villages, as well as Kothapalle in Telangana, Mahatwar in Uttar Pradesh, and Siresandra in Karnataka).

Nevertheless, the picture that emerges clearly is that educational attainment – reflected in literacy rates for the population aged 7 years and above – was distinctly lower for persons belonging to small farmer households as compared to persons in large farmer households. This was true for both females and males.

In respect of other indicators of educational achievement, such as median (completed) years of schooling among males and females aged 16 years and above, and the proportion of persons who had completed 10 years of schooling in the 25-plus age category, small farmers again fared poorly relative to large farmers. The situation was better, of course, among males than females even within the small farmer category. Educational attainment was better also in the more agriculturally advanced (irrigated) villages than other villages, as seen in the two villages of Maharashtra, for example. Further, there was large intra-village inequality, with landlord and capitalist farmer households constituting a class of their own in terms of educational attainment.

Overall, the evidence is consistent with the observations of Kautsky and Lenin, cited in Chapter 1 of this volume, on the levels of deprivation that small farmers suffer as they desperately try to survive in the face of advancing capitalism. In India, the post-1991 decline in state support to agriculture, and the processes of privatisation of health care and education that have led to increased costs of education and health have serious implications for the levels of educational deprivation among small farmer families.

The authors of this chapter are grateful to Venkatesh Athreya for his inputs.

REFERENCE

Foundation for Agrarian Studies (FAS) (2014), *Child Well-Being, Schooling and Living Standards: A Summary Report*, Bengaluru.

11

Living Standards of Small Farmers in India

Shamsher Singh

INTRODUCTION

Housing and access to basic amenities of a given population is necessary for ensuring a dignified life, better health outcomes, and increased human productivity. There are, of course, other variables, including levels of education and indicators of health, that must also be factored in while discussing the standard of living of small farmers in India.

This chapter discusses the condition of housing and access of households to selected basic amenities, namely, drinking water supply, electricity for domestic use, and toilets, in a select group of villages across India based on data canvassed from households as part of the Project on Agrarian Relations in India (PARI). In addition, we discuss the types of fuel used for cooking. The condition of housing and access to basic amenities, it may be argued, can act as a proxy for standard of living.

Access to adequate housing and basic household amenities plays an important role in countering historic exclusion and deprivation faced by socially and economically marginalised and disadvantaged sections of a society. Ensuring access to these components of modern life should therefore be an important part of the agenda to deliver social justice and inclusive growth. Indeed, adequate housing, water, and sanitation have been declared as human rights by the United Nations.[1]

---

[1] Article 25 of the Universal Declaration of Human Rights states, "Everyone has the right to a standard of living adequate for the health and well-being of himself and of his family, including food, clothing, housing." (UN 1949).The term 'adequate housing' was first used by the International Covenant on Economic, Social and Cultural Rights (ICESCR), 1966. The Committee in its Article 11.1 says, "The right of everyone to an adequate standard of living for himself and his family, including adequate food, clothing and housing, and to the continuous improvement of living conditions" (UN 2009, p. 11). It was only in 1991 that the United Nations declared right to adequate housing as a basic human right. Right to water and sanitation were recognised as basic human rights by the United Nations General Assembly in 2010.

A BRIEF REVIEW OF VILLAGE STUDIES LITERATURE

Historically, the literature on village studies in India has not focussed on the issue of housing and the access of households to basic household amenities. As a result of this neglect, little detailed information is available on various aspects of rural housing and the availability of basic household amenities across different sections of village society. Classical village studies carried out by social scientists, while they have studied various other aspects of village life in India in great detail, make only a passing mention of the condition of housing, sanitation, and other basic amenities. This extends also to the presence of rural infrastructure such as roads, transport, communication, education, and health services (Dube 1958; Srinivas 1962, 1976; Beteille 1965, 1996; Sharma 1970; Freeman 1979; Epstein *et al.* 1998; Wiser and Wiser 2001).

Wiser and Wiser's *Behind Mud Walls: Seventy-five Years in a North Indian Village*[2] is one among a few village studies that discusses, in detail, the living conditions in the village and the state of its amenities. The authors describe housing in terms of the construction material used, the condition of structures, ventilation, and the location of residences of "untouchables" (Scheduled Castes or Dalits were formerly known as "untouchables") (Wiser and Wiser 2001, p. 3).

Epstein *et al.* wrote that their study village in Karnataka did not have electricity and other basic amenities, in 1954 (Epstein *et al.* 1998, pp. 23–24). Instead, glass-covered kerosene lanterns were used for lighting (*ibid.*, p. 40). They further noted that peasant households lived in houses made of hand-made tiled roofs and mud walls. Adi-Karnataka (Scheduled Caste) households lived in thatched roof structures (*ibid.*, p. 41).

K. L. Sharma describes housing and living conditions in his study villages in Rajasthan. He finds differences among upper-caste and upper-class households, and lower and intermediate castes and classes, in respect of housing and lifestyle. Sharma observed a huge difference in the living conditions and cultural lifestyle of the privileged minority and non-privileged majority (Sharma 1970, reprinted in Jodhka 2012, p. 88).

M. N. Srinivas, in his famous book, *The Remembered Village*, describes different aspects of rural dwellings, including the use of spaces for different activities in the house he lived in during his study of the village (Srinivas 1976, pp. 11–12). He highlights the settlement patterns in the village, and caste practices relating to drinking water, sanitation, and bathing (*ibid.*, pp. 15–17).

---

[2] Originally published in 1930.

Access to Water and Caste Discrimination in Rural India

The literature on village life in India describes the practice of untouchability and other forms of pervasive discrimination against the Scheduled Castes (Dalits) in accessing sources of water. A document prepared by the Indian Council for Social Science Research (ICSSR 1973) notes, "Drinking water has been the most critical domain of the practice of untouchability" (cited in Tiwary and Phansalkar 2007, pp. 45–46).[3] Sociological studies of villages have long pointed to the role of caste and social identity-based restrictions and discrimination in the use of common resources and public infrastructure, including sources of water.[4] Indeed, the continuation of caste discrimination – including untouchability – in rural India has been the focus of several studies in recent decades: Joshi and Fawcett (2001); Krishnan *et al.* (2003); Shah *et al.* (2006); Soni (2006); Bhatia (2006); Tiwary and Phansalkar (2007); Thorat (2009); Acharya (2010); Jodhka and Shah (2010); Ramachandran, Rawal, and Swaminathan (2010); Ramachandran (2014); Swaminathan and Singh (2014); Singh (2015).[5]

While village studies have described housing conditions and standards of living while highlighting the practices of untouchability and discrimination against Scheduled Castes and other marginalised social groups in accessing social and common infrastructure in villages, only a few of them have actually tried to quantify the deprivation, unequal access, and disparities among different socio-economic groups and classes in Indian villages. There exists a large gap in research on the issue of access to housing and basic household amenities across different strata of the peasantry, farmer households, and other occupational and socio-economic classes in the Indian countryside. This chapter makes an attempt to fill this gap through a presentation of the actual attainments across different strata of cultivating classes in a wide and diverse range of villages.

---

[3] This study was conducted by Professor I. P. Desai in 1971 in Gujarat, to study the practices of untouchability against Dalits in accessing water sources.

[4] For more details, see Ghurye (1932, 1957); Dube (1958); Srinivas (1962, 1976); Desai (1973, 2013); Freeman (1979); Beteille (1996).

[5] Other important and useful sources which provide quantitative information on rural housing and amenities are official data agencies like the National Sample Survey Office (NSSO), the Census of India, and the National Family Health Survey (NFHS); some of the limitations of these sources have been discussed in Singh, Swaminathan, and Ramachandran (2013).

CONDITION OF HOUSING AND ACCESS TO BASIC HOUSEHOLD AMENITIES

This section discusses various aspects of the condition of housing and access to basic household amenities for all households, but with a specific focus on the category of small farmers. We pose the question: "How do small farmer households fare in absolute and relative terms in respect of access to housing and amenities, in comparison to households belonging to other agrarian classes in the study villages?" In the course of this analysis, we also study these disparities in the light of the social and caste identity or affiliation of small farmer households.

Ownership Rights over Housing

Having ownership rights over the place of residence is an important aspect of the housing question. Lack of ownership rights over the dwelling leads to insecurity and threat of eviction. The international literature on housing, including guidelines of the United Nations, lays emphasis on the importance of security of housing tenure. The guidelines say that housing is not adequate if the occupants do not have tenurial rights and security. It is this that guarantees legal protection against forced evictions, harassment, and other threats (UN 1991a).[6]

Findings from our village-level data show that an overwhelming majority of households in the study villages across socio-economic classes, including small farmers, had their own houses. Appendix Table 1 shows that 96 per cent of all small farmer households owned their dwellings. This proportion was 100 per cent among landlord and capitalist farmer households, and 85 per cent among artisan and manual labour households. The proportion of house ownership was noticeably low among small farmer households in two Andhra Pradesh villages, one Telangana village, and one village of Uttar Pradesh. In these villages, many of the small farmer households were tenants, and they resided in shelters provided by relatives or others.

The issue of house ownership is linked with ownership of homestead land, which again is linked with ownership of agricultural land. Households that owned agricultural land also owned their dwellings.[7]

---

[6] Of the seven guidelines suggested by the United Nations Committee on Economic, Social and Cultural Rights on adequate housing (UN 1991b), the first one is security of tenure. Similarly, UN-HABITAT and OHCHR (2002), which are the UN housing monitoring agencies, use proportion of homeless persons and security of tenure of habitants as important indicators of housing rights.

[7] This has been discussed in detail in Singh (2014).

Structural Type and Quality of Housing

A house structure has three integral components, namely, the roof, floor, and walls. The materials used in constructing these three components determine the type and quality of the structure. They also determine the strength, durability, comfort, protection, and safety provided to the residents of the structure in all weather conditions.

Official data sources in India, while defining type of house structure, take into consideration the materials used in the construction of its walls and roof. Construction materials are categorised into two: *pucca*/permanent and *katcha*/temporary.[8] This categorisation is used to define types of structures.[9] However, this categorisation is based only on durability and does not take into account basic norms of housing.[10] For example, construction materials such as tin, metal, and asbestos sheets are classified as permanent material even though structures made with them do not provide protection from extreme weather conditions, and pose health and safety hazards. Another serious problem with this definition is that construction materials used for the floor are not included in the definition of the structure. The exclusion of floor material in the definition of type of structure is a violation of the international norms of adequate housing.[11] This omission also has an important gender dimension, as the burden of maintaining a *katcha* (mud) floor falls on the women of a household (see discussion below).

---

[8] The Census of India includes stone packed with mortar, GI/metal/asbestos sheets, burnt brick, and concrete in permanent materials. Temporary materials refer to grass/thatch/bamboo, plastic/polythene, mud/unburnt brick, wood, stone not packed with mortar. See Census of India (2001). According to the NSSO, *pucca* (equivalent to permanent) material includes: cement, concrete, oven-burnt bricks, hollow cement/ash bricks, stone, stone blocks, jack boards (cement-plastered reeds), iron, zinc or other metal sheets, timber, tiles, slate, corrugated iron, asbestos cement sheet, veneer, plywood, artificial wood of synthetic material and polyvinyl chloride (PVC) material. NSSO *katcha* (equivalent to temporary) material includes: unburnt bricks, bamboo, mud, grass, leaves, reeds, thatch, etc. See NSSO (2010).

[9] Permanent/*pucca* house: house with wall and roof made of permanent materials.
Semi-permanent/*pucca* house: either the wall or the roof is made of permanent material (and the other is made of temporary material).
Temporary/*katcha* house: house with wall and roof made of temporary materials.
Serviceable temporary/*katcha*: the wall is made of mud, unburnt bricks, or wood. Non-serviceable temporary/*katcha*: the wall is made of grass, thatch, bamboo, plastic or polythene.

[10] The Workers' Housing Recommendations of the International Labour Organisation (ILO) mention that a dwelling should provide "appropriate protection against heat, cold, damp, noise, fire, and disease-carrying animals, in particular, insects" (ILO 1961).

[11] "Dirt floors not only make domestic hygiene difficult, but may harbor helminthes" (WHO 1989, p. 8).

In order to avoid this weakness of the official definition of a dwelling, we propose an alternative definition of a permanent/*pucca* structure as one where all three parts of the structure – that is, roof, walls, and floor – are made of permanent or *pucca* materials. Our results from the PARI study villages (see Appendix Table 2) show that 82 per cent of landlord and capitalist farmer households and 64 per cent of large farmer households across all the villages lived in fully *pucca* structures, while only 41 per cent of small farmer households did so. The proportion of small farmer households living in fully *pucca* structures across the study villages ranged from 1 per cent in Kalmandasguri (Koch Bihar district, West Bengal) to 82 per cent in Alabujanahalli (Mandya district, Karnataka). Of a total of 16 study villages which are discussed here, in five villages, less than one-fourth of small farmer households lived in fully *pucca* structures. Residents of all villages that had an exceptionally low proportion of permanent houses reported using construction materials such as paddy straw and jute straw, which were locally available and inexpensive.

In three villages of Karnataka – Alabujanahalli, Siresandra (Kolar district), and Zhapur (Kalaburagi district) – a very high proportion of small farmer households lived in fully *pucca* houses. Use of locally made burnt tiles, called Mangalore tiles, which were less expensive than conventional permanent construction materials such as brick, cement, and concrete, have improved roofing across all social groups. In Siresandra, a majority of the households used metal and asbestos sheets for constructing roofs, and, as a result, were classified as residents of houses with *pucca* roofs. Asbestos sheets are under criticism, however, for their harmful effects on the health of residents. In Zhapur, the high proportion of *pucca* houses was on account of easy availability of stone, as this village has many stone quarries nearby.

Further analysis of the village data shows that more than 70 per cent households, except landlord and capitalist farmer households, lived in structures that had floors made of mud and dung. Female respondents from the study villages reported that maintaining a mud floor was highly labour-intensive, and that they had to devote a large amount of time and energy to this task. What follows is a description of the experience of a woman from Ananthavaram village in Guntur district, Andhra Pradesh, who lives in a house with mud floors.

Anjamma, 24, lives in a house of one room and a kitchen. The household is landless and leases in 2.5 acres of land. Anjamma and her husband, Rosayya, 32, both work as manual labourers, apart from their work on the leased-in land. Their room is a semi-*pucca*/permanent structure. The walls of their dwelling are made of brick and mud, and for the roof they have used paddy straw and dry maize plants, which they get as their share of by-products from leased-in land every season. The couple have two daughters aged four and one-and-a-half. Anjamma said that she has to frequently skip work on their land, as well as wage employment, just to wipe the floor with mud-and-dung paste. The mud floor requires multiple coatings every couple of days. During the monsoon her burden increases even further. She said that while she does her household chores, she has to carry her younger daughter, as she cannot leave the toddler on the mud floor to play because of the health hazards.

It is clear from the analysis that an overwhelming majority of households live in houses with mud floors. By not including the floor component in the definition of a structure, the official data sources hugely underestimate the number and proportion of non-*pucca* houses in rural India. This misrepresentation of the condition of structures has serious policy implications for housing. The official definition of type of structure needs to be broadened by including flooring material so that it reflects the actual condition of dwellings.

Discussions with village residents reveal that there are several factors that influence a household's choice of construction material – cost, local availability, suitability of the material to the local weather conditions, and cultural and traditional preferences. Further analysis of the data shows that households used multiple combinations of construction materials. For example, commonly used material combinations for roofs were thatch/paddy straw and metal sheets; brick/tile and cement; brick and mud; metal sheet, bamboo and mud; concrete, tile, and wood; stone, brick, and wood; concrete and thatch; stone and wood. In the construction of walls, fewer combinations were used. Walls were mainly made of brick and mud, and, in many cases, cement plaster. Depending on the circumstances of the families, they either cement-plastered the brick and mud walls entirely, or applied cement only on the joints.

It was also seen that rich households used relatively costly material combinations such as brick and cement, stone and concrete, only brick, only cement and iron. Households that could not afford costly materials used relatively cheap materials such as unburnt bricks and tiles, thatch, metal and asbestos sheet, wood and bamboo, polythene, etc.

The urban influence on house construction, especially among richer households, was clearly visible in the study villages. This was reflected in construction materials, construction technologies, layout and space management. This reflects the expansion of markets to the rural areas, and improvements in the skills of local masons and their adaptation of modern methods of construction.

Table 1 *Proportion of households living in fully pucca/permanent houses, by social group among small farmer class, all study villages* in number and per cent

Social group	No. of households	Per cent
Scheduled Caste (SC)	93	22
Scheduled Tribe (ST)	10	23
Muslim	4	7
Other social groups (all remaining)	518	52
All	625	41

Note: The classification of materials into *pucca*/permanent and *katcha*/temporary categories is similar to that in the Census and NSSO. 'All' includes all households belonging to the small farmer class across all the study villages.
Source: PARI survey data.

The analysis here shows that locally available construction materials played an important role in the construction of both permanent and temporary dwellings, and especially for poor households.

Table 1 shows differences across various social groups among small farmers with respect to having fully *pucca* structures as dwellings. While 41 per cent of small farmer households lived in fully *pucca* structures, only around one-fifth of all Scheduled Caste (SC) households among small farmers did so. The situation was similar for Scheduled Tribe (ST) households. There were very few Muslim households in the small farmer class, and of them a very small proportion, only 7 per cent, lived in fully *pucca* structures. Differences across social groups among the small farmer class were very evident, as more than half of the households belonging to other social groups lived in fully *pucca* structures.

Availability of Living Space

The availability of living space in a dwelling is an important indicator of the quality of housing, crucial as it is for organising daily life and routine by its residents according to their needs. For example, an ill person requires adequate separation from other household members, schoolgoing children require a separate space to study, and women and married couples need private spaces and separation. Lack of space and privacy can cause serious inconvenience to the inhabitants of a household, and can have serious implications for the domestic environment, psychological well-being, and interpersonal relationships of household members.[12]

[12] Overcrowding, particularly in conjunction with poverty and inadequate facilities, has been shown to increase the transmission rates of such communicable diseases as tuberculosis, pneumonia, bronchitis, and gastrointestinal infections (WHO 1989, p. 8). Also see UN-HABITAT and OHCHR 2002, p. 5.

Living space, defined as built-up area, is used for living and sleeping purposes. In some cases the space is used for sleeping as well as for other purposes such as storage, or activities related to home manufacturing excluding cooking. The space where cooking happens is not included in our definition of living space.

The adequacy of living space in a dwelling can be measured in terms of floor area, cubic volume, or size and number of rooms (for details, see UN-HABITAT 2006, p. 82). Here we use the room criterion.

Appendix Table 3 shows that a little more than one-fourth of all small farmer households across the study villages lived in highly congested, single-room structures. This proportion was only 6 per cent among large farmer households, and no landlord and capitalist farmer household lived in a single-room structure. The proportion of small farmer households living in single-room structures ranged from 1 per cent in Gharsondi (Gwalior district, Madhya Pradesh) to more than half, i.e. 51 per cent, in Ananthavaram (Guntur district, Andhra Pradesh). In six out of all the study villages, more than one-third of small farmer households lived in single-room houses. The analysis shows that large farmer households lived in houses with more living space. Only in two study villages did 20 per cent or more of large farmer households live in single-room dwellings. In a majority of the villages, this proportion was 0 or less than 5 per cent. Having adequate living space is closely linked to the availability of homestead land. For example, in Rewasi village (Sikar district, Rajasthan) and to some extent in Gharsondi village, where a large number of small farmer households have constructed their houses in fields outside the main village settlements, the proportion of single-room dwellings was lower than in other villages. The data show that a large proportion of artisan and manual labour households, of which a majority were landless, lived in single-room houses.

A close look at the data show that in Ananthavaram, Bukkacherla (Anantapur district, Andhra Pradesh), Kothapalle (Karimnagar district, Telangana), and Harevli (Bijnor district, Uttar Pradesh), where the proportion of house ownership was relatively lower among small farmer households, the proportion of single-room structures was high. Small farmer households in these villages lived in single-room dwellings provided by others. Another interesting fact regarding the quality of housing comes from an analysis of two villages, Siresandra and Zhapur (both in Karnataka, in Kolar and Kalaburagi districts, respectively). Here the proportion of small farmer households living in permanent structures was relatively high (79 and 77 per cent, respectively), and the proportion of small farmer households living in single-room structures was also relatively high (35 and 48 per cent, respectively). This shows that

Table 2 *Proportion of households living in single-room house, by social group among small farmer class, all study villages* in number and per cent

Social group	No. of households	Per cent
Scheduled Caste (SC)	135	32
Scheduled Tribe (ST)	24	54
Muslim	14	26
Others (all remaining)	225	23
All	398	26

Note: Rooms that were used for living and sleeping purposes were recorded. "All" includes all households belonging to the small farmer class across all the study villages.
Source: PARI survey data.

even when quality in terms of construction material improved, living space remained a problem.

Internationally, two persons per room is a recommended norm (UN 2003). During household interviews in the study villages, we observed that dwellings of small farmer households in most cases accommodated four to five persons per room, and this number went up to seven to eight members in some cases, mostly in single-room houses. We came across households where family members spanning three generations shared a single room, accommodating married couples, children, and other members of the household. Another issue with the living space arrangements of small farmer households was the location of cattle.[13] Animal husbandry is an important activity and component of the small farmer household economy. As discussed above, small farmer households lack access to enough homestead and resources to expand their dwellings. The problem of crowding becomes more severe when these households have to accommodate cattle in already insufficient living space.

Table 2 shows that there were high levels of differentiation on the basis of social identity among small farmers with respect to rooms. Almost one-third of SC and more than half of ST households lived in single-room dwellings, whereas this proportion was only 23 per cent among "others."

> Santosh, aged 41, who belongs to a Dalit caste from Zhapur village, owns 1 acre of unirrigated agricultural land and leases in another 2 acres. There are a total of six members in the household, including his wife Ramabai, aged 35; two daughters, Laxmi and Sama, 20 and 15 years old respectively; their son Vijay, who is 10 years old; and Santosh's mother Rathnamma, aged 70.

[13] According to international standards of adequate housing, there should be a suitable separation of living rooms from quarters for animals (UN-HABITAT 2006, p. 82).

Other than working on their own and leased-in land, Santosh, Ramabai, and their elder daughter Laxmi work in stone quarries as daily labourers. Their other daughter and son study in class 8 and class 5, respectively. Santosh has a homestead of 15 x 10 feet that he inherited from his father when he separated from him a few years ago. There is one living room of 10 x 10 feet dimension, and the remaining area is covered by an asbestos sheet and used as a veranda. The veranda is also used for cooking. Santosh says that during summers he and Vijay sleep in the veranda, and all the female members of the household sleep in the room. The household owns a pair of bullocks used for their own cultivation, which they also rent out for work on others' land. They keep the bullocks in the public street in front of their house. In winters, all six family members sleep in the room and the bullocks are kept in the veranda. Within the room they have made a temporary partition by putting up a sheet made of empty fertilizer bags.

Ramabai says that Sama has failed twice in class 8 and Vijay also does very poorly in studies as they are unable to study properly at home due to lack of space in the house. Laxmi too had to give up her studies after failing class 8 despite several attempts.

Provision of Kitchen

The provision of a separate space for cooking is an important aspect of the quality of housing. Separation of the kitchen from living areas is important for hygienic cooking and for maintaining a healthy environment inside the house. While a separate kitchen is of great importance for the well-being of all members of a household, its absence affects the women in a household more adversely as they are the ones who do the cooking in most cases. A separate kitchen and the type of fuel used for cooking are closely interlinked. Lack of a kitchen exposes women to more health hazards than men in the same household. While collecting data on housing in our village surveys, we defined the kitchen as "a separate room used for either cooking or storing cooking material or both, and not for any other purpose."

Appendix Table 4 shows that there were very significant disparities with respect to the presence of kitchens among different classes of farming households. On the one hand, almost all the landlord and capitalist farmer households, and more than 80 per cent of the large farmer households, had kitchens in their houses. On the other hand, only a little more than 60 per cent of small farmer households had this provision. This means that in a large number of households belonging to small farmers, cooking was done in unhygienic conditions – either in the open or in the living area. As a result, the health of the residents of these dwellings was compromised.

If we look at figures village-wise, we find that a large proportion of small farmer households in two Karnataka villages – Alabujanahalli (Mandya district) and Siresandra (Kolar district) – reported having a separate kitchen. Some interesting facts emerged when this issue was further probed. Residents from these two villages told us that the provision of a separate space for the kitchen had been a part of their cultural heritage and traditional house layout, and that they had continued with this traditional component of housing while adopting modern house construction formats. We observed this to be the case more among households belonging to the dominant castes.

Similarly, three-fourths of all small farmer households in Warwat Khanderao village (Buldhana district, Maharashtra) reported having a kitchen. Focus group discussions with the residents of the village revealed that the type of fuel used for cooking played an important role in this. Cotton being a major crop of the area, cotton bushes served as the main fuel for cooking along with other crop residues. In Kalmandasguri village (Koch Bihar district, West Bengal), where 96 per cent of small farmer households had a kitchen, jute was a major crop and jute sticks were used as fuel for cooking. In these two villages mentioned above, after harvesting the crops, households stored cotton bushes and jute sticks in the kitchen room. The requirement of storing fuel separately encouraged persons in the household to make a separate room. It was also noticed that small farmer households in both these villages owned substantial homestead plots, which made it possible for households to construct a room for this purpose.

A relatively small proportion of small farmer households in Bukkacherla (Anantapur district, Andhra Pradesh), Kothapalle (Karimnagar district, Telangana), Harevli (Bijnor district, Uttar Pradesh), and Mahtwar (Ballia district, Uttar Pradesh) villages had a kitchen. It is interesting to note these are the same villages (with the exception of Mahatwar) where a high proportion of small farmer households lived in single-room accommodations.

Though most of the residents in the survey villages reported having a kitchen, we noticed that a majority of these households used the kitchen rooms to store firewood, utensils, and groceries, and the actual cooking activity happened in the courtyard of the house or in other available open spaces. The main reason for cooking in the open was the use of solid and smoke-generating fuel by a majority of the village residents.

Type of Fuel Used for Cooking

The type of fuel used for cooking is important in terms of the individual's health as well as the health of the broader environment. Using smoke-generating fuel for cooking can cause serious health problems, especially when the cooking happens indoors. While the type of fuel used for cooking affects the well-being and health of all household members, women and children are particularly vulnerable.

> There is growing evidence that exposure to indoor smoke can cause serious respiratory and other adverse health effects. There is compelling evidence linking indoor smoke to acute respiratory infections in children and chronic obstructive pulmonary disease (COPD) or chronic bronchitis in women. (Mishra 2004, p. 2, cited in IIPS and Marco International 2007, p. 412)

In our discussion here, we use the National Family Health Survey–3 (NFHS–3) classification of smoke-generating fuels, namely, wood, animal dung, crop residues/grasses, coal, and charcoal under solid fuels (*ibid.*). For non-smoke-generating fuels such as biogas, LPG, and electricity, we use the term "improved" fuels. Table 3 shows the type of fuel used by farmer households in the study villages.

The table shows that the picture regarding the type of fuel used for cooking was an alarming one. Findings from the surveys show that an overwhelming majority of households in the study villages used solid (smoke-generating) fuel for cooking. The proportion of households using solid fuel among small farmers was 97 per cent and among large farmers, 91 per cent. In almost half the villages for which data are available, not a single household in the small farmer class used improved (non-smoke-generating) fuel for cooking. The situation was equally bad in large farmer households in the study villages.

The data reveal that while only a negligible proportion of households had provision of LPG, which is an improved fuel due to its high cost, they used it sparingly, for tasks like making tea and boiling milk, and only on occasion for preparation of meals. On a regular basis, food was prepared by using firewood, dung-cakes, and crop residue.

Table 4 shows the situation across different social groups among small farmers with respect to type of fuel used for cooking. It shows that even though an overwhelming majority of households belonging to the small farmer class used solid fuel for cooking, there were still disparities among social groups in this regard. While 4 per cent of small farmer households belonging to "other" social groups used improved fuel for cooking, this proportion was 1 per cent

Table 3 *Proportion of households using solid fuel for cooking, by selected socio-economic classes in study villages in number and per cent*

Village	State	Small farmers		Large farmers		All	
		No. of households	Per cent	No. of households	Per cent	No. of households	Per cent
Rewasi	Rajasthan	105	98	37	100	214	99
Alabujanahalli	Karnataka	101	89	19	76	217	90
Siresandra	Karnataka	56	100	3	75	78	99
Zhapur	Karnataka	26	100	9	100	105	100
Amarsinghi	West Bengal	53	98	–	–	125	98
Kalmandasguri	West Bengal	68	100	–	–	146	99
Panahar	West Bengal	141	99	1	100	241	98
Total		550	97	69	91	1126	97

Notes: In the PARI surveys, this variable (source of energy for cooking) was added from the Karnataka round (2009) onwards.
– not applicable.
All = all households of a village.
Source: PARI survey data.

Table 4 *Proportion of households using solid fuel for cooking, by social group among small farmer class, all study villages* in number and per cent

Social group	No. of households	Per cent
Scheduled Caste (SC)	156	99
Scheduled Tribe (ST)	34	100
Muslim	20	100
Others (all remaining)	340	96
All	550	97

Note: "All" includes all households belonging to the small farmer class across all study villages.
Source: PARI survey data.

among SC households, and there was no household from ST and Muslim groups that used an improved fuel for cooking in the study villages.

ACCESS TO ELECTRICITY, DRINKING WATER, AND SANITATION

Access to Electricity

Provision of electricity in a dwelling is one of the important amenities of modern life. Though the uses of electricity in rural households in India are diverse, and depend on the electrical equipment and assets the household owns, one major use of electricity in rural India across households is lighting. Not having access to this amenity can cause isolation, backwardness, and exclusion in our fast-changing world.

A striking feature that emerges from the PARI survey data is that not a single village among the study villages was 100 per cent electrified (Appendix Table 5). The data show that one-fifth of all small farmer households across the study villages did not have access to an electricity connection (authorised or unauthorised) for domestic use. Across all farmer households in the study villages, small farmer households lagged behind. There were five villages where more than 40 per cent of small farmer households did not have access to an electricity connection. The proportion of small farmer households without an electricity connection ranged from 2 per cent in Kothapalle (Karimnagar district, Telangana) and Siresandra (Kolar district, Karnataka), to 96 per cent in Kalmandasguri village (Koch Bihar district, West Bengal).

It is important to note that in 2010, at the time of the survey of Kalmandasguri, the village was not electrified. Four households, situated on the boundary of the village and near another village, had taken connections from the distribution line of that neighbouring village. During our resurvey of the village in 2015 we found that the village was electrified. The second highest proportion of small farmer households not having electricity was in Harevli (Bijnor district, Uttar Pradesh). In this village, the entire Dalit settlement, which comprised small farmers, tenants, and manual wage labourers, was segregated from the main village settlement, which comprised upper-caste and middle-caste households. The Dalit settlement was not only segregated spatially, but also in respect of electricity and piped water supply, which the rest of the village was provided with.

Another feature of access to electricity in the study villages was unauthorised electricity connections. A significant proportion of households in some of the villages had such connections. Our data show that of all small farmer households, 9 per cent households in Panahar (Bankura district, West Bengal), 16 per cent in Nimshirgaon (Kolhapur district, Maharashtra), and 40 per cent in Harevli were accessing electricity through unauthorised means. This involved drawing electricity through a wire directly from the main distribution pole or a nearby distributing line, and accessing electricity from other households without an authorised metre. In some cases, bribes were paid to local electricity department employees who facilitated unauthorised access to these households. These practices were prevalent in other study villages as well, but at a negligible level. Our discussions with residents of the above-mentioned villages revealed the reasons behind such practices. Residents of Harevli, where the proportion of households having unauthorised connections was the highest, said that electricity supply in the village was very poor and erratic. The village got power supply only for a couple of hours in the entire day and that too at very low voltage. In addition, the households of the village got very high electricity bills every month. The residents felt that it was not worth taking an authorised electricity connection for such poor power supply. Some residents complained about administrative hurdles, and the lengthy and bureaucratic procedure of getting an authorised connection.

The pattern of disparities among different social groups in the small farmer class is well reflected in Table 5. It shows that one-fifth of all small farmer households did not have an electricity connection. Of these, one-fourth of SC households, and more than 40 per cent of ST and Muslim households were deprived of this amenity. A lower proportion, 16 per cent, of "other" households lacked an electricity connection in the study village among all small farmer households.

Table 5 *Proportion of households not having a domestic electricity connection, by social group among small farmer class, all study villages* in number and per cent

Social group	No. of households	Per cent
Scheduled Caste (SC)	105	25
Scheduled Tribe (ST)	19	43
Muslim	23	43
Others (all remaining)	160	16
All	307	20

Note: "All" includes all households belonging to the small farmer class across all the study villages. Having an electricity connection includes both authorised and unauthorised connections.
Source: PARI survey data.

Access to Drinking Water

There are two important aspects to the issue of access to drinking water: first, the quality or type of source/s of water used for drinking purposes; secondly, the location or distance of the source from the dwelling.

Type of drinking water source

During the PARI village surveys, information on the quality of source of drinking water was not collected due to various technical limitations. Here, in order to define the quality of the water, we have divided sources of water into two categories: covered and open sources. Covered water sources include taps, hand pumps, tube wells, and bore wells, where water is supplied through pipes. Sources such as rivers, ponds, lakes, open tanks, wells, and streams come under open sources.[14]

Appendix Table 6 discusses village-wise proportion of households across different classes of farmers having access to open sources of drinking water. The table shows that the proportion of small farmer households using open sources of water for drinking purposes was double those belonging to landlord and capitalist farmer households, i.e. 10 per cent and 5 per cent respectively, across all the study villages. Though this proportion was zero or negligible in most of the villages, in a few villages it was significant: more than one-fourth of small farmer households in Kothapalle (Karimnagar district, Telangana), more than one-third in Nimshirgaon (Kolhapur district, Maharashtra), 13 per cent in Rewasi (Sikar district, Rajasthan), and 14 per cent households in Gharsondi (Gwalior district, Madhya Pradesh) reported using open sources of

---

[14] Official data sources in India use covered or piped sources of water as a proxy for safe or improved sources of drinking water, and open sources as unsafe. However, this categorisation has its own problems as covered sources are not necessarily always safe and vice versa.

water. In most of these villages there was piped water supply, but the residents claimed that tap water was not good enough to drink. Therefore households in these villages used water from various community wells. Across the study villages, handpumps were a major source of drinking water.

Location of drinking water source

The distance of a source of water from the dwelling has a direct impact on the quantity of water collected by the household, and influences hygiene and health outcomes (Howard and Bartram, cited in WHO 2003). The location and distance of source of water has a very important gender dimension too, as the responsibility of collecting water is invariably a task performed by women and girls of the household. In the absence of a source that is either inside or close to the dwelling, female members of the household end up investing a substantial amount of their time and energy in fetching water. Having to travel far to fetch water also makes women vulnerable to physical and sexual abuse.

Appendix Table 7 shows the proportion of households having their source of drinking water within the premises of the dwelling. Differences across farmer classes were stark in this respect as well. On the one hand, more than 80 per cent of landlord and capitalist farmer households, and around 60 per cent of large farmer households had sources of drinking water within their house premises; on the other hand, only 38 per cent of small farmer households had this amenity. The provision of a source of drinking water within the homestead varied a great deal across the study villages. While in three out of 15 study villages more than 70 per cent of small farmer households had a source of drinking water within their homestead, in the remaining study villages this proportion was far below one-fifth of all households.

Further analysis shows that for a little more than half or 55 per cent of small farmer households, the source of water was within a distance of 200 metres, while 2 per cent of the households fetched water from a distance of more than 1 kilometre.

Kalmandasguri (Koch Bihar district, West Bengal), Rewasi (Sikar district, Rajasthan), and Harevli (Bijnor district, Uttar Pradesh) were all villages with a relatively high proportion of small farmer households having a source of drinking water within the homestead. In two of these three villages, namely Rewasi and Harevli, there were panchayat-operated water supply mechanisms, and individual household connections were provided on payment of installation and monthly charges. Personal handpumps and borewells were also present in high numbers in these villages. In Kalmandasguri, most of the households had received handpumps from the government. Though some of the other study villages, like Bukkacherla (Anantapur district, Andhra Pradesh), Warwat Khanderao (Buldhana district, Maharashtra), Siresandra (Kolar district, Karnataka), and Zhapur (Kalaburagi district, Karnataka), also had panchayat-operated water supply systems, water was supplied through common stand-posts rather than individual household connections.

High one-time installation costs in addition to monthly charges pushed economically poor households out of piped water supply systems. For example, in Nimshirgaon village, a World Bank-funded water supply scheme had replaced the already existing panchayat-operated system. Under the new scheme, installation charges and monthly tariffs were hiked multiple times in comparison to the earlier scheme, and public stand-posts were removed. This actually deprived a substantial section of poor households of access to drinking water.

By contrast, the experience of Warwat Khanderao village showed that imaginately designed schemes could reverse prevailing exclusion of poor and marginalised sections. In this village, a Left candidate, Gopal Galkar, was elected as village *sarpanch* (council head) three times in a row. With the help of the residents of the village and people's mass organisations, he pressurised the district administration to release funds for water supply to the village. His efforts and initiative bore fruit. The village water supply was strengthened and expanded. Earlier there were only a few public stand-posts in Warwat Khanderao. The village panchayat under Gopal Galkar's leadership installed two new borewells and households were given individual connections at very reasonable charges. The panchayat also increased the number of public stand-posts so that those who could not take individual connections could access water from these outlets.

Discrimination against Scheduled Caste households
Indian villages have been sites of discrimination and brutal forms of untouchability against people of certain groups on the basis of social origin, and these practices are particularly prominent when it comes to access to water sources. Though we did not study these practices in detail during our village surveys, we did observe blatant discrimination against Scheduled Caste (SC) households in terms of water-sharing and access to water.

First, our data show that in the small farmer class, the proportion of SC households that used open sources of water was lower than among large farmers. On the face of it, this looked like a positive attainment. However, this was not the case. Rather, it was on account of the inability of SC households to access common wells in some villages. For example, in Gharsondi (Gwalior district, Madhya Pradesh) and Zhapur (Kalaburagi district, Karnataka), where well water was used for drinking purposes, Backward Class (BC) and upper caste households did not allow households from SC communities to access common wells, claiming that these "belonged" to their communities.

Secondly, if we look at the results from the surveys regarding location of the source of drinking water across social groups, we find that overall only one-third of SC households had a source of drinking water within their dwellings, whereas this proportion was 40 per cent for others. In some villages – for example, Harevli (Bijnor district, Uttar Pradesh), Mahatwar (Ballia district, Uttar Pradesh), Ananthavaram (Guntur district, Andhra Pradesh), and Kothapalle (Karimnagar district, Telangana) – a higher proportion of SC households had sources of drinking water within their homestead. This was on account of the discrimination faced by these communities in terms of access to common water sources. Respondents belonging to SC groups in these villages told us that it was because of the discrimination they faced that priority was given to them in respect of installing a water supply system in the homestead, although they had to raise money for this by investing their savings and often by borrowing.

It was also noticed during our field visits that either SC habitations were completely deprived of piped water supply or the common sources that had been installed were very few in number. Often, the common sources were not attended to or repaired by the local authorities when they malfunctioned (for details, see Singh 2015). Generally, SC households across most of the study villages had to depend on multiple sources for their water requirements.

Access to Toilets

Access to clean sanitation facilities is crucial for better health in society. Open defecation, a dismal feature of the Indian countryside, compromises the individual's dignity, safety, and health. It pollutes the environment and thus affects the well-being of society at large. Here we discuss the situation with respect to sanitation among small farmer households in the PARI study villages.

Our data (Appendix Table 8) show that the inmates of more than half the households in the study villages did not have access to a lavatory and were forced to defecate in the open. This was the norm for a majority (58 per cent) of small farmer households in these villages. We found that not all households – even those belonging to the richest and most resourceful strata, like landlords and capitalist farmers – in the study villages had access to a lavatory, though this proportion was low (15 per cent) in comparison to other households in the villages. There were very high disparities in this regard across different farmer classes. Around one-third of "other" farmer households were without a lavatory.

In nine out of 15 study villages, more than 60 per cent of small farmer households defecated in the open. Zhapur stood out in terms of this practice. There was only one household in this village, out of a total of 106 households, that had access to a lavatory. In Mahatwar (Uttar Pradesh) and Siresandra (Karnataka), more than 90 per cent of small farmer households defecated in the open.

There were some exceptions to this situation with some study villages showing better coverage of sanitation facilities. For example, in Nimshirgaon (Maharashtra), three-fourths of small farmer households had access to a lavatory. Improved sanitation in this village was due to the active efforts of the village panchayat. In an interview, the panchayat *sarpanch* (panchayat head) of the village said:

> While going for a morning walk towards my fields I used to come across so many men, women, and children of all ages defecating in the fields along the village road. This was my everyday routine, and I felt really ashamed of this situation. These encounters used to cause a lot of inconvenience to me and others, especially to women. After my retirement from service, I decided to change this situation. I decided to run for the post of panchayat *sarpanch* in the elections and I got elected. I took the initiative to pass a resolution in the panchayat meeting to construct common lavatories from the village funds on common land near the panchayat office. I also got special assistance from the district authorities for constructing toilets. The problem of open defecation was most severe among the Scheduled Caste households of the village, as these households did not have the resources to build their personal lavatories. We constructed a complex of around fifteen lavatories and assigned one lavatory to three households. These lavatories were locked to ensure proper usage and each assigned household was given the key of their lavatory. The panchayat appointed cleaning staff to clean the lavatories. Other households in the village also got motivated and constructed their individual lavatories. Though there are still households that do not have access to sanitation facilities, this initiative changed the sanitation situation in the village.

In Kalmandasguri village (West Bengal), around 70 per cent of all households and 85 per cent of small farmer households reported having access to toilets (Appendix Table 8). Further investigation into the improved sanitation in this village revealed two main reasons. First, due to the location of the village in a high rainfall zone, it had an intensive cropping pattern. With either paddy or jute under cultivation all the year round the fields were flooded with water throughout the year, making it impossible to use the fields for open defecation. Secondly, households in this village had significant homestead land which

Table 6 *Proportion of households not having access to a lavatory, by social group among small farmer households, all study villages* in number and per cent

Social group	No. of households	Per cent
Scheduled Caste (SC)	316	74
Scheduled Tribe (ST)	37	84
Muslim	18	35
Others (all remaining)	512	51
All	882	58

Note: "All" includes all households belonging to the small farmer class across the study villages.
Source: PARI survey data.

they had received from the Left Front government of the State during land reforms. This enabled homesteads to construct lavatories.

Table 6 shows that the problem of open defecation was not common to all small farmer households. There were stark differences among different social groups within the small farmer class in respect of access to lavatories, with SC and ST small farmer households lagging far behind others. Almost three-fourths of SC and 84 per cent of ST households had to defecate in the open, while the proportion was 51 per cent among others. Across all social groups, Muslims had better access to lavatories. It is important to note here that most Muslim households in the small farmer class were from Kalmandasguri village in West Bengal, which in general had better lavatory coverage (as discussed above).

Lack of access to a lavatory affects women the most. Open defecation is a threat to women's safety and dignity. Incidents of violence and sexual assaults against women while defecating in the open have been widely reported and highlighted in the national and international media in the recent past. Women respondents from Mahatwar village (Uttar Pradesh), where 94 per cent of the small farmer households did not have access to a lavatory, narrated the ordeal of open defecation to us:

> In practice there are specific locations that men and women from the village use for relieving themselves. But women and girls have to walk long distances early in the morning or after sunset under the cover of dark for defecation. If they are late in the morning, they face extreme inconvenience and embarrassment as the movement of villagers increases in the fields. It is impossible for women to go alone in the dark for defecation so they always go in a group. During the rainy season, and when there are crops on the field, it becomes even more difficult to access open places and fields, as landowners start guarding the fields.

The experiences of women from marginalised and landless households are even more horrifying and traumatising.

AN INTEGRATED VIEW OF HOUSING

A broad definition of housing is required to understand the condition of housing in rural India. A serious flaw in the official understanding of the housing question in rural areas is that housing is seen only in terms of a structure, without taking into consideration the provision of basic household amenities (see Singh, Swaminathan, and Ramachandran 2013). There are very detailed norms and standards on housing recommended by various international agencies and forums. One of the first of these was the International Labour Organisation (ILO) Workers' Housing Recommendation, 1961. It recommended that housing should have adequate space per person or per family, supply of safe water, adequate sewage and garbage disposal systems, appropriate protection against heat, cold, damp, noise, etc., adequate sanitary and washing facilities, ventilation, cooking and storage facilities, among other things (for details see UN-HABITAT 2006, p. 82).[15]

India has ratified these basic norms of housing structure and amenities. In our study, we define these norms as integrated housing. Our concept of integrated housing includes a house with: (1) a pucca roof, walls, and floors; (2) at least two rooms; (3) a source of water inside the house; (4) an electricity connection (authorised or unauthorised); and (5) a functioning lavatory. Based on this definition of housing, the results from the PARI village surveys are reported in Appendix Table 9.

The village survey data show a picture of large-scale and generalised deprivation with respect to the basic norms of housing, which were, in any case, far below the standards of housing to which India is committed internationally. Among all households residing in the study villages, 87 per cent fell short of achieving the specified criteria. The results are surprising as just 50 per cent of even those households belonging to the richest and wealthiest (landlords and capitalist farmers) class of village residents had housing that conformed to the required standards. The proportion was highest among the class of "all farmers" in the study villages. Of large farmer households, 28 per cent had houses that met the standards of integrated housing. The housing conditions of small farmer households were the worst among all classes of farmers. Only one-tenth of small farmer households across the study villages lived in houses

[15] Various international agencies recommend that housing should be defined as more than just a structure with a roof and four walls. The Commission on Human Settlements and the Global Strategy for Shelter to the Year 2000 states, "Adequate shelter means . . . adequate privacy, adequate space, adequate security, adequate lighting and ventilation, adequate basic infrastructure and adequate location with regard to work and basic facilities – all at a reasonable cost" (UN-HABITAT 2006, p. 13). (Also see UN 1949; ILO 1961; WHO 1989; UN 1991a; UN-HABITAT and OHCHR 2002; UN 2009; UN 2010.)

Table 7 *Proportion of households having access to integrated housing, by social group among small farmer households, all study villages* in number and per cent

Social group	No. of households	Per cent
Scheduled Caste (SC)	8	2
Scheduled Tribe (ST)	1	2
Muslim	0	0
Others (all remaining)	151	15
All	160	11

Note: "All" includes all households belonging to the small farmer class across the study villages.
Source: PARI survey data.

that fulfilled the minimal norms of housing. In five out of a total of 15 study villages, not a single small farmer household met these minimal housing norms. The highest proportion, i.e. one-third, was in Alabujanahalli (Mandya district, Karnataka). Two other villages where a relatively high proportion of small farmer households had access to minimum housing were Rewasi (Sikar district, Rajasthan) with 21 per cent and Nimshirgaon (Kolhapur district, Maharashtra) with 20 per cent.

Table 7 shows a similar pattern of disparities across different social groups among small farmer households that have been observed in the analysis of earlier variables related to housing and access to basic amenities. In other words, the results shown in Table 7 are combined outcomes of existing disparities across different social groups with respect to housing and basic amenities. While the overall levels of deprivation with regard to basic housing are deplorable and alarming, the situation among SC, ST, and Muslim households was much worse than among households belonging to other social groups. If we put together all the SC, ST, and Muslim households of small farmers across the study villages, we find that only 2 per cent of SC and ST households met the basic criteria of integrated housing. There was not a single Muslim household that lived in what we have defined as integrated housing. On the other hand, this proportion was 15 per cent for households belonging to "other" groups.

CONCLUSIONS

This chapter has examined the conditions and quality of housing, and the access of households to basic household amenities such as type of fuel used for cooking, electricity for domestic use, source of drinking water, and lavatories in the study villages. While the main focus of discussion is the situation of small farmer households, the situation of "other" classes of farmers is also discussed. We highlighted the disparities and inequalities across different social groups among small farmers. The discussion also brought international norms, standards, and ·guidelines put forward by the United Nations and its various agencies relating to adequate housing, water, and sanitation.

Results from the village surveys show that homelessness was non-existent in the study villages, and an overwhelming proportion of households across all socio-economic classes owned their dwellings. In our analysis we gave *pucca/* permanent structures an alternative definition: structures where the roof, walls, and floor are all made of *pucca/*permanent materials. This, we believe, is a better and improved definition as compared to the one used by various official data sources and government departments in India. We showed that the official definition, which considers only construction materials used in roofs and walls, and ignores the floor component, is inadequate given the international norms of adequate housing, and falls short in terms of assuring the health, safety, and comfort of the inhabitants.

Results from the survey data show that a large majority of small farmer households (59 per cent) across the study villages lived in non-*pucca/*permanent structures. There were large variations in the condition of and types of structure across the study villages. Use of different types of construction materials and combinations of materials were special features of housing construction, influenced as they were by socio-economic, cultural, and local factors such as availability of materials and their suitability to local weather conditions. Within the small farmer class, a much higher proportion of SC, ST, and Muslim households lived in non-*pucca/*permanent housing as compared to other farmer households.

Availability of living space in terms of number of rooms was a serious issue among all households in general, and small farmer households in particular. More than one-fourth of all small farmer households lived in single-room structures. Among small farmers, almost one-third of SC households and more than one-half of ST households (though their absolute numbers were very small) lived in single-room dwellings. Through a case study of a household living in a single-room dwelling, we showed how crowding and lack of adequate space for women and young girls, married couples, the elderly and the ill, and schoolgoing children affected the well-being of different members of the household and that of the family as a whole.

Around 40 per cent of small farmer households in the study villages did not have·a separate kitchen in their dwellings. In these households, cooking was done in the living space of the dwelling or in the open. The situation with regard to the type of fuel used for cooking was alarming throughout the study villages and across all socio-economic classes. For 97 per cent of all small farmer households, solid (smoke-generating) fuels such as firewood, dung cakes, and crop residue were the main sources of cooking fuel. Given that cooking was mainly done by women in the households, they bore the brunt of the negative consequences of solid fuel use, which exposed them to serious health hazards.

In terms of access to basic household amenities, we found that in no study village were all the households electrified. Among small farmer households, one-fifth did not have a domestic electricity connection. Kalmandasguri (West Bengal) was entirely unelectrified in 2010, at the time of our first survey, but had got electricity coverage by 2015. Among the electrified study villages, Harevli (Uttar Pradesh) had the highest proportion of small farmer households not having an electricity connection. A majority of these were SC households who lived in a separate settlement that was not covered under the electricity distribution mechanism of the village. A significant proportion of small farmer households in Panahar (West Bengal), Nimshirgaon (Maharashtra), and Harevli (Uttar Pradesh) had unauthorised access to electricity.

In our analysis of the type of drinking water source/s, we found that a significant proportion (10 per cent) of all small farmer households in the study villages was using open sources of water. Under the open source category wells were the most used source, and in some villages wells were preferred for drinking purpose as, according to residents, the quality and taste of piped drinking water was not good. Among covered sources handpumps/borewells were the most used sources across all the study villages.

Availability of a source of water within a short distance or inside the housing premises can reduce the time and energy spent in fetching water, besides preventing incidents of discrimination, violence, and physical abuse, all of which are very common especially towards women belonging to the Scheduled Castes and other marginalised groups. Our study found that only 38 per cent of all small farmer households in the study villages had a source of drinking water within their housing premises. In five of the study villages, this proportion was less than 10 per cent. In some villages we found that SC households were discriminated against in terms of access to drinking water. Case studies of two Maharashtra villages showed that while a profit-oriented water supply scheme excluded poor households, another scheme that had minimal installation and monthly charges, and focused on providing more common water outlets, benefited more people.

Households' access to lavatories in the study villages was shockingly limited, and open defecation was the norm for a majority of the households. This was more so among small farmer and manual labourer households. More than half of all small farmer households in the study villages did not have access to a lavatory and its members defecated in the open. Not a single small farmer household in Zhapur village (Karnataka) had access to lavatories. In nine out of all 15 study villages, more than 60 per cent of small farmer households defecated in the open. A much higher proportion of SC and ST households than other households among small farmers lacked access to lavatories. A village case study showed that women in general, and SC and landless households in particular, faced huge inconvenience and trauma while using open spaces for defecation.

While international standards of adequate housing are very detailed and include a wide range of indicators, we take the minimal norms to be a fully *pucca* structure with at least two rooms and basic household amenities, such as water within the housing premises, an electricity connection, and a functional lavatory. We call this integrated housing. In all the study villages and across all socio-economic groups, our data showed high levels of deprivation with respect to access to integrated housing. Only 13 per cent of all households living in the 16 study villages fulfilled the criteria mentioned above.[16] This proportion was only one-tenth among small farmer households. Even among landlord and capitalist farmer households, who belong to the richest and wealthiest strata, this proportion was only a little more than 50 per cent. In five out of all the study villages, there was not a single small farmer household that lived in integrated housing as defined above.

Levels of deprivation and disparities were much higher among SC, ST, and Muslim small farmer households on the one hand, and households belonging to other groups on the other. While 15 per cent of other households belonging to the small farmer class lived in structures that fulfilled the criteria of integrated housing, this proportion was merely 2 per cent among SC and ST households, and 0 per cent among Muslim households.

It is noteworthy that though the Government of India has in principle agreed to the right to adequate housing, it does not take these norms into consideration in its actual policies and estimates of housing shortage in rural India. The government has reduced the comprehensive concept of housing to merely a structure with four walls and a roof, without including basic amenities.

Our analysis shows that the status of artisan/manual labourer households is at the bottom of the ladder with respect to housing structure and access to basic household amenities among all socio-economic groups in the study villages. While small farmer households performed a little better, an overwhelming majority of these continued to live in sub-standard, crowded dwellings that lacked a source of drinking water, electricity, and basic sanitation facilities.

---

[16] For this calculation, 25F Gulabewala was also included.

On the basis of these findings it could be suggested that government policy initiatives relating to housing, drinking water, electricity, and sanitation should bring small farmer households within the ambit of schemes to improve their living conditions.

While small farmer households are the worst off among the peasantry, there exist disparities and differences within the class of small farmers on the basis of social identity. The analysis presented in this chapter shows that SC, ST, and Muslim households among small farmers are far more deprived in terms of housing and access to basic household amenities than households belonging to other social groups. This points to the fact that in Indian society, and more so in rural society, deprivation is not merely economic but social as well. Even though a uniform criterion was used to define small farmer households, we find that higher levels of deprivation among SC, ST, and Muslim households are an outcome of the historical exclusion and accumulated disadvantages faced and inherited by these social groups. Continued practices of untouchability, physical and residential segregation, and isolation shape current outcomes for these groups.

REFERENCES

Acharya, Sanghmitra S. (2010), "Caste and Patterns of Discrimination in Rural Public Health Care Services," in Sukhdeo Thorat (ed.), *Blocked by Castes: Economic Discrimination and Social Exclusion in Modern India*, Oxford University Press, New Delhi, pp. 208–09.

Beteille, Andre (1996), *Caste, Class and Power: Changing Patterns of Stratification in a Tanjore Village*, Oxford University Press, New Delhi.

Beteille, Andre (2004), "The Distribution of Power," in Vandana Madan (ed.), *The Village in India*, Oxford University Press, New Delhi, pp. 227–37.

Bhatia, Bela (2006), "Dalit Rebellion against Untouchability in Chakwada, Rajasthan," *Contributions to Indian Sociology*, vol. 40, no. 1, pp. 29–61.

Census of India (2001), *Census of India, 2001: H-Series, Tables on Census House, Household Amenities and Assets*, Office of the Registrar General and Census Commissioner, Ministry of Home Affairs, Government of India.

Census of India (2011), *Census of India, 2011: H-Series, Tables on Census House, Household Amenities and Assets*, Office of the Registrar General and Census Commissioner, Ministry of Home Affairs, Government of India.

Desai, I. P. (1973), *Water Facilities for Untouchables in Rural Gujarat*, Indian Council of Social Science Research, New Delhi.

Desai, I. P. (2013), *Rural Sociology in India*, Popular Prakashan, Mumbai.

Dube, S. C. (2004), "A Rural Development Project in Action," in Vandana Madan (ed.), *The Village in India*, Oxford University Press, New Delhi, pp. 351–72.

Epstein, T. Scarlet, Suryanarayana, A. P., Thimmegowda, T. (1998), *Village Voices: Forty Years of Rural Transformation in South India*, Sage Publications, New Delhi.

Freeman, James M. (1979), *Untouchable: An Indian Life History*, George Allen and Unwin, London.

Ghurye, G. S. (1932), *Caste and Race in India*, K. Paul, Trench, Trubner and Co. Ltd., London.

Ghurye, G. S. (1957), *Caste and Class in India*, Popular Book Depot Company, Bombay.

Indian Council for Social Science Research (ICSSR) (1973), "Water Facilities for the Untouchables in Rural Gujarat," *Occasional Monograph Number 8*, Indian Council for Social Science Research (ICSSR), New Delhi.

Indian Institute of Population Sciences (IIPS) and Macro International (2007), *National Family Health Survey (NFHS-3), 2005–06, India: Volume 1*, Mumbai.

International Labour Organisation (ILO) (1961), *Workers' Housing Recommendation, 1961 (No. 115)*, Recommendation Concerning Workers' Housing, 45th ILC session (28 June 1961), Geneva, available at http://www.ilo.org/dyn/normlex/en/f?p=1000:12 100:::NO:12100:P12100_INSTRUMENT_ID:312453, viewed on 16 January 2107.

Jodhka, Surinder S. (ed.) (2012), *Village Society*, Orient Black Swan, New Delhi.

Jodhka, Surinder S., and Shah, Ghanshyam (2010), "Comparative Contexts of Discrimination: Caste and Untouchability in South Asia," Working Paper Series, vol. 3, Indian Institute of Dalit Studies, New Delhi.

Joshi, Deepa, and Fawcett, Ben (2001), "Water, Hindu Mythology and an Unequal Social Order in India," paper presented at the Second Conference of the International Water History Association, Bergen, Norway, 10–12 August.

Krishnan, Rekha, Bhadwal, Suruchi, Javed, Akram, Singhal, Shaleen, and Sreekesh, S. (2003), "Water Stress in Indian Villages," *Economic and Political Weekly*, vol. 38, no. 37, pp. 3879–84.

National Sample Survey Organisation (NSSO) (2010), *Housing Conditions and Amenities in India, 2008–09, Report No. 535, NSS 65th round, July 2008–June 2009*, National Sample Survey Office, Ministry of Statistics and Programme Implementation, Government of India, New Delhi.

Ramachandran, V. K. (2014), "Introduction," in V. K. Ramachandran and Madhura Swaminathan (eds.), *Dalit Households in Village Economies*, Tulika Books, New Delhi, pp. 1–13.

Ramachandran, V. K., Rawal, Vikas, and Swaminathan, Madhura (eds.) (2010), *Socio-Economic Surveys of Three Villages in Andhra Pradesh*, Tulika Books, New Delhi.

Sangameswaran, Priya (2010), "Rural Drinking Water Reforms in Maharashtra: The Role of Neoliberalism," *Economic and Political Weekly*, vol. 44, no. 50, pp. 47–54.

Shah, Ghanshyam, Mander, Harsh, Thorat, Sukhdeo, Deshpande, Satish, and Baviskar, Amita (2006), *Untouchability in Rural India*, Sage Publications, New Delhi.

Sharma, K. L. (1976), *The Remembered Village*, Oxford University Press, New Delhi.

Sharma, K. L. (2012), "Modernisation and Rural Stratification: An Application at the Micro-Level," in S. Jodhka (ed.), *Village Society: Essays from Economic and Political Weekly*, pp. 82–92.

Singh, Shamsher (2014), "Access to Basic Amenities: A Sociological Study of Villages in Selected States of India," unpublished PhD thesis, University of Calcutta, Kolkata.

Singh, Shamsher (2015), "Residential Segregation and Access to Basic Amenities: A Village-Level Case Study," *Review of Agrarian Studies*, vol. 5, no. 2, available at http://

ras.org.in/residential_segregation_and_access_to_basic_amenities, viewed on 16 January 2017.

Singh, Shamsher, Swaminathan, Madhura, and Ramachandran, V. K. (2013), "Housing Shortages in Rural India," *Review of Agrarian Studies,* vol. 3, no. 2. pp. 54–72, available at http://ras.org.in/housing_shortages_in_rural_india, viewed on 16 January 2017.

Soni, Jayshree (2006), "Water Accessibility and Marginalisation of Dalits (Scheduled Caste): Some Observations of Rural Gujarat," paper presented at the workshop on Water, Law and the Commons, International Environmental Law Research Centre (IELRC), New Delhi.

Srinivas, M. N. (1962), *Caste in Modern India and Other Essays,* Asia Publishing House, Bombay.

Swaminathan, Madhura, and Singh, Shamsher (2014), "Exclusion in Access to Basic Civic Amenities," in V. K. Ramachandran and Madhura Swaminathan (eds.), *Dalit Households in Village Economies,* Tulika Books, New Delhi, pp. 305–32.

Thorat, Sukhadeo (2009), "Access to Civic Amenities: Housing, Water and Electricity," in Sukhadeo Thorat (ed.), *Dalits in India: Search for a Common Destiny,* Sage Publications, New Delhi.

Tiwary, Rakesh, and Phansalkar, Sanjiv J. (2007), "Dalits' Access to Water: Patterns of Deprivation and Discrimination," *International Journal of Rural Management,* vol. 3, no. 1, pp. 43–67.

United Nations (UN) (1949), "United Nations Universal Declaration of Human Rights 1948," available at http://www.jus.uio.no/lm/un.universal.declaration.of.human. rights.1948/sisu_manifest.html, viewed on 16 January 2017.

United Nations (UN) (1991a), "General Comment No. 4: The Right to Adequate Housing (Art. 11 (1) of the Covenant)," Committee on Economic Social and Cultural Rights, available at www.ohchr.org/Documents/.../TB/HRI-GEN-1-REV-9-VOL-I_ en.doc, viewed on 16 January 2017.

United Nations (UN) (1991b), "The Right to Adequate Housing (1991): Art. 11 (1) of the Covenant on Economic, Social and Cultural Rights (ICESCR)," United Nations Human Rights Commission, available at http://www.unhchr.ch/tbs/doc.nsf/0/469f4d 91a9378221c12563ed0053547e?Opendocument, viewed on 16 January 2017.

United Nations (UN) (2003), *The Habitat Agenda Goals and Principles: Commitments and the Global Plan of Action,* available at http://www.unhabitat.org/declarations/habitat_ agenda.htm, viewed on 16 January 2017.

United Nations (UN) (2009), *The Right to Adequate Housing: Fact Sheet No. 21/Rev.1,* Office of the United Nations High Commissioner for Human Rights, available at http://www.ohchr.org/Documents/Publications/FS21_rev_1_Housing_en.pdf, viewed on 16 January 2017.

United Nations (UN) (2010), *Economic and Social Council Committee on Economic, Social and Cultural Rights Statement on the Right to Sanitation,* 45th Session, available at www2.ohchr.org/english/bodies/cescr/.../statements/E-C-12-2010-1.doc, viewed on 16 January 2017.

United Nations General Assembly (UNGA) (2012), "Report of the Working Group on the Universal Periodic Review," Agenda Item 6, Universal Periodic Review, Human Rights Council, 21st Session, available at http://www.ohchr.org/Documents/

HRBodies/HRCouncil/RegularSession/Session21/A-HRC-21-10_en.pdf, viewed on 16 January 2017.

United Nations Human Settlements Programme (UN-HABITAT) (2006), *Compilation of Selected United Nations Documents on Housing Rights*, Report No. 6, Nairobi.

United Nations Human Settlements Programme (UN-HABITAT) and Office of the High Commissioner for Human Rights (OHCHR) (2002), "Human Rights Legislations: Review of International and National Legal Instruments," *United Nations Human Rights Programme Report No. 1*, Nairobi, available at http://www.ohchr.org/Documents/Publications/HousingRightsen.pdf, viewed on 16 January 2017.

Wiser, William, and Wiser, Charlotte (2001), *Behind Mud Walls: Seventy-Five Years in a North Indian Village*, University of California Press, Berkeley and Los Angeles.

World Health Organisation (WHO) (1989), *Health Principles of Housing*, Geneva, available at http://apps.who.int/iris/bitstream/10665/39847/1/9241561270_eng.pdf, viewed on 16 January 2017.

Appendix Table 1 *Proportion of households owning their house, selected socio-economic classes, study villages in number and per cent*

Village	State	Landlords and capitalist farmers		Small farmers		Artisans and manual wage labourers		All owned	
		No. of households	Per cent	No. of households	Per cent	No. of households	Per cent	No. of households	Per cent
Ananthavaram	Andhra Pradesh	11	100	207	93	137	72	534	82
Bukkacherla	Andhra Pradesh	10	100	92	86	45	71	244	84
Kothapalle	Telangana	5	100	90	92	127	76	313	84
Harevli	Uttar Pradesh	3	100	42	93	26	93	104	95
Mahatwar	Uttar Pradesh	2	100	69	100	39	100	154	100
Warwat Khanderao	Maharashtra	3	100	104	99	74	97	237	95
Nimshirgaon	Maharashtra	3	100	230	97	255	86	701	93
Rewasi	Rajasthan	8	100	106	99	42	100	215	99
Gharsondi	Madhya Pradesh	12	100	71	100	63	91	249	95
Alabujanahalli	Karnataka	–	–	110	99	61	91	224	97
Siresandra	Karnataka	–	–	56	100	11	85	77	97
Zhapur	Karnataka	3	100	24	100	43	86	95	91
Amarsinghi	West Bengal	–	–	54	100	48	96	123	97
Kalmandasguri	West Bengal	–	–	67	100	44	100	132	99
Panahar	West Bengal	7	100	140	99	59	97	240	99
		87	100	1464	96	1164	85	3818	91

Notes: – not applicable.
"All" = all households of a village.
Source: PARI survey data.

Appendix Table 2 *Proportion of households living in fully pucca/permanent houses,* * selected socio-economic classes, study villages in number and per cent*

Village	State	Landlords and capitalist farmers		Small farmers		Large farmers		All	
		No. of households	Per cent	No. of households	Per cent	No. of households	Per cent	No. of households	Per cent
Ananthavaram	Andhra Pradesh	9	90	52	23	30	88	233	36
Bukkacherla	Andhra Pradesh	6	60	42	40	18	43	137	49
Kothapalle	Telangana	5	100	65	69	4	100	244	67
Harevli	Uttar Pradesh	1	33	5	11	3	16	14	13
Mahatwar	Uttar Pradesh	0	0	2	3	0	0	3	2
Warwat Khanderao	Maharashtra	3	100	33	32	8	62	79	32
Nimshirgaon	Maharashtra	3	100	116	49	27	52	328	44
Rewasi	Rajasthan	8	100	87	81	34	92	183	85
Gharsondi	Madhya Pradesh	12	100	28	39	34	58	105	40
Alabujanahalli	Karnataka	–	–	93	82	24	96	180	75
Siresandra	Karnataka	–	–	44	79	4	100	57	72
Zhapur	Karnataka	2	100	20	77	7	70	78	73
Amarsinghi	West Bengal	–	–	20	37	–	–	25	20
Kalmandasguri	West Bengal	–	–	1	1	–	–	3	2
Panahar	West Bengal	5	71	16	11	0	0	31	13
Total		70	82	625	41	216	64	1754	42

Notes: * Classification of materials into pucca/permanent and katcha/temporary categories is the same as in the Census and NSS.
Rooms used for living and sleeping purposes were recorded.
– not applicable.
"All" = all households of a village.
Source: PARI survey data.

Appendix Table 3 *Proportion of households living in single-room houses, selected socio-economic classes, study villages in number and per cent*

Village	State	Small farmers		Large farmers		Artisans and manual wage labourers		All	
		No. of households	Per cent	No. of households	Per cent	No. of households	Per cent	No. of households	Per cent
Ananthavaram	Andhra Pradesh	115	51	0	0	129	67	326	50
Bukkacherla	Andhra Pradesh	42	39	12	25	21	33	89	31
Kothapalle	Telangana	28	28	0	0	56	35	103	28
Harevli	Uttar Pradesh	16	36	1	5	20	83	40	39
Mahatwar	Uttar Pradesh	8	12	0	0	9	24	22	15
Warwat Khanderao	Maharashtra	25	24	0	0	30	41	70	29
Nimshirgaon	Maharashtra	47	20	0	0	131	45	203	27
Rewasi	Rajasthan	8	7	1	3	8	19	20	9
Gharsondi	Madhya Pradesh	1	1	2	3	17	24	24	9
Alabujanahalli	Karnataka	11	10	1	4	25	36	43	18
Siresandra	Karnataka	19	35	0	0	4	40	24	33
Zhapur	Karnataka	12	48	2	20	31	62	50	49
Amarsinghi	West Bengal	11	20	–	–	23	47	41	33
Kalmandasguri	West Bengal	23	34	–	–	35	64	70	48
Panahar	West Bengal	33	23	0	0	28	44	69	28
Total		398	26	19	6	597	44	1225	29

Notes: – not applicable.
"All" = all households of a village.
Rooms used for living and sleeping purposes were recorded.
Source: PARI survey data.

Appendix Table 4 *Proportion of households having a separate kitchen, selected socio-economic classes, study villages in number and per cent*

Village	State	Landlords and capitalist farmers		Small farmers		Large farmers		All	
		No. of households	Per cent	No. of households	Per cent	No. of households	Per cent	No. of households	Per cent
Bukkacherla	Andhra Pradesh	10	100	36	40	27	64	126	49
Kothapalle	Telangana	5	100	24	31	0	0	87	30
Harevli	Uttar Pradesh	3	100	21	48	14	70	48	47
Mahatwar	Uttar Pradesh	2	100	32	46	1	100	57	38
Warwat Khanderao	Maharashtra	2	100	80	75	13	100	170	68
Nimshirgaon	Maharashtra	3	100	145	61	48	100	348	48
Rewasi	Rajasthan	8	100	80	76	29	76	156	73
Gharsondi	Madhya Pradesh	11	100	36	52	46	79	133	53
Alabujanahalli	Karnataka	–	–	97	86	25	100	201	85
Siresandra	Karnataka	–	–	48	87	3	75	62	79
Zhapur	Karnataka	1	50	13	52	6	60	50	49
Amarsinghi	West Bengal	–	–	37	70	–	–	72	61
Kalmandasguri	West Bengal	–	–	64	96	–	–	134	92
Panahar	West Bengal	7	100	69	54	1	100	106	49
Total		71	99	782	63	245	82	1839	55

Notes: – not applicable.
"All" = all households of a village.
Source: PARI survey data.

Appendix Table 5 *Proportion of households not having an electricity connection, selected socio-economic classes, study villages in number and per cent*

Village	State	Landlords and capitalist farmers		Small farmers		Large farmers		All	
		No. of households	Per cent	No. of households	Per cent	No. of households	Per cent	No. of households	Per cent
Ananthavaram	Andhra Pradesh	0	0	8	4	0	0	82	12
Bukkacherla	Andhra Pradesh	0	0	9	8	12	25	33	11
Kothapalle	Telangana	2	39	2	2	0	0	29	8
Harevli	Uttar Pradesh	0	0	33	77	5	24	68	66
Mahatwar	Uttar Pradesh	0	0	32	46	0	0	75	49
Warwat Khanderao	Maharashtra	0	0	15	15	0	0	50	21
Nimshirgaon	Maharashtra	0	0	22	9	10	19	180	24
Rewasi	Rajasthan	2	25	47	45	16	42	100	46
Gharsondi	Madhya Pradesh	0	0	5	7	1	2	26	10
Alabujanahalli	Karnataka	–	–	5	4	0	0	19	8
Siresandra	Karnataka	–	–	1	2	0	0	1	1
Zhapur	Karnataka	0	0	3	12	1	10	13	12
Amarsinghi	West Bengal	–	–	23	43	–	–	80	63
Kalmandasguri	West Bengal	–	–	65	96	–	–	143	97
Panahar	West Bengal	0	0	37	26	0	0	71	29
		4	5	307	20	46	13	1017	24

Notes: – not applicable.
"All" = all households of a village.
Having an electricity connection includes both authorised and unauthorised connections.
Source: PARI survey data.

Appendix Table 6 *Proportion of households using open sources of drinking water, selected socio-economic classes, study villages in number and per cent*

Village	State	Landlords and capitalist farmers		Small farmers		Large farmers		All	
		No. of households	Per cent	No. of households	Per cent	No. of households	Per cent	No. of households	Per cent
Ananthavaram	Andhra Pradesh	1	9	0	0	0	0	5	1
Bukkacherla	Andhra Pradesh	0	0	3	3	0	0	3	1
Kothapalle	Telangana	2	39	25	26	0	0	71	19
Harevli	Uttar Pradesh	0	0	0	0	0	0	0	0
Mahatwar	Uttar Pradesh	0	0	0	0	0	0	0	0
Warwat Khanderao	Maharashtra	0	0	1	1	0	0	1	0
Nimshirgaon	Maharashtra	0	0	87	37	9	18	260	35
Rewasi	Rajasthan	0	0	14	13	7	18	26	12
Gharsondi	Madhya Pradesh	1	8	10	14	5	8	26	10
Alabujanahalli	Karnataka	–	–	3	3	2	8	7	3
Siresandra	Karnataka	–	–	0	0	0	0	0	0
Zhapur	Karnataka	0	0	2	8	5	50	27	25
Amarsinghi	West Bengal	–	–	0	0	–	–	0	0
Kalmandasguri	West Bengal	–	–	0	0	–	–	0	0
Panahar	West Bengal	0	0	1	1	0	0	1	0
Total		4	5	147	10	29	9	440	10

Notes: Open sources of drinking water include river, wells, pond, open tank, canal, lake, and stream.
– not applicable.
"All" = all households of a village.
Source: PARI survey data.

Appendix Table 7 *Proportion of households with source of drinking water within homestead, selected socio-economic classes, study villages in number and per cent*

Village	State	Landlords and capitalist farmers		Small farmers		Large farmers		All	
		No. of households	Per cent	No. of households	Per cent	No. of households	Per cent	No. of households	Per cent
Ananthavaram	Andhra Pradesh	10	91	108	48	25	74	317	49
Bukkacherla	Andhra Pradesh	3	30	9	8	9	20	33	11
Kothapalle	Telangana	3	61	49	50	4	100	166	45
Warwat Khanderao	Maharashtra	0	0	3	3	0	0	10	4
Nimshirgaon	Maharashtra	3	100	110	46	39	76	335	45
Harevli	Uttar Pradesh	3	100	32	71	19	100	78	73
Mahatwar	Uttar Pradesh	2	100	31	46	1	100	65	43
Rewasi	Rajasthan	7	88	83	78	35	92	176	81
Gharsondi	Madhya Pradesh	11	92	13	18	12	21	48	19
Alabujanahalli	Karnataka	–	–	50	45	17	68	89	37
Siresandra	Karnataka	–	–	3	6	0	0	3	4
Zhapur	Karnataka	0	0	0	0	1	10	1	1
Amarsinghi	West Bengal	–	–	1	2	–	–	2	2
Kalmandasguri	West Bengal	–	–	57	85	–	–	122	84
Panahar	West Bengal	7	100	32	22	1	100	52	21
Total		69	81	582	38	198	59	1625	39

Note: – not applicable.
"All" = all households of a village.
Source: PARI survey data.

Appendix Table 8 *Proportion of households not having access to a lavatory, selected socio-economic classes, study villages in number and per cent*

Village	State	Landlords and capitalist farmers		Small farmers		Large farmers		All	
		No. of households	Per cent	No. of households	Per cent	No. of households	Per cent	No. of households	Per cent
Ananthavaram	Andhra Pradesh	0	0	144	63	4	12	354	54
Bukkacherla	Andhra Pradesh	6	60	89	83	36	75	229	78
Kothapalle	Telangana	2	39	57	58	0	0	207	56
Harevli	Uttar Pradesh	0	0	28	65	7	33	66	63
Mahatwar	Uttar Pradesh	0	0	65	94	0	0	144	94
Warwat Khanderao	Maharashtra	0	0	59	56	3	23	134	54
Nimshirgaon	Maharashtra	0	0	76	32	3	6	225	30
Rewasi	Rajasthan	1	13	66	62	25	69	137	64
Gharsondi	Madhya Pradesh	2	17	44	65	18	31	146	56
Alabujanahalli	Karnataka	–	–	46	41	2	8	115	48
Siresandra	Karnataka	–	–	50	91	2	67	71	92
Zhapur	Karnataka	2	100	26	100	10	100	105	99
Amarsinghi	West Bengal	–	–	22	41	–	–	73	58
Kalmandasguri	West Bengal	–	–	10	15	–	–	42	29
Panahar	West Bengal	0	0	101	71	0	0	175	72
		13	15	882	58	111	32	2253	54

Note: – not applicable.

"All" = all households of a village.

Source: PARI survey data.

Appendix Table 9 *Proportion of households with access to integrated housing, selected socio-economic classes, study villages in number and per cent*

Village	State	Landlords and capitalist farmers		Small farmers		Large farmers		All	
		No. of households	Per cent	No. of households	Per cent	No. of households	Per cent	No. of households	Per cent
Ananthavaram	Andhra Pradesh	8	80	24	11	23	72	146	23
Bukkacherla	Andhra Pradesh	0	0	0	0	3	7	9	3
Kothapalle	Telangana	1	21	12	12	4	100	56	15
Harevli	Uttar Pradesh	1	33	3	7	2	10	6	6
Mahatwar	Uttar Pradesh	0	0	0	0	0	0	0	0
Warwat Khanderao	Maharashtra	0	0	0	0	0	0	4	2
Nimshirgaon	Maharashtra	3	100	48	20	14	29	134	18
Rewasi	Rajasthan	5	63	22	21	5	14	44	21
Gharsondi	Madhya Pradesh	9	75	4	6	4	7	25	10
Alabujanahalli	Karnataka	–	–	37	33	14	58	63	26
Siresandra	Karnataka	–	–	1	2	0	0	1	1
Zhapur	Karnataka	0	0	0	0	0	0	0	0
Amarsinghi	West Bengal	–	–	1	2	–	–	1	1
Kalmandasguri	West Bengal	–	–	0	0	–	–	0	0
Panahar	West Bengal	5	71	7	5	0	0	16	6
Total		47	57	160	11	92	28	555	13

Note: – Not applicable.
"All" = all households of a village.
Source: PARI survey data.

12

How Do Small Farmers Fare?

Madhura Swaminathan

At the end of this analysis of evidence from village surveys, we return to the question we posed at the beginning of this book: how do small farmers – that section of the peasantry whose holdings of 2 hectares or less than 2 hectares constitute more than 80 per cent of the total number of farms in India – fare in the country's fast-changing agrarian sector? This volume is our answer. In the preceding chapters we have examined the socio-economic characteristics and viability of small farmers and small farming in different agro-ecological regions of India, locating farms and farmers in the broader context of capitalist development in Indian agriculture. As in many other developing countries, agriculture in India has historically been dominated by peasant farming, with a little less than 50 per cent of total operated area under smallholdings. Given their numerical importance, small farmers are an important analytical category in any study of agrarian India.

This volume draws largely on empirical material on the Indian countryside collected by researchers at the Foundation for Agrarian Studies (FAS) as part of our ongoing Project on Agrarian Relations in India (PARI). Detailed questionnaires canvassed from households in 25 villages (and counting) located in varied agro-economic regions across 11 States in India have given us a unique database on many features of change in agrarian society across India.

For the purposes of this volume, namely, to analyse the status of small farmers in village economies in contemporary India, we have examined detailed data at the level of the farm household in 17 villages across nine States. The analysis includes, but is not restricted to, production, production systems and livelihoods, and socio-economic characteristics of different strata of the rural population.

The idea of a book on small-scale farming in India arose partially as a response to some of the theories and assumptions in the Indian and international literature on this subject. On the one hand, we have the influential viewpoint that Venkatesh Athreya, Deepak Kumar, R. Ramakumar, and Biplab Sarkar, in their introductory chapter, refer to as "romanticisation" of small-scale farming systems. This viewpoint considers small farms and small farmers as the solution to the problems of efficiency, equity, food security, and environmental sustainability.

Small-scale farming has come to be accepted by some as a one-size-fits-all solution to problems in the countryside across much of the underdeveloped world. The support for and advocacy of small-scale farming come from many (seemingly) antagonistic sources. While it is promoted by the World Bank, the United Nations and many of its constitutive organisations, and various international and domestic NGOs, it also finds sympathisers amongst individuals and organisations such as the popular Via Campesina, which oppose globalisation and the influence of transnational organisations.

However, as a response to this, a forward-looking agrarian movement should not reject the reality of small-scale farming, Athreya *et al.* warn. Rather, it should recognise and address the problems of small farmers. While there may not be much ground for arguing that small-scale farming embodies the set of virtues claimed for it by its romantic advocates, it is important to recognise that any democratic agrarian policy/perspective must reckon with the fact that small farmers account for a substantial proportion of the rural/ agrarian population, and that they require concrete policy support to stay viable in the context of the hostile assault on their livelihoods by neoliberal capitalist globalisation. The empirical analysis of small-scale farming in this volume substantiates this argument.

The book begins with a brief review of the contemporary literature on small farms and small farmers. The authors also review the relevance of the discussion on small peasant agriculture in the context of the historical advance of capitalism in agriculture, a matter that engaged the attention of Marx and Engels, and, in the case of early twentieth-century agriculture in Russia, of Lenin and Kautsky. They examine the relevance of that historical debate to the current Indian situation where the status of the small farmer and small-scale farming is determined among other things, *inter alia*, by the advance of capitalism in agriculture in India.

Our analysis begins with a socio-economic classification of households in the PARI study villages (Ramachandran, Rawal, and Swaminathan 2006). This classification is based on four criteria: ownership of the means of production, occupation, the use of family and hired labour, and the surplus that a household is able to generate within a working year (Ramachandran 2011). Based on these general criteria, households in the study villages were categorised into the following five classes: landlords; capitalist farmers; peasants; manual workers; and households dependent on business, salaries or other sources of income. Within each village, the peasantry was further subdivided based on the specific conditions of the village.

There is an important caveat in respect of the present study that must be stated at the outset. We have deliberately chosen a definition of small farmer that is narrower than the classical Marxist definition of the peasant. We use the extent of operational holding to identify the category of small farmer so as to establish some commonality with policy definitions in India and elsewhere. Based on our socio-economic categorisation, we first selected households belonging to the peasantry as a whole. Among peasants, we then identified small farmer households as *those with an operational holding of less than 2 hectares of irrigated land or 6 hectares of unirrigated land, or any combination thereof.* This also differs from the official Indian policy definition of small farmer, which does not distinguish between irrigated and unirrigated land held by small farmers. Large farmers are defined by us as all peasant households with more than 2 hectares of irrigated land or 6 hectares of unirrigated land. Manual worker households and households dependent *primarily* on incomes from business or salaries or remittances, whether or not they had operational holdings, were excluded from the analysis. Landlords and capitalist farmers, who invariably owned more than 2 hectares of irrigated land or its equivalent, are included in our discussion only in order to illustrate intra-village differences and inequalities among farm households.

Chapter 2 provides a brief description of the study villages, including features of their agrarian structure, agrarian relations, social composition, and production conditions. While each village is unique and offers insights into specific processes, there is also a broad typology of villages covered by the present study. In the 17 villages taken up for detailed analysis, small farmers were numerically significant in all but two villages (25F Gulabewala in Sri Ganganagar district, Rajasthan, and Tehang in Jalandhar district, Punjab). Landlords and capitalist farmers primarily controlled agricultural production in these two villages. In terms of the weight of small farmers in the village economy and in crop production, there was a great deal of diversity among the study villages. Small farmers constituted more than 80 per cent of cultivators and accounted for more than 50 per cent of gross cropped area in eight villages. Amarsinghi (Malda district, West Bengal), Kalmandasguri (Koch Bihar district, West Bengal), Alabujanahalli (Mandya district, Karnataka), Mahatwar (Ballia district, Uttar Pradesh), Warwat Khanderao (Buldhana district, Maharashtra), Siresandra (Kolar district, Karnataka), Panahar (Bankura district, West Bengal), and Kothapalle (Karimnagar district, Telangana) belong to this group of villages. Of this group, five villages – three in West Bengal, Siresandra in Karnataka, and Kothapalle in Telangana – can be termed small farmer-dominated villages.

Another clear distinction that can be made is between rainfed villages and irrigated villages. Bukkacherla in Anantapur district of Andhra Pradesh, Warwat Khanderao in Buldhana district of Vidarbha, Zhapur in Kalaburagi district in north Karnataka, and Rewasi in Sikar district in Rajasthan are all rainfed or dry villages. The canal-irrigated villages (with surface and groundwater irrigation) were Ananthavaram in Guntur district (Andhra Pradesh), Harevli in Bijnor district (Uttar Pradesh), Nimshirgaon in Kolhapur district (Maharashtra), Alabujanahalli in Mandya district (Karnataka), 25F Gulabewala in Sri Ganganagar district (Rajasthan), and Tehang in Jalandhar district (Punjab). Tank and groundwater irrigation were important in Kothapalle (Karimnagar district, Telangana). In Siresandra, groundwater irrigation was important. The three villages of West Bengal had a mix of surface and groundwater irrigation. While Mahatwar village in Ballia district, Uttar Pradesh, and Gharsondi in Gwalior district, Madhya Pradesh were canal-irrigated as per revenue records and the official classification, in our survey year there was very little water for irrigation in these two villages.

Certain broad conclusions emerge from the chapters of this book.

LOW AND INADEQUATE INCOMES

Our study establishes that there is a *crisis facing small farmers and small farming* in terms of the inability of these households to generate adequate incomes for maintaining a minimum standard of living. The average income of small farmers was inadequate to provide for investment or requirements other than daily consumption needs.

In Chapter 5, Aparajita Bakshi and Tapas Modak show the extent of income deprivation among small farmer households, using data from the PARI database. The method used to collect this detailed and disaggregated information is now available in an online manual. As the PARI surveys do not collect information on consumption expenditure, Bakshi uses minimum wages (as per the Official Minimum Wages Act in each State) as a point of reference. The ratio of mean (as also median) incomes per household to minimum wages among small farmers was close to one in most villages, implying that the average small farmer's income barely covered expenses on basic necessities.

There were, of course, variations across villages. The ratio of median income to minimum wage was around two only in Tehang village in Punjab and 25F Gulabewala in Rajasthan, where there were very few small farmers. The ratio was less than one in rainfed villages such as Bukkacherla (Anantapur district, Andhra Pradesh) and Zhapur (Kalaburagi district, Karnataka), as well as the three villages of West Bengal (where small farmers had suffered losses in the survey year on account of a crash in potato prices).

The absolute returns would be even lower if the imputed value of family labour were to be included in the cost of production. Our definition of net returns from farming is gross value of output minus paid-out costs (or Cost A2) excluding the cost of family labour (FL). As the extent of family labour used in crop production is inevitably higher for small farmers than for large farmers, if we had included the imputed cost of family labour (Cost A2 + FL), it would have reduced net incomes from farming even more sharply for small farmers as compared to large farmers.

The inadequacy of income from farming is reflected in what can be termed as "pluriactivity." Almost all small farmer households across the study villages engaged in multiple activities and reported multiple sources of income. On average, small farmers reported three to five separate sources of income. The fact that net household per capita incomes were nonetheless very low and close to the minimum wage for small farmer households would suggest that pluriactivity was a survival strategy for most small farmer households, and not evidence of aspiration-driven choice.

Although crop production and livestock-rearing continued to be the two most important sources of income for small farmer households, agricultural wages and incomes from non-agricultural sources also contributed significantly to household incomes, a process that could be characterised as proletarianisation without depeasantisation. More than 30 per cent of small farmer households engaged in manual wage work, and significant sections of small farmer households were also involved in petty trade or other activities in the non-agricultural sector. These income-generating channels, however, did not (and could not) provide a regular and sustained income flow to rural households. This was perhaps owing to the fact that only a small section of small farmer households received salary incomes, while the majority participated in wage employment and non-farm enterprises.

Why, then, do small farmers not exit farming altogether, Aparajita Bakshi asks. The answer perhaps lies in the fact that for them, access to non-agricultural employment and incomes is limited and uncertain.

Proletarianisation and Self-exploitation

In Chapter 3, Niladri Sekhar Dhar and Subhajit Patra discuss the extent to which small farmer households participated in the market for hired labour. A striking feature of the PARI data is that they provide information on the extent of participation of small farmer households in wage labour or sale of labour power. Indeed, the number of days of labouring-out were often more than the number of days spent working on the family farm.

The days of wage employment obtained by a small farmer household was high in some villages. In Mahatwar in eastern Uttar Pradesh (a village where men specialised in well-digging), wage employment accounted for around 240 days, and in Zhapur, Kalaburagi district (where employment was available in stone quarries), it was 263 days. These are both villages with access to non-agricultural employment. It was relatively low in West Bengal: 60 days in Amarsinghi (Malda district) and 125 days in Kalmandasguri (Koch Bihar district). In these last two villages, family labour was used intensively for crop production and there were no major sources of non-agricultural employment. In Kalmandasguri, family labour accounted for 57 per cent of total labour use in crop production.

A second feature of small farmer households is self-exploitation, as indicated by the high worker to non-worker ratio. Family labour use was widespread in the study villages except in the two villages in the northwest of the country, 25F Gulabewala and Tehang. In most villages, women from small farmer households participated in work, both on their own farms and on others' farms, though not as much in non-farm, non-agricultural employment. The unfortunate reality of child labour was evident in some of the villages, particularly among small farmer households in irrigated villages (Ananthavaram, Harevli, and Nimshirgaon), where children contributed to labour on the household's operational holding.

Thirdly, the total number of days of hired labour far outstripped the total number of days of family labour utilised in crop production. Although the ratio of family labour to hired labour was without exception higher for small farmers than large farmers and capitalist farmers, even small farmers employed hired labour for crop production. There was also variation in this ratio across villages. In the majority of the villages, the ratio was higher than one among small farmers, implying that family labour exceeded hired labour in crop production. However, the ratio was less than one among small farmers in six villages. In these villages, small farmers usually cultivated labour-intensive crops such as cotton or sugarcane, and used hired labour for several operations.

Interestingly, the expansion of piece rate-based contract labour in many villages has led to a situation where even small farmers hired contract workers for specific operations, even though contract labour was more expensive than daily paid casual labour (and family labour).

There is no simple relationship between hiring in labour, labouring out of the family farm, and family labour among small farmer households in contemporary India. All three forms can be found in small farmer families.

While the balance between hired labour and family labour on a farmer's own holdings depends primarily on production conditions (cropping pattern, crop intensity, degree of mechanisation, timeliness of operations, and so on), participation by members of small farmer households in wage labour reflects their poverty and the insufficiency of incomes from farming. In general, it is clear from the PARI data that the "rural poor, particularly manual workers and poor and middle peasants, continue to be the great reserve army of labour of capitalism in India" (Ramachandran 2016).

INEQUALITIES AND DIFFERENTIATION

Another important set of findings from this study and earlier PARI-based research relates to the prevalence of high inequalities in contemporary rural India. There are three kinds of inequalities that the authors of this volume highlight: inequalities between small farmers in a village; inequalities between small farmers in a village on the one hand, and large farmers and capitalist farmers on the other; and inequalities between small farmers in different agro-ecological regions.

Let me illustrate, in reverse order.

Small farmers are not a homogenous category across India. There was substantial variation in net crop incomes and overall household incomes of small farmers across the study villages and, correspondingly, across different agro-ecological regimes. In Chapter 4, Arindam Das and Madhura Swaminathan show that net incomes from farming were very low for the average small farmer in rainfed and drought-prone villages. At 2010–11 prices, annual net income from crop production per hectare was less than Rs 5,000 in the villages of Bukkacherla (Anantapur district, Andhra Pradesh), Rewasi (Sikar district, Rajasthan), and Zhapur (Kalaburagi district, Karnataka). Access to irrigation, as expected, tended to raise crop incomes on average. Incomes from crop production exceeded Rs 40,000 per hectare in Nimshirgaon (Kolhapur district, Maharashtra), Alabujanahalli (Mandya district, Karnataka), and Tehang (Jalandhar district, Punjab). Irrigation alone, however, was not a guarantee of higher returns, as incomes from farming depend on a host of natural and market factors. A good example is that of the irrigated village of Panahar in Bankura district, West Bengal, where average income from crop production was low in the survey year on account of a steep post-harvest crash in potato prices.

In terms of household income, the contrast across villages depended on an even larger set of factors, including factors such as connectivity and access to non-agricultural employment. For example, the annual household income per capita of a small farmer household was Rs 7,000 in Alabujanahalli in Karnataka's sugar belt, an irrigated agricultural village, as compared to Rs 22,000 in Rewasi, a dry village with large-scale out-migration and remittance inflow.

Secondly, there was variation in incomes among small farmers in a village. The coefficient of variation of crop income per hectare across small farmer households was greater than 2 in seven villages (Chapter 4). In all but one village (25F Gulabewala), some small farmers made losses in crop production in the survey year – that is, their estimated income from crop production was negative. There were five villages where the proportion of households making net losses in farming was 20 per cent or more. Of these five, three were dry villages – Bukkacherla in Anantapur district, Rewasi in Sikar district, and Zhapur in Kalaburagi district. However, two were irrigated villages – Ananthavaram in Guntur district, where small farmer tenants incurred losses on account of high rent payments, and Panahar in Bankura district, where small farmers incurred losses from potato cultivation. Surprisingly, there were only a handful of small farmers across villages who reported having received crop insurance payments.

The third inequality, and one that speaks to a large literature in agricultural economics, is that between small farmers and large farmers. Net crop incomes per hectare were lower for small farmers than for large farmers, on average, in every village. The difference was statistically significant in the irrigated villages of Ananthavaram, Harevli, Tehang, and Rewasi (we excluded six villages that were dominated by small farmers from the statistical test, as well as 25F Gulabewala with only two small farmers). The differences in gross and net incomes between small farmers and large farmers were on account of factors such as differences in cropping pattern, access to irrigation, and cost of inputs. In no village was an inverse relationship between farm size and profitability per unit area statistically significant. No differences were found between small farmers and large farmers in fertilizer use efficiency (or the ratio of output to fertilizer input). Chapter 7 concluded that there was no conclusive evidence to suggest a higher efficiency in usage of fertilizers by small farmers vis-à-vis large farmers.

Further, if we compare small farmers with capitalist farmers and landlords in a village, then the difference in net crop incomes was invariably much higher. In most villages, the ratio of net crop income per hectare of a landlord or capitalist farmer was four to six times that of a small farmer. This is no surprise, of course, given that landlords and capitalist farmers have access to the best and highest quality and extent of land in every village (including irrigated land), and to other means of production.

Turning to aggregate household incomes, the gap between small farmers and large farmers, particularly capitalist farmers, was wide. Not only was the average level of household income of a small farmer much lower than that of a big capitalist farmer, there were also significant differences in the sources of income between the two categories. There were differences in the composition of incomes between the relatively rich and poor among small farmer households. For instance, while poorer small farmer households depended on manual wage incomes to supplement incomes from farming, richer households had a higher share of "other incomes," including transfers and remittances.

ACCESS TO INPUTS, CREDIT, AND MARKETS

While Chapters 3 to 5 deal with the question of "how small farmers fare" in terms of labour deployment, incomes from farming, and aggregate household incomes, the next three chapters examine some features of input use and market access among small farmers.

The overall picture that emerges from these analyses of input use, fertilizer use, and credit is that small farmers are subject to severe constraints in the process of crop production and allied activities. Of the total paid-out costs of production examined in Chapter 6 by Arindam Das *et al.*, the cost of hired labour constituted one of the largest components in most villages. The proportion of cost of hired labour to paid-out costs was as high as 40 per cent in some villages, particularly villages in which labour-intensive crops such as jute and sugarcane were grown. Secondly, irrigation was a major cost for small farmers in many villages. The cost of irrigation varied with the type of irrigation, being lowest for canal irrigation and highest for private, tubewell-based irrigation. As ownership of irrigation equipment was less among small farmers than large farmers, the former depended more on water markets, which resulted in pushing up the costs of irrigation further. Thirdly, the costs of machine labour were higher for small farmers than for large farmers as the ownership of machines was concentrated in the hands of the latter. Small farmers rarely owned tractors, power tillers or threshers in the study villages. Fourthly, the cost of seed varied a great deal across villages in absolute terms and as a share of total paid-out costs. Expenditure on seeds was higher among small farmers than large farmers in paddy-growing villages. Differences across farmer categories were not, however, apparent in wheat-growing villages. Fifthly, in villages with a high degree of tenancy, rent of leased-in land was extremely high for small tenant farmers.

Chapter 6 shows that the burden of input costs was heavier for small farmers than for large farmers. A preliminary exercise with village data showed the limited reach of agricultural extension systems among small farmers, a serious concern both for technological development, and control over the quality and costs of production.

In Chapter 7, Kamal Murari, T. Jayaraman, and Sanjukta Chakravarty examine in some detail the application of fertilizers in the study villages. Official statistics invariably show that the application of fertilizers is not very different as between small and large farmers. The authors' first and critical finding from the PARI data contradicts this conclusion. The ratio of fertilizer cost to total input cost was higher for small farmers than large farmers in most villages, though the difference in median values was not statistically significant. Secondly, they found that both small and large farmers in the study villages paid a higher price for fertilizer per unit of operated area than reported by the Commission on Agricultural Costs and Prices (CACP). They argue that the calculation of minimum support prices based on CACP data may be underestimating the actual costs of fertilizer use among small farmers. Thirdly, they found imbalances in the use of NPK (nitrogen–phosphorous–potassium) fertilizers across crops and villages. In villages where neither wheat nor paddy was grown, nutrient use, including the use of nitrogen, was below recommended levels. By contrast, in wheat and paddy areas there was excessive use of nitrogen and very low application of potassium. Inappropriate application of fertilizers by small farmers is a concern for soil health and the sustainability of agriculture.

Based on a detailed review of banking statistics, Pallavi Chavan shows, in Chapter 8, that the growth of credit to small farmers has decelerated and the share of credit to small farmers in total agricultural credit has declined after 1991. This decline was arrested in 2005 when policies of financial inclusion began to be introduced, and there was some increase after 2010. Nevertheless, data from the study villages from the late 2000s paint a very grim picture. The overall finding is of a general inadequacy of and limited access to formal credit for small farmer households. Only one-third to one-quarter of these households had access to fresh loans from the formal banking sector in the year of the PARI survey in the village. This was also reflected in the relatively high interest costs for loans taken by small farmer households. Further, in no village did credit from the formal sector account for more than one-half of total credit. And in no village was actual credit from the banking sector adequate even in the limited sense of corresponding to the prevailing scale of finance for the particular crop. Ultimately, small farmers relied on informal sources to meet their agricultural credit needs more than large farmer households did.

Regression analysis showed two variables to be statistically significant in explaining access to formal credit: the extent of operational holding and the social background of the household. The larger the extent of land, the higher was the probability of gaining access to formal credit; and the more oppressed the social group to which the household belonged, the lower was the access to formal credit.

Moving from costs to prices, evidence presented in Chapter 6 shows that the official minimum support price or MSP acted as a floor price in all the study villages for wheat but not for paddy. The exception was Tehang village in Punjab, where small farmers received prices higher than MSP for both wheat and paddy. As is well known, price support and procurement systems are well established in Punjab. For crops and regions without an effective MSP, existing market structures and marketing channels played a central role. While marketing is not discussed in depth in this book, the example of a potato-growing village in West Bengal brings out the risks faced by small farmers who sold their produce to local traders. In short, except for a few crops, small farmers were at the mercy of markets for sale of their produce.

CLIMATE CHANGE

Agriculture is a risky enterprise. The risks emanate not only from natural and climatic factors, but also from institutional and market factors. T. Jayaraman makes a forceful argument in Chapter 9 on the seriousness of the impact of climate variability on small farmers, a group more vulnerable than large farmers to climate change. Small farmers' susceptibility to climate shocks arises from their specific socio-economic conditions. Jayaraman argues that technological solutions to mitigate the impact of climate change alone will not work unless accompanied by a policy shift that implements programmes to reduce the gross inequalities in land and resources in the agrarian sector.

LIVING STANDARDS

We have noted that small farmers in our study villages received incomes that were a bare minimum for the requirements of a household. This has grave implications for living conditions. In Chapter 11, Shamsher Singh describes the quality of housing of small farmer households. A majority of these houses were non-permanent structures and provided inadequate space per person. Most dwellings lacked a separate kitchen and access to even basic amenities such as electricity, safe drinking water, and toilets. One-half of all small farmer families defecated in the open. Within the category of small farmers, a disproportionately high proportion of Scheduled Caste, Scheduled Tribe and

Muslim households lived in houses that were not fully *pucca* structures and lacked adequate basic amenities.

Educational Deprivation

A very important component of living standards is educational attainment. The village data bring out the situation of adult rural men and women in respect of basic literacy and years of school education (Chapter 10). In every village, literacy rates among men and women in small farmer families were lower than in large farmer families. The gap was even larger in relation to capitalist farmer households. To illustrate, in Harevli and Mahatwar villages of Uttar Pradesh, the median number of years of schooling among women above the age of 16 in small farmer families was zero. The corresponding numbers were 12 and 11 for women from capitalist farmer and landlord families. In nine out of the 17 villages, the median number of years of schooling for women was zero. The situation was slightly better among men. Educational attainments among men were better in irrigated villages than in rainfed villages.

Child labour was prevalent in small farmer families, particularly among those in irrigated villages, where children worked on family operational holdings and did not attend school. The ongoing process of privatisation and decline in state support to education is likely to accentuate the problem of education deficit among small farmers, and has serious long-term implications for children and adults in small farmer families.

SMALL FARMERS AND SMALL FARMING[1]

The evidence put together in this book does not support many of the popular arguments in the literature, including on subsistence production by small farmers, relative efficiency of small farmers, equity or social justice in small-scale farming, and sustainability.

An interesting finding from the data is that small farmers tend to have relatively more land under intercropping than large farmers, and this was particularly notable in rainfed and drought-prone villages. This choice can be interpreted as a risk-reducing strategy in an uncertain agricultural environment. Small farmers also had relatively more area under foodgrains and less under purely commercial crops like sugarcane than large farmers. This, however, cannot be taken as an indicator of subsistence production or lack of market participation by small farmers.

[1] One issue that is not discussed here is biodiversity conservation and the role of small farmers in conservation.

Small farmers participate in markets, including markets for their produce, for purchase of inputs, and for sale of labour power.[2] As in other market economies, however, participation is constrained by ownership of resources (primarily land, in this case), as well as by specific production relations and caste and other social barriers.

The evidence from 17 villages presented in this book shows that while the gross value of output per hectare did not vary much as between small farmers and large farmers, there were significant differences in the net incomes of small farmers and large farmers, particularly in irrigated villages. In other words, we do not find support for an inverse relationship between farm size and profitability. On the contrary, the relationship was positive; that is, small farmers received a lower return per unit of operational holding than larger farmers.

Moreover, it is clear that the crisis of small farmers is not so much of productivity (we did not find much difference in productivity of paddy and wheat as between small farmers and large farmers within a village), but of income, on account of the price squeeze they experienced in terms of high costs and low and variable prices.

As discussed in an earlier section, there was high variability and inequality in crop incomes among small farmers in a village, between small farmers in different agro-ecological regions, and between small and large farmers.

POLICY SUPPORT

Small farmers account for a substantial proportion of the rural population today and require urgent policy support to survive in the context of neo-liberal capitalist globalisation. Small farmers in India operate in a capitalist market economy where a small section of households (of landlords and capitalist farmers) have control over the bulk of the means of production. In this context, how can public support measures deliver economies of scale to small farmers?

Public policy needs to address a range of problems in order to ensure minimum living standards for small farmers. Concrete measures are needed to address the problems of low productivity, credit, timely availability of quality inputs at reasonable prices, and assured prices for output. Some of the policy issues that emerge from this book include providing access to appropriate practices with regard to fertilizer application by strengthening public extension

---

[2] It should be pointed out here that no tribal village is included in the list of 17 villages. Market orientation is distinctly less and different in tribal villages (for an example, see Ramachandran 2010 on Dungariya village of Udaipur district, Rajasthan).

networks and services. Institutional mechanisms are needed to lower the costs of inputs such as machine labour and irrigation for small farmers, as they lack ownership of machines and irrigation equipment, and are dependent on those who do. Collective or group ownership of machines and implements, more public investment in irrigation, and public regulation of private water markets are required. With growing tenancy in many parts of the country, careful regulation of tenant contracts is needed.[3] Extension systems need to be revamped to ensure the delivery of modern and appropriate agricultural knowledge and information to small farmers. Crop insurance must reach every farmer. The national agricultural research system (NARS) needs to be reoriented and strengthened to serve small farmers.

An overhaul of rural credit policy is required to ensure adequate and timely availability of formal credit to small farmers. This could include measures such as removing the requirement of land as collateral for bank loans, ensuring that crop credit corresponds to realistic scales of finance, and extending credit to small farmers for consumption-related needs as well.

On the output front, instability of agricultural commodity prices is not a new phenomenon. For raising absolute levels of income as well as ensuring stability in incomes for small producers, interventions to stabilise the prices of produce are required. The system of minimum support price (MSP) accompanied by active procurement has worked well for a few crops in a few regions. Policy must build on these successes and MSP policy should be extended to all regions. MSP should be fixed so as to provide a reasonable return to small farmers. The C2[4] plus 50 per cent formula suggested by the National Commission on Farmers is the most important recent policy recommendation in this regard.[5] To ensure implementation of MSP for new crops and regions, the policy can be given a statutory basis, as recommended by the High Level Committee on Food Grain Policy (2003) and, more recently, the Karnataka Agricultural Prices Commission. A price stabilisation fund

[3] The suggestions of a recent report by the Niti Ayog, however, propose leaving the terms of tenancy to bargaining between lessors and lessees.

[4] C2 cost of production includes all costs including imputed costs of family labour, owned capital, and rental on land. It is one of the various cost concepts used by the Commission for Agricultural Costs and Prices (CACP) in India to determine crop incomes. See Sarkar, Ramachandran, and Swaminathan (2014, pp. 393–95).

[5] The National Commission on Farmers (NCF) appointed by the Government of India in 2004 argued that "implementation of MSP had to be improved for crops other than paddy and wheat; the commission recommended that MSP should be at least 50 per cent more than the weighted average of costs of production C2. In other words, the costs of all major producing regions would need to be considered in estimating C2, and MSP should give a return over C2 (Sarkar, Ramachandran, and Swaminathan (2014, p. 395)." See Swaminathan (2016), for a commentary on Ten Years of the National Policy for Farmers, drafted by the NCF.

should be established for crops subject to world market price fluctuations. At the same time, small farmers need more and better market information, and advice on how to lower marketing costs.

Since household incomes of small farmers include a significant wage component, ensuring the economic viability of small farmer households requires employment guarantee schemes with reasonable minimum wages that contribute to the strengthening of the production base of small farmers by improving irrigation facilities and efficiency of water use.

Lastly, public support needs to be extended to small farmer households (and other sections of the rural poor) to improve conditions of housing and ensure provision of all basic amenities. For the next generation in small farmer households, access to quality education and the completion of 10 years of schooling must be ensured.

India is home to hundreds of millions of small farmers. They cannot be left in conditions, to use Kautsky's phrase, of "overwork, undernourishment, and ignorance." To enhance the well-being of small farmers at this historical juncture, we require strong and effective public policy support. This must go with the recognition that, in the long run, policy has to empower small farmers by giving them the power of scale through social and technological measures.

I am very grateful to V. K. Ramachandran for his comments.

REFERENCES

Ramachandran, V. K. (2011), "The State of Agrarian Relations in India Today," *The Marxist*, vol. 27, nos. 1–2, January–June.

Ramachandran, V. K., Rawal, Vikas, and Swaminathan, Madhura (eds.) (2010), *Socio-Economic Surveys of Three Villages in Andhra Pradesh: A Study of Agrarian Relations*, Tulika Books, New Delhi.

Sarkar, Biplab, Swaminathan, Madhura, and Ramachandran, V. K. (2014), "Aspects of the Political Economy of Crop Incomes in India," *World Review of Political Economy*, vol. 5, no. 3, Fall, pp. 392–413.

Swaminathan, M. S. (2016), "National Policy for Farmers: Ten Years Later," *Review of Agrarian Studies*, vol. 6, no. 1, available at http://ras.org.in/national_policy_for_farmers_ten_years_later, viewed on 6 September 2017.

Contributors

VENKATESH ATHREYA, Consultant, M. S. Swaminathan Research Foundation, Chennai

APARAJITA BAKSHI, Assistant Professor, School of Development Studies, Tata Institute of Social Sciences, Mumbai

SANDIPAN BAKSI, Programme Coordinator, Foundation for Agrarian Studies, Bengaluru

SANJUKTA CHAKRABORTY, Data Analyst, Foundation for Agrarian Studies, Bengaluru

PALLAVI CHAVAN, Researcher on rural credit based in Mumbai

ARINDAM DAS, Senior Programme Manager, Foundation for Agrarian Studies, Bengaluru

NILADRI SEKHAR DHAR, Assistant Professor, School of Rural Development, Tata Institute of Social Sciences, Tuljapur

RITAM DUTTA, Data Analyst, Foundation for Agrarian Studies, Bengaluru

T. JAYARAMAN, Professor, School of Habitat Studies, Tata Institute of Social Sciences, Mumbai

DEEPAK KUMAR, PhD scholar, Yokohama National University, Yokohama, Japan

VIJAY KUMAR, Data Analyst, Foundation for Agrarian Studies, Bengaluru

RAKESH KUMAR MAHATO, Data Analyst, Foundation for Agrarian Studies, Bengaluru

TAPAS SINGH MODAK, Senior Data Analyst, Foundation for Agrarian Studies, Bengaluru

KAMAL KUMAR MURARI, Assistant Professor, School of Habitat Studies, Tata Institute of Social Sciences, Mumbai

SUBHAJIT PATRA, Data Analyst, Foundation for Agrarian Studies, Bengaluru

R. RAMAKUMAR, Professor and Dean, School of Development Studies, Tata Institute of Social Sciences, Mumbai

A. BHEEMESHWAR REDDY, Assistant Professor, Department of Economics and Finance, Birla Institute of Technology & Science (BITS) Pilani (Hyderabad Campus)

BIPLAB SARKAR, Assistant Professor, Centre for Developmental Studies, PES University, Bengaluru

SHAMSHER SINGH, Postdoctoral Fellow, Indian Institute of Management, Ahmedabad

T. SIVAMURUGAN, Programme Manager, Foundation for Agrarian Studies, Bengaluru

MADHURA SWAMINATHAN, Professor and Head, Economic Analysis Unit, Indian Statistical Institute, Bengaluru